D1416140

Comparative Studies
in
Science and Society

Edited by

Sal P. Restivo
The University of Hartford

Christopher K. Vanderpool
Michigan State University

Charles E. Merrill Publishing Company
A Bell & Howell Company
Columbus, Ohio

12.95

Merrill Sociology Series

Under the editorship of
Richard L. Simpson
University of North Carolina, Chapel Hill
and
Paul E. Mott

Published by
Charles E. Merrill Publishing Company
A Bell & Howell Company
Columbus, Ohio 43216

ISBN: 0-675-08907-7

Library of Congress Catalog Card Number: 73-81553

1 2 3 4 5 6—79 78 77 76 75 74

Printed in the United States of America

For
Lorraine, Valerie,
Aaron, David, Daniel, and Eric

Preface

Our objectives in preparing this book were (1) to contribute to the development of a comparative, macrosociological, and historical perspective on science, and (2) to link this perspective to the critical problems of science as a social activity and social process. We have selected articles which confront the social problems of science and at the same time contribute to the identification of theoretical problems and perspectives, methods, and variables in the comparative sociology of science. Most of the selections are unedited reprints of research papers and essays written by recognized authorities in the sociology, political science, and history of science. With the exception of an essay by Edward Shils originally published in 1957, the previously published materials originally appeared between 1960 and 1970. New material includes a previously unpublished paper by Karp and Restivo on science in China and Western Europe, and our own contributions, drawn from our dissertations, based on a survey of visiting foreign scientists at several midwestern United States universities.

Myths and uncritically idealistic conceptions of science are rapidly giving ground to realistic views of science as a human activity subject to the functional and dysfunctional consequences of human psychology and social organization. The emergence of a more realistic conception of science will hopefully lead to a better appreciation of science as a manifestation of human creative and critical intelligence, and also as a world-unifying force of friendship among human beings. We view the comparative sociology of science as an important part of the foundation for such an appreciation of science.

We wish to acknowledge our appreciation for the encouragement, guidance, criticisms, suggestions, and friendship of our mentors, Professors John and Ruth Hill Useem. We also wish to thank the following persons for their many helpful criticisms and suggestions regarding our dissertations: William H. Form, Charles Loomis, James B. McKee, Vincent Salvo, and Frederick Waisanen. Susan Asch, Gurdyp Aurora, Arnold Holden, Florence McCarthy, and Shelby Stewman provided critical stimulation for and evaluation of many of the ideas we discuss in this book. Primary support for our dissertation research was provided by a National Science Foundation Dissertation Grant (GS2326), and fellowships from the National Science Foundation and the National Institute of Mental Health.

We are grateful to our editor, Roger Ratliff, for his patient encouragement and assistance during all phases of the writing, editing, and production of this book. Susan Ziegler, our production editor, was extremely helpful during the final stages of production, especially in her careful, critical reading of the manuscript. Finally, we thank Marilyn Lovall, Sharon Oakley, Ellen Ritter, Jean Schanz, Susan Roggelin, and Joy Forgwe for their assistance in typing and editing portions of the manuscript.

Contents

Science: Social Activity and Social Process

Robert K. Merton defined the sociology of science as the study of "the dynamic interdependence between science, as an ongoing social activity giving rise to cultural and civilizational products, and the environing social structure" (1957, p. 531). The relationships between science and society are reciprocal. But more than a decade after Merton deplored the "uneven attention" accorded the impact of society on science, this subject continues to be virtually ignored. Students of the sociology of science emphasize the impact of science on society and focus on the internal structure of the social system of science. For example, Storer analyzes the scientific process in terms of an autonomous system of exchanges governed by rewards and sanctions (1966); and there is a rapidly growing literature on the study of citations and scientific communication networks (DeReuck and Knight 1967; Crane 1969; Nelson and Pollock 1970; Cole and Cole 1972).

This uneven emphasis has been accompanied by a pervasive conception of science as a "monolithic entity" (Kaplan 1964, p. 854; Rose and Rose 1970, p. 263). Such a conceptualization may reflect an implicit belief that the assumed purity of science is preserved if it is viewed as developing in a "social vacuum" (Merton 1957, p. 532). In any case, this viewpoint is not consistent with the idea of a sociology of science, and is, in fact, being countered in the newer literature (e.g., Nagi and Corwin 1972).

Science is a universal social activity, rooted "in the generic human attribute of empirical rationality" (Barber 1952). Contemporary scientific activities are linked to each other by more or less well-developed systems of transportation, communication, and exchange that cut across geographical, political, and cultural boundaries; these systems are parts of a suprasocietal system variously referred to as the international scientific community, the third-culture of science, the scientific superculture, and the scientific lateralization (Restivo 1971, pp. 187-205). Scientific activities at this level are primary links between and among societies. Historically, science has been an important linkage between and among human populations. This conception of the "dynamic interdependence" between science and society is the basis for our emphasis on the study of science in comparative, historical, macrosociological perspective.

Macrosociology is the study of large-scale social systems and interrelationships within, between and among these systems. Etzioni (1970, p. 109) refers to these social systems as societies, subsocieties, and suprasocieties. At this level, the relationship between history (conceived as a macroscopic process) and sociology is easy to establish (Etzioni 1970, p. 107). History can also be viewed as a set of social facts for testing and deriving sociological ideas, hypotheses, theories, and laws (Restivo 1970).

The logical and methodological virtues of comparative analysis have been widely noted (e.g., Durkheim 1938; Radcliffe-Brown 1952; Murdock 1957; Holt and Turner 1970). The interest in comparative sociology that has emerged and crystallized in the last decade (e.g., Andreski 1964; March 1967) has been in part a response to the parochialism of American and Western sociology, and the realities of an increasingly interdependent world. The comparative history of science links macrosociology with problems such as those posed by Needham on science in China and Western Europe, and by Joseph Ben-David on the establishment of the scientific role. The maintenance of communication among eighteenth century French and English scientists while their countries were at war with each other (DeBeer 1960), and the changes in French science wrought by the Jacobin convention in 1793 (Gillispie 1959) are examples of historical problems which should be central to the interests of sociologists of science. Weber's contributions exemplify the fruitfulness of a perspective which is at once comparative, historical and macrosociological; in the contemporary sociology of science, Ben-

David's work is the outstanding example of the nature and consequences of applying this perspective to the study of science.

An important substantive basis for emphasizing this perspective is the significance of science—logically, technologically, and ideologically —for industrialization, modernization, and social change in general at societal and global levels. Science is a crucial factor in national and international politics; it has different but high-priority functions in all contemporary nation-states; and it is an important element in the images of social change which guide politicians, intellectuals, and others who view themselves as agents of change in their societies and in the international system.

Science, as a social activity which generates cumulative increases in knowledge, closer and closer approximations to reality, increasingly sophisticated theories, and more convincing explanations and deeper understandings, is possible only under certain conditions. Among the determinants of whether or not science will be progressive (in the sense so often taken for granted by students and practitioners of science) are: (1) how scientific activity is organized, (2) the degree of organizational complexity and extent of bureaucratization, (3) the degree of institutionalization, (4) the sociocultural context, and (5) the availability of resources for scientific inquiry.

By contrast, the impression derived from the literature in the sociology of science and allied fields of inquiry is that science emerged in Western Europe, developed, and settled into a pattern characterized by evolutionary growth, continuity in norms, values, and goals, and a capacity for unaltered diffusion to other cultural settings. But the institutionalization of scientific activity beginning in sixteenth century Western Europe created a potential, as in all processes of institutionalization, for the rigidification of scientific activity, and ultimately, for the emergence and development of science as profession, bureaucracy, and ideology. The acknowledgment and investigation of such a potential have been obstructed by an idealistic view of science and scientists. From this perspective, science is a fixed approach to discovery which evolves independent of internal or external changes.

The sociology of science has never been so narrowly defined or so systematically cultivated by sociologists as to discourage contributions by non-sociologists, or to make such contributions inconsequential. Approximately half the articles in Kaplan's *Science and Society* (1965) were by men originally trained in physical and natural sciences; and while few contributors to Barber and Hirsch's *The Sociology of Science* (1962) were not social scientists, the variety of their professional training reflects the diverse fields of inquiry out of which the sociology of science has been fashioned.

The diversity of contributors to the study of science and society must be credited with stimulating an awareness of, and generating excitement about, problems in the sociology of science. Unfortunately, this diversity, in conjunction with the relative neglect of science by sociologists, and the underdevelopment of general sociological theory, has not significantly advanced the development of a sociological theory of science. One need not be an advocate of disciplinary imperialism or a radical differentiation of subject matters to recognize the legitimacy of our concern for the special contribution sociologists—or, more accurately, the sociological perspective—can make to the study of science and society. We have attempted to select articles for this volume which indicate this special contribution.

PROBLEMS IN THE SOCIOLOGY OF SCIENCE

Our organization of the text has been influenced by a set of basic problems in the sociology of science: (1) What do the norms of science tell us about science and scientists? (2) What is the relationship between science and ideology? (3) What are the consequences of professionalization and bureaucratization for "progress" in science? (4) What are the sources of heterogeneity in the social system of scientific activities? and (5) What is the "international scientific community"? The first three questions will be discussed together; a discussion of conditions of work as sources of heterogeneity in science follows; finally, we discuss the concept of an international scientific community, and introduce the concept "third-culture of science."

PROFESSIONALIZATION, BUREAUCRATIZATION, IDEOLOGY, AND THE NORMS OF SCIENCE

Storer (1966, pp. 78-80), proceeding from the works of Merton and Barber, identified three orientation and three directive norms in science: the orientation norms are universalism, disinterestedness, and rationality; the directive norms are organized scepticism, communality, and emotional neutrality. Storer argued that scientists support the norms because they are aware that such support is the basis for the proper operation of the exchange system in science. To the extent that the individual scientist wishes to maximize his creative potential, he must be concerned with the maintenance of the social structure of science. Only within science can he be assured of receiving "honest, competent response from others;" this "need" of the scientist gives him a "personal stake" in supporting the norms of science (Storer 1966, p. 86). But this is inadequate if we wish to understand, empirically and theoretically, the nature and functions of the norms, especially in terms of the reciprocal relations between science and society.

When originally formulated by Merton (1957, pp. 551-61), universalism, communism, disinterestedness, and organized scepticism were elements of an "ethos" of science, what Merton defined as an "affectively toned complex of values and norms which is held to be binding on the man of science." Norms-values were also defined as "institutional imperatives," a term Merton used synonymously with "mores" (1957, p. 552). These imperatives, or mores, were conceived to have a moral as well as a technical dimension.

Bernard Barber (1952, pp. 122-34), viewing science as a "moral enterprise," identified a set of "moral values" common to science and to "liberal society": rationality, emotional neutrality, universalism, and individualism. In addition, Barber identified communality (a politically and ideologically "clean" rendition of Merton's communism) and disinterestedness, or other-orientation (following Parsons 1949, pp. 343-45), as the "moral ideals" of science.

In the third major contribution to the study of norms in science, Parsons discussed three levels of norms (Kaplan 1964, pp. 855-56): (1) technical norms—empirical validity, logical clarity, logical consistency and generality of principles.(Parsons 1951, p. 335); (2) the set of pattern variables associated with the occupational role of the scientist—universalism, affective neutrality, specificity, achievement orientation, and collectivity orientation (1951, p. 343); and (3) research norms—tentativeness and "an *obligation* . . . to accept the validity of scientific findings and theories which have been adequately demonstrated" (1951, p. 353).

Two problems are associated with these attempts to identify normative-evaluative orientations in science. Kaplan noted one in his excellent review article for the *Handbook of Modern Sociology* (1964, p. 857): "Whether their analysis is correct or not, the point to be stressed here is that the values posited by Merton and Barber and Parsons have been fully accepted as those which prevail today, without any additional empirical verification or theoretical analysis." It is interesting to note that Kaplan chose the term "values" even though he consistently used the term "norms" in his preceding discussion. This leads us to the second problem—are these overlapping sets of concepts in Merton, Barber, Parsons, and Storer "norms," "values," "moral ideals," "institutional imperatives," or "mores"? Are the norms supposed to be a set of "shared should statements" (cf. Storer 1966, pp. 76-77)? If so, little empirical support has been offered for such an interpretation of the norms of science Storer outlines.

Sociologists of science have some recognition of the narrow, intuitive basis for Merton's influential conceptualization of the norms of science. Kaplan (1964, p. 855), for example, notes Merton's admission that the institutional imperatives were largely based on seventeenth century documents. Storer (1966, p. 77) is even more explicit in writing that Merton "was able to conceptualize the norms of science, working presumably in part through intuition and testing his ideas against what scientists have said since the seventeenth century about their work and about how scientists should behave." Idealistic conceptions of science and scientists existed before anything approaching a systematic sociology of science emerged. These conceptions have had an undue influence, relative to empirical research, on ideas about norms in science.

To argue that science has identifiable norms implies that science as a social system has been more or less clearly delineated and that the norms are operative among the members of that system. But what is the social system of science? Among what aggregates are the norms operative; do they operate for all persons who define themselves as scientists, at all times in all places? Kaplan points out, for example, that "implicit in Merton's formulation of his four institutional imperatives is the idea that these have remained relatively unchanged from the time of their early origins" (1964, p. 855). Are the norms relevant for all living Ph.D.'s in science, or for an elite? What is the intensity associated with each of the norms; does the sense of "shouldness" vary in intensity among scientists for different norms? Is a norm of science a norm if scientists show no signs of real commitment to it (Gibbs 1968, p. 210)? And to what extent are departures from the norms followed by some punishment (Homans 1950, p. 121)? What, in brief, do the norms of science tell us about the world's working scientists?

Given the fact that scientific activities have become increasingly professionalized and bureaucratized, it is reasonable to inquire whether science has become a more "rigid" enterprise than it has been in the past. Have professional scientists, concerned with status and job security, developed an ideology, and incorporated the norms of science in that ideology? To define one's research as "basic," for example, may explain and justify what amounts to an obligation to eschew responsibility as a scientist for the present and future social consequences of one's scientific work. For the idea that science can be pursued independent of all extra-scientific considerations to be defensible, it must be possible to show that science is an autonomous, self-correcting process which develops or progresses according to its own internal laws. But this is a notion applicable only in the abstract to the logic of inquiry—it does not take into account the impact of social and psychological factors on science as a process of human inquiry. Admitting the validity of a sociology of science leads to the following imperative: at the very least, the issue of purity in science must be subjected to the same forms of inquiry scientists are expected to exhibit in their own studies (Schwartz 1971, pp. 25-26; Commoner 1971, p. 178; Marcuse 1969, p. 477).

A zoologist we interviewed asserted, in response to a question on social responsibility, "My role as a scientist, insofar as society is concerned, is negligible; what I produce is negligible in its consequences." This statement is typical of the responses elicited among scientists we have interviewed. It illustrates the tenacity of the "social vacuum" assumption in science, more than a century after Marx clearly expressed the sociological view of the scientific role:

> Even when I carry out *scientific work,* etc., an activity which I can seldom conduct in direct association with other men—I perform a *social,* because *human,* act. It is not only the material of my activity—like the language itself which the thinker uses—which is given to me as a social product. My *own* existence *is* a social activity (1956, p. 77).

The emergence (beginning in sixteenth century Western Europe) of science as an autonomous, functionally differentiated social activity, and the emergence of the scientific role were preconditions for the cumulative, progressive characteristics of modern science. But this process eventually placed science in a manifest competitive relationship with other more or less autonomous institutions for the control of scarce societal resources. This competition, and the growing demand for resources in science, has promoted the development of explanations and justifications for the resource demands of scientists. Under the set of conditions associated with the emergence of modern science, it may be that the self-correcting, rational, tentative, and open nature of scientific inquiry was approximated. But even then, the operation of the norms could not be taken for granted, as Merton illustrates in his critique of Clark's conception of the significance of disinterestedness in seventeenth century English science (Merton 1957, pp. 607-8). It is naive to *assume* that such a model has a high degree of descriptive or explanatory power today. Two processes in particular require investigation in terms of their impact on science: professionalization and bureaucratization. These two processes are linked at least to the extent that they are concomitant in the modern history of industrializing societies (Jencks and Reisman 1969, p. 202).

The literature on professionals and complex organizations has traditionally stressed the conflicts inherent in linking professional and bureaucratic roles (Parsons 1954, pp. 34-49; Francis and Stone 1956, pp. 153-57; Blau and Scott 1962, pp. 60-63). "Independent professionals" in bureaucratic settings are studied as actors resisting supervisors and bureaucratic rules, rejecting bureaucratic standards, and giving conditional loyalty to the bureaucracy. Kornhauser's *Scientists in Industry-Conflict and Accommodation* (1962) is an important application of this view in the sociology of science. But Kornhauser is as much interested in the relations between profession and bureaucracy as "two institutions" as he is in relations between individuals and organizations (1962, p. 8). Professions and bureaucracies can be viewed as two institutions linked at one level by "areas of conflict" (Scott 1966). This linkage has led Scott to note a significant convergence between the professionalization of bureaucrats and the simultaneous bureaucratization of professionals (1966, p. 267).

Whatever the virtues of professionalization and bureaucratization have been in the past or during the early stages of their impact on work and occupations, the dysfunctions of these processes have become increasingly salient and predominant in the absence of counter-processes. The dysfunctional aspects of the two processes tend to converge. Bureaucratization, with its demands for reliable response and strict devotion to rules (Merton 1957, p. 200), stimulates and reinforces the tendencies in professionalization toward occupational closure and dogma. Sociologists have tended to emphasize the functional aspects of professionalization and the dysfunctional aspects of bureaucratization, thus obscuring the convergence of dysfunctions. A simple convergence has received some attention—that associated with the bureaucratization of professional associations, and of the training and educational settings for professional socialization. The links between the bureaucratic professional associations and the bureaucratic professionalizing mileux have, however, created a complex convergence with far-reaching actual and potential effects on the socialization of professionals. The medical profession may represent a paradigm for this process (e.g., Friedson 1970; and Brewer 1971, pp. 149-62).

The tendency of professionalization and bureaucratization to generate closure, i.e., an ethnocentrism of work (Jencks and Reisman 1967, pp. 201-2), may lead to a decrease in the capacity of individuals to respond to problems in creative and critical ways. According to one student of these processes, the price of professional training and precision may ultimately be the loss of objectivity (Horowitz 1970, p. 347; cf. de Grazia 1963).

To the extent that no potentially effective counter processes (e.g., "radical caucuses") arise and give root to new organizational forms, the convergence of dysfunctions can lead to the routinization of rationality. Such a possibility has been referred to in the most recent literature in the sociology of science (Rose and Rose 1970, p. 159, 179; Friedrichs

1970, p. 114; Gouldner 1970, p. 497; Haberer 1969, p. 321; see also West 1960). This suggests that a new chapter in the structure of scientific revolutions is emerging, based on a conception of science as a social process.

Kuhn notes that one of the concomitants of normal science is the narrowing and rigidifying of education in the natural sciences (1967; 1970). Scientific training, according to Kuhn, "is not well designed to produce the man who will easily discover a fresh approach;" but he optimistically adds that "so long as somebody appears with a new candidate for a paradigm—usually a young man or one new to the field —the loss due to rigidity accrues only to the individual." Kuhn also argues that "individual rigidity is compatible with a community that can switch from paradigm to paradigm when the occasion demands" (Kuhn 1962, p. 165). Optimism is a keynote of Kuhn's perspective. He refers to the "continuing evolution" of science (1962, p. 159), and to scientists as "reasonable men" (1962, p. 157). He affirms the "fact" that scientific development, like biological development, is "a unidirectional and irreversible process;" later theories are better than earlier ones "for solving puzzles in the often quite different environments to which they are applied" (1970, p. 206). This is the basis for Kuhn's statement that he is "a convinced believer in scientific progress" (1970, p. 206). But while unidirectionality and irreversibility may be attributes of the logic of inquiry, they cannot be assumed to be attributes of science as a social process. The point Kuhn ignores is that rigidity is not only an individual fact, but a social fact as well. A convergence of dysfunctions such as we have discussed should lead us to anticipate some resistance to "progress" in science. Is the supply of men who are "young" and "new," Kuhn's Bolsheviks of science, independent of social processes and changes in social structure? Is it possible that the professionalization-bureaucratization of scientific activities within corporate, university, and governmental structures, and the diffusion of science as ideology, will effectively decrease the supply of the young and new men who make scientific revolutions? Can youth and newness become increasingly unlikely and ultimately impossible as men become "standardized" and as deviation becomes not merely less likely, but more intolerable and more at the mercy of agents and agencies of social control?

At least two "damping" sources to consider in evaluating Kuhn's model can be hypothesized. The convergence of dysfunctions in professionalization-bureaucratization is one. Its effect is to lengthen the period between peaks of scientific revolution, lessen the intensity of revolutions, progressively decrease periods of crisis, and progressively decrease the probability that: (1) an individual scientist will conceptualize a revolutionary idea and (2) such a scientist will be recognized and precipitate a crisis. A second damping source is the "cost" associated with each revolution. Boulding, for example, argues that the dialectical processes accompanying scientific revolutions are costs, representing "the heat of crystallization in a process of essentially continuous change" (1970, pp. 60-61). Scientific progress, he argues, is not a function of "conflicts of theories"; science does not progress "by one theory conquering another in a revolution but rather by the slow growth of testable and tested images" (Boulding 1970, p. 63). The "heat of crystallization" analogy is noteworthy when considered in conjunction with the effects of professionalization-bureaucratization on science. Crystallization is a change from the liquid to the solid state. During this process, particles align themselves symmetrically and undergo a diminution in their "freedom" of motion. This state is generally characterized by retention of volume and shape. Boulding's reliance on this analogy suggests that while he is aware of the costs factor, he, like Kuhn, ignores the possibility that a cumulation of costs may progressively increase resistance to growth, to change in general. The costs of dialectical or revolutionary change may cumulate like fatigue products in animals during physical exercise periods—a temporary diminution of skills, power and efficiency is followed by a "recovery phase" and increase in skill, power and efficiency, i.e., progress. But such products can also cumulate in such a way as to continually decrease skills, power and efficiency, as in the life-cycle of animals. Granting the weaknesses and logical pitfalls of such an analogy, it at least suggests that a costs argument should be extended to incorporate the consequences of cumulation, to take into account the idea of a decreasing capacity in social systems to "recover," that is, to progress, having incurred certain costs.

The recognition that science is a social process open to scientific study carries with it an obligation to put aside belief and faith in progress; human activity exhibits evolutionary and devolutionary characteristics, growth and decay, uneven and multi-linear development. Progress is problematic, and not to be taken for granted or otherwise assumed to be inevitable under any and all socio-cultural conditions.

CONDITIONS OF SCIENTIFIC WORK

One important step in developing a more realistic conception of science is the identification of sources of heterogeneity in the scientific community. Numerous sources of variations have been suggested, most of which involve differentiating scientific activities on the basis of "subject matter" into physical, biological and social sciences (Davy 1967; Gullahorn and Gullahorn 1962, p. 285; Hagstrom 1964 and 1965, pp. 10, 11, 245; Hilgard and Lerner 1951; Hirsch 1961; Lazarsfeld and Thielens

1958; Meier 1951; Mills 1966, p. 39; Rose 1952, p. 148; and Weiss 1967, p. 156). The criteria for differentiation in the scientific community include the following: (1) the length of the theoretical chain linking general principles with common-sense language and experience or the extent to which mathematics is part of the theoretical structure (Frank 1958, p. 60; Oppenheimer 1958, p. 69); (2) cost and scale of research (Coser 1965, p. 299); (3) the degree to which theory and methodology are susceptible to social and political influences (Hirsch 1961); (4) the degree to which theory and methodology are "well-organized" (Menzel 1958; Price 1965, p. 107; Storer 1967); (5) the degree to which norms are specified for a concrete set of practices (Hagstrom 1965, p. 11); (6) the degree to which paradigms organize theory and methodology (Kuhn 1962 and 1963, p. 344); (7) the nature of methodology including techniques (Bucher and Straus 1961, pp. 328-30); (8) the level of development of the scientific community and sectors thereof, and associated systems of prestige-status-esteem relative to other roles, organizations, and institutions (Basalla 1967, p. 622); (9) variations in occupational role behavior (Becker and Carper 1956, pp. 288-300); and (10) differences in intellectual or cognitive styles of performance (Spencer 1966, p. 296). These criteria are aspects of the social organization of scientific work. Studies in the sociology of work, occupations, and professions support the hypothesis that conditions of work are sources of heterogeneity in science. The relationship between variations in conditions of work and variations in conceptual and activity patterns in occupations and professions has been extensively documented (Becker and Straus 1956; Berger 1964, p. 231; Bucher and Straus 1961; Caplow 1954, p. 124; Form 1946; Form and Nosow 1962, p. 441; Glaser 1960; Gottlieb 1961; Rosenberg 1957, p. 298; Wilensky 1960, p. 553). However, no general theory has been developed to logically relate specific conditions of work to specific conceptual and activity patterns.

It is sociologically naive to suppose that significant variations within science will neatly differentiate physical, biological, and social scientists. It seems more reasonable to assume that specific work conditions are systematically related to specific conceptual and activity patterns. These conditions may, of course, significantly overlap so that a particular configuration is associated with a given occupation or profession; but the appropriate independent variables are work conditions and not broad occupational or professional categories.

These questions become much more complex when we consider the idea of an international scientific community. The conception of science as an evolving, diffusing, monolithic social system rapidly loses any capacity to generate a convincing description or interpretation of worldwide scientific activities.

THE INTERNATIONAL SCIENTIFIC COMMUNITY

Philosophers, theologians, physical scientists, politicians, intellectuals, laymen, and specialists in the sociology and history of science, irrespective of their professional and personal orientation, have used the term "scientific community" to refer to a monolithic group or collectivity. Edward Shils, for example, writes that the scientific community "approximates most closely to the ideal of a body bound together by a universal devotion to a common set of standards derived from a common tradition and acknowledged by all who have passed through the discipline of scientific training" (1958, p. 15). This idea, however, has been more often asserted or assumed than subjected to systematic study.

In the opening paragraph of *The Scientific Community*, Hagstrom notes that his discussion is limited to basic, experimental sciences with "well-established theories":

> In this type of research, the scientific community is relatively autonomous, and the group of colleagues is the most important source of social influence on research (1965, p. 1).

His book is presented as a study of the scientific community; but nowhere is the reader offered an explicit definition of "scientific community," or an explanation of why it is a community or why it is referred to as one, and who the individuals are who constitute that community beyond the scientists Hagstrom has data for.

Ben-David, in a recent review article (1970, p. 12), attributes the formulation of the term "scientific community" to Michael Polanyi (1951). According to Ben-David, Polanyi used the term to describe the network of training, refereeing, and informal sanctions which promote discipline in the midst of individual freedom. In spite of the fact that Polanyi was more concerned with exercising his formidable talents as a philosopher than with contributing to the sociological study of science, Ben-David finds his construct "a perfectly adequate sociological formulation" (1970, p. 12). This might be a reasonable assertion if Polanyi's formulation is viewed as a basis for asking sociological questions, constructing exploratory and testable hypotheses, and giving sociological substance to the concept "scientific community." But sociologists of science have not pursued this course. Shils' influential usage of the concept, for example, suffers from the biases of his reflections on the "atomic scientists' movement" and is a concept which is not subject to test and refinement (1954; 1956).

More recently, Kuhn has argued that "scientific community" can be defined in terms of specialization or the sharing of a paradigm (1970, p. 177). But if a definition of scientific community must follow from Kuhn's less than rigorous definition of paradigm, then we are not significantly advanced beyond Polanyi's formulation.

In attempting to develop some guidelines for comprehending scientific activities sociologically, it would be useful to consider what sociologists mean by "community." Definitions of community have undergone complex changes. One of the main lines of development has been from a "locality" based conception of community to one which seeks to detach community from geography, territory, locality. This development is a response to the increase in scale of human activities. Useem uses the term "international community" for "any group formed of people who stem from disparate societies, who regularly interact through interpersonal contacts and communication networks, and who share mutual interests and a common ethos" (1963, p. 482). This usage, Useem notes, "detaches the sociological meaning of the community from its conventional geographical reference points." If this is accepted as a working definition of community as opposed, for example, to one such as Martindale's, "a set or system of groups sufficient to solve all of the basic problems of ordinary ways of life" (1964, p. 69), it becomes appropriate to consider the applicability of community to suprasocietal activities including, and especially, those of scientists.

The elements of community Useem includes in his definition of international community are regular interaction, interpersonal contacts, communication networks, shared mutual interests, and a common ethos. The problem is to conceptualize, describe, and analyze phenomena commensurate with their degree of complexity and structural durability. For example, the extent to which interpersonal contacts are an important part of scientific communication must be considered in evaluating the utility of "community" for describing and interpreting scientific activities. While research in scientific communication is probably the fastest growing specialty within the sociology of science, the relative incidence and functions of face-to-face interaction among scientists are still obscure. Meetings seem to be important for establishing new contacts and entering informal networks (Lin, Garvey, and Nelson 1970, pp. 31, 35). The significance of meetings, however, seems to be different for physical as opposed to social scientists. The social scientists, according to Garvey, Lin, and Nelson, have ineffectual premeeting networks, and the national meetings are important for establishing relationships with people working in given areas of specialization (1970, p. 76).

In a study of sources from which scientists acquired prepublication information for articles subsequently published, Lin, Garvey, and Nelson found that face-to-face sources were cited most often, preprint or prepublication drafts ranked second, and correspondence with authors ranked third (1970, pp. 50, 59). Interpersonal interaction may also be a function of the complexity of the message being communicated. This highly speculative idea has been formulated by Wolek as follows: "the probability that a communication will involve interpersonal interaction between source and receiver varies directly with the complexity of the message communicated" (1970, p. 233). It is reasonable to assume that, in general, the incidence and function of face-to-face interaction varies among scientists.

One final bit of evidence that illustrates the need for a more complex conceptualization of scientific activities is provided by the literature on growth patterns in science. Early periods of growth within a scientific specialty, for example, are associated with low levels of communication. As specialties grow, communication links increase, and highly coherent groups of active researchers form. As activity in a specialty declines, the groups within the specialty change in a variety of ways, some becoming loose networks, others forming into schools (Griffith and Miller 1970, p. 137; Crane 1969).

It might be useful to evaluate the vocabulary of social networks and social sets for describing and interpreting world-wide scientific activities. It should be possible to conceptualize these activities as a relational system constituted of categories, groups, quasi-groups, sets, action-sets, networks, formal organizations, core groups, cliques, factions, seminars, and caucuses. Such a system is conceived to be stratified, along various dimensions, at individual, collective, group, organizational, national, regional, and international levels. The advantage of a social network vocabulary is that it is ideally self-contained—the necessary concepts are logically related to one another (e.g., Srinivas and Beteille 1966; Mayer 1966). This seems more promising than a vocabulary of "invisible colleges" and "social circles" (Crane 1969).

The term "international" as used by Useem (1963, p. 482) refers to "disparate societies," not to "all societies," and is thus in line with standard usage in social science (e.g., Kelman 1965). When used with reference to the scientific community, however, "international" tends to connote "world-wide," or "universal." Calder, for a recent example, refers to "the international spirit of scientific research and the world-wide network of interchange and cooperation between research workers, regardless of nationality" (1970, p. 50). Conceptions of the international scientific community are rooted in statements such as accompanied the opening of the Royal Society in 1662 in which invitation was extended to "all inquisitive strangers of all countries" (Rose and Rose 1970, p. 179). The cognitive maps, and experiences of scientists, judging from our research and from studies of scientific communication are, by contrast, strictly delimited. What is needed to replace the universalistic "international" is a concept or set of concepts which can (1) differentiate bi-national, multi-national, and global situations, and (2) convey variations in centrality and peripherality among nations in terms of their control over resources, including scientific facilities and personnel.

This is an important task because the so-called "international scientific community" has been characterized as the most important suprasocietal system. It has been referred to as the basis for, or microcosm of, an emerging world community, a critical system for unifying national programs for controlled manipulation of the environment, and a crucial part of developmental processes. These ideas involve implicit and explicit conceptions of the relationships between science and society. Such conceptions can be roughly classified according to their basis in (1) intuitive and metaphysical assumptions and (2) empirically-grounded ideas and theories.

The metaphysics of science and society has its modern roots in Francis Bacon's interpretation of the prophecy of Daniel—"many shall run to and fro, and knowledge shall be increased." Bacon saw in this statement the fated destiny, by Divine Providence, of scientific advance and societal progress meeting "in the same age" (1963, p. 236). Bacon's intellectual heirs today proclaim science as an essential factor in world integration, a "savior of mankind," and the primary basis for a "true world culture" (Wagar 1963, pp. 154, 174). Recently, the Nobel laureate Sir Peter Medawar (in Greenberg, 1969, p. 1239) expressed the belief, sanctified by reference to Bacon, that "the deterioration of the environment produced by technology is a technological problem for which technology has found, is finding, and will continue to find solutions." Such uncritical expressions of faith ignore the relationship between values and other non-technological factors, and scientific and technological developments, and manifest an optimism that draws its sustenance from the ill-fated "idea of progress."

In one sense, Bacon also anticipated more critical conceptions of the relationship between science and society. Merton, for example, reiterates Bacon's intimation—but without any reference to inevitability or fate—when he notes that "the interplay between socio-economic and scientific development is scarcely problematical" (1957, p. 607). In this context, scientists are defined as significant actors in various aspects of national and world development. In economic terms they are defined as "strategic human capital" (Harbison and Myers 1964); in political terms, they are "elites" (Apter 1965); and in social psychological terms, the value-orientations of scientists are conceived to be a basis for global cooperation, and the development of a world community (Apter 1965, p. 436).

The growing awareness among scholars and laymen of the ecological unity of the earth is the source of changing conceptions about the functions of international scientific cooperation. Roger Revelle, for example, writes:

> It is by no means clear how scientific cooperation on a worldwide basis can best be used to attack [the] appalling questions of our time. But it is obvious that their solutions will be found only if science and technology are brought to bear in the broadest possible way and with urgent intensity (1963, p. 138).

The critical relationship between science, technology, and societal problems is noted elsewhere, and in more quantitative terms, in such works as Brown's *The Challenge of Man's Future* (1954) and Meier's *Science and Economic Development* (1966). It is manifest in the proliferation of international conferences on science and the new nations (e.g., UNESCO 1963, 1970; Gruber 1961; Shah 1967; Shils 1967).

That science and scientists are strategic components of large-scale social change is hardly problematic. There are, however, serious gaps in our knowledge concerning science and scientists in social change. The concept of a "third-culture of science" is designed to direct attention to and provide a framework for studying science and society in a global, developmental or social change perspective.

THE THIRD-CULTURE OF SCIENCE

The physical mobility of persons engaged in scientific activities has been a noteworthy aspect of the history of science (Dedijer 1968; Albright 1957, p. 339; Haarhof 1920, p. 241; Needham 1969, p. 243; Moskerji 1947, p. 563; Altekar 1948, p. 123, 125; Galt 1951, p. 328). From the time of the emergence of modern science in Western Europe in the 1500s until the present, an increase in the scale of "international science" and in the degree of institutionalization of scientific activities has occurred. During the twentieth century, scientists have participated in a social system developed over a period of hundreds of years. The growth of this system has made it possible for scientists to participate disproportionately in the increased movement of persons across cultures (Shils 1960; Thomas 1967 and 1968).

Scientists trained during the post-World War II decades have experienced cross-cultural mobility on a larger scale than prior generations of scientists. Geographically, they have been visitors and advisors abroad, and hosts to foreign scientists. Psychologically, they have been senders and receivers in a growing and increasingly global system of scientific communication. The increase in scale of scientific activities during the last quarter century is manifested in: (1) exchange of scholar programs, (2) institution-building and other forms of technical assistance programs in the new states, especially for higher education, (3) "brain drains," (4) the frequency of international conferences, (5) the number of individuals participating in such conferences, (6) the emergence of international scientific organizations, (7) laboratories supporting the work of scientists from different nations, and (8) the organization of international scientific "cooperatives"; on the regional level, for example, these include (a) the European Atomic Energy Community (EURA-TOM), which maintains joint research centers at Ispra (Italy), Geel (Belgium), Petten (Holland), and Karlsruhe (Germany), (b) the European Organization for Nuclear Research (CERN) in Geneva, (c) the European Space Research Organization (ESRO), which maintains several centers, the largest (the European Space Technology Center—ESTEC) located in Noordwijk, Holland, and (d) the International Rice Research Institute (IRRI) in the Philippines; on the global level, activities such as the International Geophysical Year are illustrative. These aspects of scientific activity are among the basic empirical referents for the idea of an international scientific community.

In contrast to that idea, the third-culture of science *emphasizes* science and the scientific role as links among nation-states and cultures; it underlines the relation between scientific activities and large-scale processes of social change, e.g., economic development, political modernization, and world development. Following Useem, our working definition of third-culture of science is "the cultural (including intrascientific) patterns created, shared, and learned by scientists of different societies who are in process of relating their societies or sections thereof to each other" (Useem 1963). The term has several drawbacks. It entered the literature in an entirely different context through S. A. Lakoff's critical evaluation of C. P. Snow's concept of "two-cultures" (1966). The term also reflects its roots in bi-national studies, though Useem has broadened the empirical referent for "third-culture" to include international, or multi-national linkages. The problem of definition will not be resolved here.

"Culture" is not a standardized research concept, but it is a useful "shorthand" idea. We will use "third-culture of science" to refer to those cultural patterns created out of or emergent from the interpersonal contacts of scientists from plural societies. This concept is the theme of our concluding chapter.

Introduction to the Readings

The preceding discussion of problems in the sociology of science high-lighted some of the major reasons for undertaking comparative, histori-cal, macrosociological studies of science. We are experiencing a *mal du siècle* in the counter-culture revolt against science and technology (e.g., Roszak 1969) and in the Ellul-ian nightmares of a "dictatorship of the test tubes" (Ellul 1964). But such disenchantment is founded on a very narrow conception of science, one rooted in professional roles, organiza-tions, and Nobel prizes. Science is not professional scientific activity, nor "physical science," nor "technology." It can be considered, to bor-row Jacob Bronowski's phrase, "the method of all human inquiry." In this sense, one condition for world development and world community is the proper cultivation of science as an expression of human creative and critical intelligence (cf. Maslow 1966; Siu 1957). Science, whatever else it may be, is on this view the process by which man raises the upper level of his capacity to mobilize energy and information in the adaptive process, i.e., makes progress (cf. Lenski 1970, p. 55; Mayr 1963). The idea of a third-culture of science is critical because it emphasizes the link between inquiry and progress. It can be considered the system (or sys-tems) in which "the method of human inquiry" develops and diffuses.

Progress, as just defined, depends on the ability of the human popula-tion to adapt within the global ecological system; the raising of the upper level of the capacity of human populations to mobilize energy and information is now necessarily dependent on the creation of some form of global social organization. Progress, in this sense, is dependent on diversity. We know, for example, that genetic diversity is directly related to the capacity in a population for successful adaptation to rapidly changing conditions (Lenski 1970, p. 55; Mayr 1963). This idea has been formulated by Sahlins and Service as "The Law of Evolution-ary Potential": "The more specialized and adapted a form in a given evolutionary stage, the smaller is its potential for passing to the next stage" (1960, p. 97). In other words, the potential for progress varies inversely with the degree of specialization and adaptation (cf. Dahlsten 1971, p. 202; and Durkheim, 1938, p. 140).

The significance of attending to problems of professionalization, bu-reaucratization, and homogeneity-heterogeneity in science lies in the fact that some of the social forces affecting scientific activities can be interpreted as contributing to a "loss of social diversity" (cf. Molnar 1961, p. 259). It is impossible to ignore the fact that as scientists respond to problems of control and coordination in an interdependent world by simplifying, specializing, and standardizing, the evolutionary potential of human culture may be decreased.

In order to understand changes in contemporary science, and to consider potential changes, idealized conceptions of science must be replaced by a conception of science as a human activity, directed by human values, and subject to the functional and dysfunctional consequences of human social organization. This is a precondition for fully and humanely incorporating science into world-views of social change, and for maximizing our chances for developing rational institutions organized to stimulate creative, critical, and active human living. Hopefully, a critical reading of the selections in this reader will stimulate and reinforce attempts to understand and explain science as a social activity and social process, and concomitantly, to integrate scientific activity into the search for world community.

REFERENCES

Albright, W. F. 1957. *From the Stone Age to Christianity.* New York: Doubleday-Anchor.

Allen, J., ed. 1970. *Scientists, Students, and Society.* Cambridge, Mass.: M.I.T. Press.

Altekar, A. S. 1948. *Education in Ancient India.* Benares: Kishore.

Andreski, Stanislav. 1964. *The Uses of Comparative Sociology.* Berkeley and Los Angeles: The University of California Press.

Apter, D. E. 1965. *The Politics of Modernization.* Chicago: University of Chicago Press.

Bacon, Francis. 1963. Novum Organum (1620). In *The Complete Essays of Francis Bacon,* pp. 179-264. New York: Washington Square.

Barber, Bernard. 1952. *Science and the Social Order.* Rev. ed. New York: The Free Press.

Barber, Bernard, and Walter Hirsch, eds. 1962. *The Sociology of Science.* New York: The Free Press.

Basalla, George. 1967. The Spread of Western Science. *Science* 156 (May 5): 611-22.

Becker, H. S., and James Carper. 1956. The Elements of Identification with an Occupation. *American Sociological Review* 21: 341-48.

Becker, H. S., and Anselm Strauss. 1956. Careers, Personality and Adult Socialization. *American Journal of Sociology* 66 (November): 253-63.

Ben-David, Joseph. 1970. Introduction, (special issue on the sociology of science). *International Social Science Journal* 22 (1): 7-27.

Berger, Peter. 1964. Some General Observations on the Problem of Work. In *The Human Shape of Work,* ed. Peter Berger. New York: Macmillan.

Blau, P. M., and W. R. Scott. 1962. *Formal Organization: A Comparative Approach.* San Francisco: Chandler.

Boulding, Kenneth. 1970. *A Primer on Social Dynamics.* New York: The Free Press.

Brewer, T. H. 1971. Disease and Social Class. In *The Social Responsibility of the Scientist,* ed. Martin Brown, pp. 143-62. New York: The Free Press.

Brown, Martin, ed. 1971. *The Social Responsibility of the Scientist.* New York: The Free Press.

Bucher, Rue, and Anselm Strauss. 1961. Professions in Process. *American Journal of Sociology* 66 (January): 325-34.

Calder, Nigel. 1970. *Technopolis: Social Control of the Uses of Science.* New York: Simon and Schuster.

Caplow, T. 1954. *The Sociology of Work.* New York: McGraw-Hill.

Cole, J. R. and S. Cole. 1972. The Ortega Hypothesis. *Science* 178 (October): 368-75.

Commoner, Barry. 1971. The Ecological Crisis. In *The Social Responsibility of the Scientist,* ed. Martin Brown, pp. 174-82. New York: The Free Press.

Coser, Lewis. 1965. *Men of Ideas.* New York: The Free Press.

Cotgrove, Stephen, and Steven Box. 1970. *Science, Industry, and Society.* New York: Barnes and Noble, Inc.

Cottrell, L. S., and E. B. Sheldon. 1966. Social Sciences. In *Professionalization,* ed. Howard M. Vollmer and Donald L. Mills, pp. 232-36. Englewood Cliffs, N.J.: Prentice-Hall.

Crane, Diana. 1969. Social Structure in a Group of Scientists: A Test of the "Invisible College" Hypothesis. *American Sociological Review* 34 (June, 1969): 335-52. Reprinted in *The Sociology of Sociology,* ed. L. T. and J. M. Reynolds, pp. 295-323. New York: David McKay, 1970.

Dahlsten, Donald. 1971. Pesticides. In *The Social Responsibility of the Scientist,* ed. Martin Brown, pp. 201-18. New York: The Free Press.

Davy, John. 1967. Sense in Young Britain's Swing from Science. *Observer,* October 1.

DeBeer, Gavin. 1960. *The Sciences Were Never at War.* London: Thomas Nelson and Sons, Ltd.

Dedijer, S. 1968. "Early" Migration. In *The Brain Drain,* ed. Walter Adams, pp. 9-28. New York: Macmillan.

de Grazia, Alfred, ed. 1963. The Politics of Science and Dr. Velikovsky. *The American Behavioral Scientist* 7 (September).

DeReuck, Anthony, and Julie Knight, eds. 1967. *Communication in Science: Documentation and Automation.* Boston: Little, Brown.

Durkheim, Emile. 1938. *The Rules of Sociological Method,* 8th ed. Trans. S. A. Solovay and J. H. Mueller. Chicago: University of Chicago Press. Paperback ed., New York: The Free Press, 1964. Originally published in French in 1895.

Ellul, Jacques. 1964. *The Technological Society.* Trans. John Wilkinson. New York: A. A. Knopf.

Etzioni, Amitai. 1970. Toward a Macrosociology. In *Macrosociology: Research and Theory,* ed. James S. Coleman, Amitai Etzioni, and John Porter, pp. 107-43. Boston: Allyn and Bacon, Inc.

Form, William H. 1946. Toward an Occupational Social Psychology. *Journal of Social Psychology* 24 (August): 85-99.

Form, William H., and Sigmund Nosow. 1962. Ideologies of Occupational Groups. In *Man, Work, and Society,* ed. William H. Form and Sigmund Nosow, pp. 403-7. New York: Basic Books.

Francis, R. G., and R. C. Stone. 1956. *Service and Procedure in Bureaucracy: A Case Study.* Minneapolis: University of Minnesota Press.

Frank, Philip. 1958. Contemporary Science and the World View. *Daedalus* 87: 57-66.

Freidrichs, R. W. 1970. *A Sociology of Sociology.* New York: The Free Press.

Friedson, Eliot. 1970. *Profession of Medicine.* New York: Dodd, Mead and Company.

Galt, H. S. 1951. *A History of Chinese Educational Institutions.* Vol. I. London: Probsthain.

Garvey, W. D., Nan Lin, and C. E. Nelson. 1970. Some comparisons of Communication Activities in the Physical and Social Sciences. In *Communication Among Scientists and Engineers,* ed. C. E. Nelson and D. K. Pollock, pp. 68-84. Lexington, Mass.: D. C. Heath.

Gibbs, J. P. 1968. Norms. In *International Encyclopedia of the Social Sciences,* ed. D. L. Sills, pp. 208-13. New York: Macmillan and The Free Press.

Gillispie, C. C. 1959. Science in the French Revolution. *Behavioral Science* 4 (January): 67-73.

Glaser, W. A. 1960. Doctors and Politics. *American Journal of Sociology* 66 (November): 230-45.

Gottlieb, David. 1961. Processes of Socialization in American Graduate Schools. *Social Forces* 40 (December): 124-31.

Gouldner, A. W. 1970. *The Coming Crisis of Western Sociology*. New York: Basic Books.

Greenberg, D. S. 1969. British AAS: Counterattack on Gloom About Science and Man. *Science* 165 (September 19): 1239.

Greenwood, E. 1957. Attributes of a Profession. *Social Work* 2 (July): 44-55. Reprinted in H. M. Vollmer and D. L. Mills, eds., *Professionalization*. Englewood Cliffs, New Jersey: Prentice-Hall.

Griffith, B., and A. J. Miller. 1970. Networks of Informal Communication Among Scientifically Productive Scientists. In *Communication Among Scientists and Engineers*, ed. Carnot E. Nelson and Donald K. Pollock, pp. 125-40. Lexington, Mass.: D. C. Heath.

Gullahorn, John T., and Jeanne E. Gullahorn. 1962. Visiting Fulbright Professors as Agents of Cross-Cultural Communication. *Sociology and Social Research* 46 (April): 282-93.

Haarhof, T. 1920. *Schools of Gaul*. London: Oxford.

Haberer, J. 1969. *Politics and the Community of Science*. New York: Van Nostrand Reinhold.

Hagstrom, Warren. 1964. Anomy in Scientific Communities. *Social Problems* 12 (Fall): 186-195.

_____. 1965. *The Scientific Community*. New York: Basic Books.

Harbison, F. and Charles A. Myers. 1964. *Education, Manpower and Economic Growth*. New York: McGraw-Hill.

Hilgard, E. R., and Daniel Lerner. 1951. The Person: Subject and Object of Science and Policy. In *The Policy Sciences*, ed. Daniel Lerner and H. D. Lasswell, pp. 38-43. Stanford: Stanford University Press.

Hirsch, Walter. 1961. The Autonomy of Science in Totalitarian Societies. *Social Forces* 40 (October): 15-22.

Holt, R. T., and John E. Turner. 1970. *The Methodology of Comparative Research*. New York: The Free Press.

Homans, G. 1950. *The Human Group*. New York: Harcourt, Brace and World.

Horowitz, I. L. 1970. Mainliners and Marginals: The Human Shape of Sociological Theory. In *The Sociology of Sociology*, ed. L. T. Reynolds and J. M. Reynolds, pp. 340-70. New York: David McKay Co.

Jencks, Christopher, and David Reisman. 1967. *The Academic Revolution*. New York: Doubleday-Anchor.

Kaplan, Norman. 1964. Sociology of Science. In *The Handbook of Modern Sociology*, ed. R. E. L. Faris, pp. 852-81. Chicago: Rand McNally.

_____, ed. 1965. *Science and Society*. Chicago: Rand McNally.

Kelman, H. C. 1965. *International Behavior*. New York: Holt, Rinehart and Winston.

Kornhauser, W. 1962. *Scientists in Industry*. Berkeley: University of California Press.

Kuhn, T. S. 1962. *The Structure of Scientific Revolutions*. Chicago: University of Chicago Press.

_____. 1963. The Essential Tension: Tradition and Innovation in Scientific Research. In *Scientific Creativity: Its Recognition and Development*, ed. C. W. Taylor and F. Barron, pp. 341-54. New York: Wiley.

_____. 1970. *The Structure of Scientific Revolutions*. 2nd ed. Chicago: University of Chicago Press.

Kwok, D. W. Y. 1965. *Scientism in Chinese Thought, 1900-1950*. New Haven: Yale University Press.

Lakoff, S. A. 1966. The Third-Culture of Science: Science in Social Thought. In *Knowledge and Power*, ed. Sanford A. Lakoff, pp. 1-61. New York: The Free Press.

Lazarsfeld, P. F., and W. Thielens, Jr. 1958. *The Academic Mind.* New York: The Free Press.

Lenski, Gerhard. 1970. *Human Societies.* New York: McGraw-Hill.

Lin, Nan, W. D. Garvey, and C. E. Nelson. 1970. A Study of the Communication Structure of Science. In *Communication Among Scientists and Engineers,* ed. C. E. Nelson and D. K. Pollock, pp. 23-60. Lexington, Mass.: D. C. Heath.

Marcuse, Herbert, 1969. The Responsibility of Science. In *The Responsibility of Power,* ed. Leonard Krieger and Fritz Sterm, pp. 475-81. New York: Doubleday-Anchor, 1969.

Marsh, R. M. 1967. *Comparative Sociology.* New York: Harcourt, Brace and World, Inc.

Martindale, Don. 1964. The Formation and Destruction of Communities. In *Explorations in Social Change,* ed. George K. Zollschan and Walter Hirsch, pp. 61-87. Boston: Houghton Mifflin.

Marx, Karl. 1956. *Selected Writings in Sociology and Social Philosophy,* ed. T. B. Bottomore and M. Rubel. New York: McGraw-Hill.

Maslow, Abraham. 1966. *The Psychology of Science.* Chicago: Henry Regnery Company.

Mayer, Adrian C. 1966. The Significance of Quasi-Groups in the Study of Complex Societies. In *The Social Anthropology of Complex Societies,* ed. Michael Banton, pp. 97-122. London: Tavistock.

Mayr, E. 1963. *Animal Species and Evolution.* Cambridge, Mass.: Harvard University Press.

Meier, R. L. 1951. The Origins of the Scientific Species. *Bulletin of the Atomic Scientists* 7 (June): 169-73.

_____. 1966. *Science and Economic Development.* 2nd ed. Cambridge, Mass.: Massachusetts Institute of Technology Press.

Menzel, Herbert. 1958. The Flow of Information Among Scientists: Problems, Opportunities, and Research Questions. New York: Columbia University Bureau of Applied Social Research (mimeographed).

Merton, R. K. 1957. Science and Economy of 17th Century England. In *Social Theory and Social Structure.* Rev. and enlarged. ed., pp. 607-27. New York: The Free Press.

Mills, T. J. 1966. Scientific Personnel and the Professions. *The Annals* 367 (September).

Molnar, Thomas. 1961. *The Decline of the Intellectual.* Cleveland: World Publishing Co.

Moskerji, R. K. 1947. *Ancient Indian Education.* London: Macmillan.

Murdock, G. P. 1957. Anthropology as Comparative Science. *Behavioral Science* 2 (October): 249-54.

Needham, Joseph. 1954. *Science and Civilization in China* (successive volumes). Cambridge, England: Cambridge University Press.

_____. 1969. *The Grand Titration.* London: Allen and Unwin, 1969.

Nelson, C. E., and D. K. Pollock, eds. 1970. *Communication Among Scientists and Engineers.* Lexington, Mass.: D. C. Heath.

Oppenheimer, R. 1958. The Growth of Science and the Structure of Culture, Comments on Dr. Frank's Paper. *Daedalus* 87 (Winter).

Parson, T. 1951. *The Social System.* New York: The Free Press.

_____. 1954. *Essays in Sociological Theory.* Rev. ed. New York: The Free Press.

Polanyi, Michael. 1951. *The Logic of Liberty.* Chicago: University of Chicago Press.

Prandy, Kenneth. 1965. *Professional Employees—A Study of Scientists and Engineers.* London: Faber and Faber, Ltd.

Price, Don K. 1965. *The Scientific Estate.* Cambridge, Mass.: Harvard University Press.

Radcliffe-Brown, A. R. 1952. The Comparative Method in Social Anthropology. *The Journal of the Royal Anthropological Institute of Great Britain and Ireland* 81 (Parts I and II): 15-22.

Restivo, S. P. 1970. Sociology and History: Notes on Rapproachement. *The Kansas Journal of Sociology* 6 (Fall and Winter): 134-42.

_____. 1971. Visiting Foreign Scientists at American Universities: A Study in the Third Culture of Science. Unpublished Ph.D. thesis. East Lansing, Michigan: Michigan State University.

Revelle, Roger. 1963. International Cooperation and the Two Faces of Science. In *Cultural Affairs and Foreign Relations,* ed. Robert Blum, pp. 122-38. Englewood Cliffs, N.J.: Prentice-Hall Inc.

Roe, Ann. 1952. *The Making of a Scientist.* New York: Dodd, Mead and Company.

Rose, H., and S. Rose. 1970. *Science and Society.* Baltimore: Penguin Books.

Rosenberg, Morris. 1957. *Occupations and Values.* New York: The Free Press.

Roszak, T. 1969. *The Making of a Counter-Culture.* New York: Doubleday-Anchor.

Sahlins, M. D., and E. R. Service. 1960. *Evolution and Culture.* Ann Arbor: University of Michigan Press.

Schwartz, Charles. 1971. Professional Organization. In *The Social Responsibility of the Scientist,* ed. Martin Brown, pp. 19-34. New York: The Free Press.

Scott, W. R. 1966. Professionals in Bureaucracies—Areas of Conflict. In *Professionalization,* ed. H. M. Vollmer and D. L. Mills, pp. 265-75. Englewood Cliffs, N.J.: Prentice-Hall.

Shah, A. B. ed. 1967. *Education, Scientific Policy and Developing Societies.* Bombay: Masskalas.

Shapere, Dudley. 1971. The Paradigm Concept (Book Review). *Science* 172 (May 14): 706-709.

Shils, Edward. 1954. Scientific Community: Thoughts after Hamburg. *Bulletin of the Atomic Scientists* 10 (May): 151-55.

_____. 1956. *Torment of Secrecy.* New York: The Free Press.

_____. 1958. The Intellectuals and the Powers: Some Perspectives for Comparative Analysis. *Comparative Studies in Society and History* I (October): 15-22.

_____. 1960. The Traditions of Intellectual Life: Their Conditions of Existence and Growth in Contemporary Societies. *International Journal of Comparative Sociology* I: 177-194.

_____. 1967. On the Improvement of Indian Higher Education. In *Education, Scientific Policy and Developing Societies,* ed. A. B. Shah, pp. 475-99. Bombay: Massakalas.

Siu, R. G. H. 1957. *The Tao of Science.* Cambridge, Mass.: M.I.T. Press.

Spencer, Metta. 1966. Professional, Scientific, and Intellectual Students in India. *Comparative Education Review* 10 (June): 296-305.

Srinivas, M. N., and Andre Beteille. 1964. Networks in Indian Social Structure. *Man* (November–December): 165-68.

Storer, Norman. 1966. *The Social System of Science.* New York: Holt, Rinehart and Winston.

_____. 1967. The Hard Sciences and the Soft: Some Sociological Observations. *Bulletin of the American Library Association* 55 (January): 75-84.

Thomas, Brinley. 1967. The International Circulation of Human Capital. *Minerva* 5 (Summer): 479-506.

_____. 1968. "Modern" Migration. In *The Brain Drain,* ed. Walter Adams, pp. 29-49. New York: Macmillan.

UNESCO. 1963. *Science and Technology for Development, Report on the United Nations Conference on the Application of Science and Technology for the Benefit of the Less Developed Areas.* In several volumes. New York: UNESCO.

_____. 1970. *Science and Technology in Asian Development.* Paris: UNESCO.

Useem, John. 1963. The Community of Man: A Study in the Third-Culture. *Centennial Review* 7 (Fall): 481-98.

Vucinich, A. 1963, 1970. *Science in Russian Culture.* 2 vols. Stanford: Stanford University Press.

Wager, W. W. 1963. *The City of Man.* Boston: Houghton Mifflin. Paperback ed., Baltimore, Md.: Penguin, 1967.

Weiss, Paul. 1967. Science in the University. In *The Contemporary University: U.S.A.*, ed. Robert S. Morison, pp. 152-85. Boston: Beacon.

West, S. S. 1960. The Ideology of Academic Scientists. *IRE Transactions of Engineering Management.* EM-7: 54-62.

Wilensky, H. L. 1960. Work, Careers and Social Integration. *International Social Science Journal* 12 (Fall): 543-60.

Wolek, F. W. 1970. The Complexity of Messages in Science and Engineering: An Influence on Patterns of Communication. In *Communication Among Scientists and Engineers,* ed. C. E. Nelson and D. K. Pollock, pp. 233-65. Lexington, Mass.: D. C. Heath.

PART I

Science and Social Structure

The selections in this part deal with the relationships between scientific activities and sociocultural milieux. Cardwell and Ben-David discuss scientific growth in the modern states of Western Europe, Britain, and the United States. In noting a correlation between quantity and quality in higher education in chemistry, and the development of advanced industrial techniques, Cardwell shows England to have been a poor second to Germany in the development of systematic applied science. He offers evidence that Bacon's goal of inventing the method of invention was achieved first in Germany's coal tar color industry. English scientific education was not significantly influenced by this industry in which science was most readily applicable. This selection provides some background for Ben-David's study of the shift of the scientific center away from Germany. Of special note are Cardwell's concluding paragraphs, in which he raises questions concerning the functions and dysfunctions of specialization. We need to know more about the relation between specialization in individuals and collectivities, and adaptability, and to answer these questions: At what point and under what conditions does specialization cease to foster adaptability?; and To what extent does our highly competitive society encourage specialization without at the same time encouraging interdependence of specialities, a process which could lower our collective potential for responding adaptively to rapid social and environmental changes?

Two aspects of how Joseph Ben-David approaches comparative analysis are worth noting. First, the Germany-U.S. comparison is strengthened methodologically, theoretically, and substantively by references to scientific activities in Britain and in Western Europe. Second, the comparisons detail a two-way flow of diffusion of innovations between the U.S. and Germany. Ben-David argues that the growth in eminence of the U.S. and Britain (relative to Germany) as centers of science, especially in the newer fields of study, was mainly due to the creation of "departments" in the universities. The establishment of the graduate schools is viewed as a "crucial step" in the process of importing the European university model into the United States. This was a primary factor in stimulating the organization of research in the universities. The growth of new disciplines as a consequence of the differentiation of higher education into the undergraduate college, graduate, and professional school is illustrated in a brief case study of statistics.

The general receptivity to innovations in America is attributed to (1) decentralization of the American academic system, and (2) the fact that no significant monopolies impinged on that system. Negative aspects of the American system are discussed, but cursorily, and in such a way as to reinforce a basically positive perspective on scientific activities in American universities. Here, as in his recent book (Ben-David 1971), Ben-David's analysis of sources of conflict and threats to the institutional autonomy of science and the creative autonomy of individual scientists is weak. The dysfunctions of institutionalization, professionalization, and bureaucratization rarely interfere with Ben-David's somewhat idealistic conception of contemporary science. Actual and potential alienative and exploitative consequences of social organization in science for working scientists are not discussed. Neither these nor any other criticisms can diminish the significance of Ben-David's contributions to the comparative history of science in a macro-institutional perspective.

There has been a pervasive Western bias in the study of science as a socio-cultural phenomenon. In the past quarter century we have begun to construct a more realistic picture of science in human culture, thanks in great part to the work of Joseph Needham. His multi-volume study of science in China is, though still incomplete, already a "classic." An essay by Needham on science in China and in the West follows a reprint of Alexander Vucinich's paper on mathematics in Russian culture. Though by no means as ambitious as Needham's work, Vucinich's study of science in Russian culture bridges an important gap in our knowledge and understanding of science and culture in comparative perspective. His paper alerts us to some of the social factors which promoted the rapid development of mathematics in Russia during the eighteenth, nineteenth, and early twentieth centuries. He argues that mathematics developed more rapidly than the other sciences because it was less ideologically and politically "sensitive." It was also less susceptible than other sciences to the influence of the metaphysical schools. Personality plays a key role in Vucinich's analysis. He notes, for example, that Peter I recognized the significance of mathematics for his modernization programs and promoted the introduction of Western mathematics into Russia. Leonard Euler is credited with establishing the reputation of the St. Petersburg Academy of Sciences among mathematicians and other scientists. The first part of Vucinich's paper is a compendium of "important persons": in addition to Peter I and Euler, Mikhail Lomonosov, T. F. Osipovskii, D. M. Perevoshchikov, M. V. Ostrogrodskii, P. Chebyshev; and Euler's students, Vucinich argues, were responsible for enhancing the role of mathematics in Russia. Thus, while the paper provides some basic information on science and Russian culture, it can only hint at the broader social structural conditions which were necessary for generating and institutionalizing the activities of individuals and groups. A more fully developed institutional analysis is undertaken in Vucinich's *Science in Russian Culture* (1963; 1970).

Needham's voluminous study of science and technology in Chinese civilization (1954–) began with the problem of why modern science emerged in Western Europe and not in China or elsewhere. An equally important question also occurred to him early: why was China "much *more* efficient than occidental (civilization) in applying human natural knowledge to practical human needs?" In the essay we reprint, Needham outlines his basic framework for approaching and answering these complex questions. He develops a rationale for a socio-economic, socio-historic analysis of variations in the history of scientific activity in European and Chinese civilizations. He argues that "Asian 'bureaucratic feudalism' at first favoured the growth of natural knowledge and its application to technology for human benefit, while later on it inhibited the rise of modern capitalism and of modern science". . . . His interpretation of Chinese history is provocative, even heretical. He argues that China was fundamentally a more rational society than Europe; he documents the scientific and technological preeminence of China over a period of fifteen hundred years; and he suggests that the diffusion of Chinese science and technology was a significant factor in the emergence of modern science in Western Europe. In part, this essay is a plea for a sociology of the scientific revolution to replace "pure chance" or "racial" explanations. Needham fails to devote sufficient energy to a sociological definition of modern science. In the following selection, Karp and Restivo attempt to formulate such a definition and to develop Needham's theses in an ecological perspective.

Karp and Restivo focus attention on the societal conditions necessary for the institutionalization of scientific activity, and argue that the progressive characteristics of modern science were dependent on the differentiation and autonomous development of science as a social activity. Using China as the critical comparative case, they sketch a theoretical orientation which emphasizes the societal bases for changes in the structure of scientific activities. They challenge reliance on "utilitarian" and "reciprocal institutional influences" hypotheses developed in the works of Weber, Merton, and Ben-David. Their work draws critically on the contributions of Ben-David and Needham. This is a working paper. The complexity of the questions Karp and Restivo attempt to answer requires a much more critical and detailed analysis of social structure in China and Western Europe than they provide or rely on in terms of secondary sources.

A hypothesis proposed by numerous students of science and society (including Vucinich and Ben-David in the papers reprinted in this section)—that the more "abstract" sciences (e.g., pure mathematics) are least likely to be subject to political interference—is examined in the paper by Walter Hirsch on scientific autonomy in totalitarian societies. Hirsch examines two general working hypotheses: (1) the *natural* sciences are more autonomous, i.e., less subject to political controls than the social sciences, and (2) the *applied* aspects of science, including *technology,* are more autonomous than the *pure* or *theoretical.* The hypotheses are discussed in the context of data on scientific activities in Nazi Germany and Soviet Russia. The strength of the article lies primarily in the questions it raises concerning the relationships between science and the polity, and the use of an historical context to test and illuminate ideas about how science as a process is affected by the nature of its links with the political system. Hirsch's paper emphasizes the importance of examining variations in the societal context of scientific activities, and normative and organizational variations among the sciences.

The final selection in part one is from Kenneth Prandy's *Professional Employees—A Study of Scientists and Engineers* (1965). In this study of British "technologists" (the term is used for both scientists and engineers), Prandy is concerned with how the various occupational groups in science have developed, and how they have become involved in political processes and "the system of power." Prandy asks whether scientists "wield power," or "even exercise very much influence" economically and politically. And how representative of the "scientific community" are those scientists who are politically active? Comparisons with other industrialized countries, including the United States and Germany, raise questions regarding the social effects of the technologists' work, and the requisites for their fully responsible participation in society. Prandy's emphasis is on manpower needs in relation to maintaining or improving material standards in society. He thus takes for granted the value an industrial society places on "growth" and a rise in standards of living. There is no serious attempt to consider conflicts and contradictions within industrial society, and potentials for radical changes in the structure of society and in the roles of scientists. The vision of what man may become is not an expansive one. But Prandy has made a major contribution to this area of inquiry in Britain and elsewhere. This selection provides a link to part two, which focuses on science and industry.

The omission of Robert K. Merton from this section requires a brief comment. Merton's 1938 monograph on science in seventeenth century England was a seminal contribution in the sociology of science, and a model for the type of sociological analysis of science and society represented by the selections in this book. The monograph is noteworthy for its emphasis on "institutional interchange" and on science as a developing, or evolving institution. Excerpts from the monograph have been widely reprinted, and the monograph itself is now available (1971); this was one reason for not including Merton in this volume. Neither have we felt any necessity for going beyond our general criticisms of Merton's contributions toward a fully developed critique of his sociology of science as a "school of thought"; such an approach has recently been taken by critics of Merton's "functional," "normative" sociology of science (e.g., Barnes and Dolby 1970; King 1971; Mulkay 1969). We have simply sought to draw attention to the importance of the institutional perspective on science and society established by Merton and to flaws in that perspective which tend to support an idealistic, unproblematic view of scientific activity.

REFERENCES

Barnes, S. B. and R. G. A. Dolby. 1970. The Scientific Ethos: A Deviant Viewpoint. *Archives Europeenes de Sociologie* 11: 3-25.

Ben-David, Joseph. 1971. *The Scientist's Role in Society.* Englewood Cliffs, New Jersey: Prentice-Hall Inc.

King, M. D. 1971. Reason, Tradition, and the Progressiveness of Science. *History and Theory* 10: 3-32.

Merton, Robert K. 1971. *Science, Technology, and Society in Seventeenth Century England.* New York: Harper and Row.

Mulkay, Michael. 1969. Some Aspects of Cultural Growth in the Natural Sciences. *Social Research* 36: 22-52.

Needham, Joseph. 1954. *Science and Civilization in China* (successive volumes). Cambridge University Press.

Prandy, Kenneth. 1965. *Professional Employees—A Study of Scientists and Engineers.* London: Faber and Faber Ltd.

Vucinich A. 1963, 1970. *Science in Russian Culture.* 2 vols. Stanford: Stanford University Press.

D. S. L. CARDWELL

The Development of Scientific Research in Modern Universities: A Comparative Study of Motives and Opportunities

Francis Bacon, the quatercentenary of whose birth fell this year [1961], was the first sociologist of science; not only by virtue of his descriptions of the social and psychological hindrances to the pursuit of science, the Idols of the Theatre, etc., but also by virtue of his clear analysis of the modes of innovation. The mariner's compass, firearms and the printing press are, he says, the inventions which established the supremacy of Europe. An acute observation, no doubt, but Bacon goes further and points out that whereas the introductions of the compass and of firearms depended on the prior discoveries of the properties of the lodestone and of gunpowder—on science, if you like—there was no such precondition for the invention of the printing press. There was, in his view, no reason in principle why the ancient Egyptians or the Greeks could not have invented it.

This distinction between invention which depends on prior science, or discovery, on the one hand, and "straightforward" invention[1] on the

D. S. L. Cardwell, "The Development of Scientific Research in Modern Universities." Chapter 22 of *Scientific Change* edited by Alistair Crombie, © 1963 by Heinemann Educational Books Ltd., Basic Books, Inc., Publishers, New York.

other, is one which runs through the history of technology from ancient times to the present day: it may be commended as a useful framework for prospective historians of technology. In the Baconian scheme the relationship between science and inventions based on science is one good reason for encouraging the advancement of science. Certainly Bacon's ideas helped to inspire the foundation of the Royal Society; but, as I understand him, he went even further in hoping that science, both pure and applied, would become a broadly based national and social institution. This did not begin to be the case until the end of the nineteenth century.

Bacon has often been criticized for his suggestion that quite ordinary people should be able to undertake scientific research. If we think of science as a succession of "big names"—Galileo, Descartes, Newton— it is easy to believe that Bacon was wrong. But he was not wrong; he was clearly right: every large modern research laboratory is a confirmatory instance. In fact, the great discovery of the nineteenth century that ordinary talents can be effectively harnessed for the process of discovery is a vindication of Bacon's judgment in this matter.

It was not unfitting that the nineteenth century opened in England with the foundation of the, apparently quite Baconian, Royal Institution. The intention was that it should further scientific investigations and seek to apply them to useful ends,[2] should disseminate scientific and other useful knowledge, and should have a library and museum of models. But it soon became clear that these intentions were not to be fully realized. Davy and Faraday established a tradition of research but no fertilizing school of science was set up,[3] the lectures were not given to industrial technicians but to the wealthy and fashionable, the museum was never started and the great problems of relating science to industry were not tackled. It would be difficult, therefore, to sustain any claim that the Royal Institution was the first of the modern teaching and research laboratories.

The Royal Institution reflected the temper and traditions of English science: those of the amateur or devotee. In much the same way the Royal Society was, at that time, an association both of men actively and of men passively interested in science. If there were relatively few really professional scientists in the Society, that was because there were few indeed in the country; and if some of the Fellows were influential non-scientists, that was not necessarily to be deplored.

Correlative with the amateurism of English science, the universities took it to be their duty to conserve and to transmit the established liberal education based on classical literature and Newtonian natural philosophy. Interpreted by a man like Whewell this philosophy of education could be persuasive, and criticism was, in fact, usually restricted either to matters of standards and syllabus or to the exclusiveness of English university education. The intending scientist would be advised, on this philosophy, to postpone his researches until his liberal education was completed, when the clarity of his thought and the soundness of his judgment would be fully developed. Accordingly, university reform in England during the first half of the nineteenth century was devoted to securing the inclusion of modern sciences in the syllabuses and ending unfair discriminations. The roles of scholarship and research in the university were hardly considered at all.

It was different in Germany. From the beginning of the century and more particularly from the foundation of Berlin University, free research was regarded as the ideal means of higher education. Wilhelm von Humboldt, a Kantian humanist, established the doctrine that the autonomy of reason was fundamental to the nature of the university. The university professor is an original scholar whose students help him in his work, acquiring their education and love of learning in the process. Beneath these academic ideals there were strong social currents: in times of defeat and political fragmentation the German universities were the acknowledged strongholds of aspirations for political and social unity: symbols of a common national heritage. This gave them an honoured place in the centre of German public life. The geography and history of England demanded no such role of the English universities. There were enough separate German states to ensure the existence of a number of competing, non-centralized universities. The ideal of free research was from the beginning impartial and non-utilitarian: it applied equally to classical philology and chemistry, to history, law, philosophy and physics, indeed to all branches of systematic knowledge. Engineering and technology were excluded from the university.

The German system was set up in very conscious opposition to the French. In 1808 Napoleon had unified all higher education in France under one central university. This in the event was unfortunate: it established the scientific domination of Paris over the provincial cities, prevented the development of autonomous "schools" of science and, thanks to the parsimony and indifference of the state, led to the neglect and underendowment of research and scholarship.[4] Much later, after 1870, Pasteur attributed many of France's political ills to her persistent neglect of science.[5]

France, in fact, has as good a claim as any country to have pioneered the systematic application of science to industry. The advance of French mathematics in the eighteenth century had been accompanied by advances in application of mathematical engineering—civil and military— to hydraulics and shipbuilding. These were really the main applied sciences of the eighteenth century. During the Revolution heroic efforts were made to develop new industries and to put the older ones on a scientific basis. In 1794 the Ecole Polytechnique was established; here, among other things, training in practical chemistry was given and students were allowed to carry out their own experiments and investigations. This practice was, in fact, copied from the much admired mining college at Schemnitz (Stiavnica, in Czechoslovakia) founded in 1760 during the reign of Maria Theresa—the first of its kind.

Great though the achievements of French and British science were during the nineteenth century, neither country, and this is especially true of Britain, proved able to turn out research students—the rank and file of science—in the numbers and of the quality that Germany achieved. T. H. Huxley put it aptly when he likened British science to an army consisting solely of officers.

In the new century systematic experimental science and later systematic applied science were pioneered in Germany. Early in the century Strohmeyer at Göttingen and Gmelin at Heidelberg had teaching laboratories, but it was really the opening, in 1825, of Liebig's famous laboratory at Giessen that marked the start of organized scientific research as well as the emancipation of chemistry from medicine. Not only was there a constant stream of highly trained scientists from this laboratory, and from that of Wöhler at Göttingen, but new sciences were to be developed: organic chemistry, biochemistry, agricultural chemistry (interest in organic chemistry was much stimulated by its practical possibilities in agriculture and medicine). Following the successes of these laboratories, others were instituted at Marburg (1840), Leipzig (1843 and 1868), and elsewhere in the 1850's. In the 1860's extremely well equipped laboratories were opened at the universities of Berlin and Bonn. The latter had accommodation for over sixty students.[6]

But the development of the German schools of chemistry may not have been so smoothly effected as a simple chronology of events might suggest. As late as 1840, Liebig himself made a sharp attack on the neglect of experimental chemistry by the six Prussian universities; they offered, he said, no place for the training of the teacher of experimental science, the way was blocked by "an overgrown humanism."[7]

The other sciences followed chemistry in setting up research and teaching laboratories from the late 1830's onwards. (In a sense, of course, physical and biological laboratories have always existed in the forms of observatories, botanical and zoological gardens, etc. And learned academies often carried out specified researches—Victor Regnault's investigations into the properties of steam are a classical example. But in the first case the institutions are very specialized and limited in number; in the second, as soon as the particular inquiry has been completed all research has stopped.) The development of biological laboratories has been closely tied to the medical faculty in which, at the beginning of the nineteenth century, the dominant scientific disciplines were anatomy and botany. But the influence of Liebig and other chemists together with the tremendous development of medicine—first in France and then in Germany[8]—led to the foundation of systematic laboratories for physiology and other sciences related to medicine. Prominent among these were those of Purkinje and Claude Bernard. The Pasteur Institute was opened in 1888.

It was not until the sixties and seventies that physics laboratories were established more or less simultaneously in most universities. It is not clear why physics should have lagged some forty years behind chemistry in this respect. Maxwell gives the rather negative reason: "it will take a good deal of effort to make Exp. physics bite into our university system which is so continuous and complete without it."[9] It might be that while the heart of physics—Newtonian mechanics—held little scope for experimental investigations, and while subjects like heat and electricity were regarded as branches of chemistry, the development of the physics laboratory would necessarily be retarded.[10]

II. A Particular Instance

It will be clear that the development of university laboratories represents a complex and far ranging subject. From this point, then, I must limit my discussion to one important topic only, and I must apologize for the serious omissions this will necessitate.

Only if the admission of students to a laboratory is recognized as a permanent practice can we talk of the laboratory as a systematic research institution. Judged by this standard, systematic research chemistry was taught for the first time in England when, in 1845, the Royal College of Chemistry was opened in London. Public recognition of the importance of Liebig's work for agriculture and medicine inspired the foundation of this institution, and the first principal, A. W. Hofmann, was in fact a nominee of Liebig's. For the first twenty-five years of its existence the College was small—numbers varied between thirty and fifty-two, indicating that the "need" was not so great as its founders may have hoped—but in that time some 140 original papers were published.

The study of organic chemistry was not being pursued all over Europe. In France, a chair in the subject was created in Paris in 1853, and men like Chevreul, Dumas and others worked in collateral subjects at about that time or earlier; but the main centres of research were in Germany. It was therefore almost an accident when, in 1856, W. H. Perkin discovered the first of the aniline dyes. Perkin, a student of Hofmann's at the Royal College, had been working on his own in the Easter vacation, attempting to synthesize quinine, when he made this discovery. Realizing its significance and encouraged by the report of a firm of dyers, Perkin, his father and brother launched into manufacture of the new dyestuff. Technologically the moment was propitious: Mansfield, a student of Hofmann's, had just discovered a process for separating benzene from coal tar, Zinin, a student of Liebig's, had reduced nitrobenzene to aniline and Béchamp greatly improved the latter process. But all this hardly detracts from the credit due to Perkin for what Sir Robert Robinson calls "his active, forceful pioneering" in the new manufacturing techniques of coal tar colours.[11] For seventeen years Perkin's firm prospered until, in 1874, he sold it and returned to chemical research.[12]

Thus a number of conspicuous threads from Giessen were woven into the pattern of an important scientific industry. And the new enterprise was not limited to England: factories were set up very quickly in Germany, Switzerland and especially in France where important contributions were made by Verguin, Girard, De Laire and others, and where there was a great tradition not only in scientific chemistry but also in the art of dyeing. In Germany one important firm which manufactured the new aniline dyes had formerly made natural dyestuffs;[13] curiously, in England, no old dye works seem to have taken up the manufacture of the new dyes. Firms were either founded specially to make the new dyestuffs, like Ivan Levinsteins, started in 1864, or else they were like Read Hollidays of Huddersfield, tar distillers started in 1830 who took up aniline dyes in 1860.

Levinsteins and Hollidays were, by the end of the century, the largest and most active manufacturers of aniline dyes in Britain. There were very good reasons why they should be: they were located near their markets, in the textile areas of Lancashire and Yorkshire, raw material —coal tar—was readily available from gas works, and in both areas there was a traditional interest in scientific industry.[14] To these assets we should add the commercial experience of these localities, the availability of capital and, in the case of Levinsteins, who were near Manchester, proximity to the largest and most active school of chemistry in England. It is therefore the more surprising that, by the end of the century, the aniline dyes industry had migrated to Germany, where about 90 per cent of the world's manufacture was carried out. In England the growth of the industry had been very slow indeed.

To revert to the question of systematic scientific education in England: the University Commissions of 1850 and 1854 paid little attention to research training in science, and the ideal continued to be a liberal education. The committee set up by London University in 1858 to consider the inauguration of science degrees similarly paid little attention to research: "science in education" meant knowledge about science rather than the art and practice of science. An emphasis reinforced by the nature of the written examination system.

The change began in the late sixties and early seventies, and it came through two distinct, although related, movements. Grove, in his Presidential address to the British Association in 1866, asked for Government endowment of research, and this plea was taken up by others in the following years. The aim seems to have been that men of science should have available public research laboratories in much the same way that scholars were provided for by the British Museum Reading Room. This suggests that in the early seventies the typical British scientist was still thought to be an amateur; but now, thanks to the growing complexity and national importance[15] of science, his work was to be officially helped. The second movement began with the criticisms of conventional liberal education and its instrument the written examination by Matthew Arnold and Mark Pattison. An added impetus was given to this movement by a young philosopher, C. E. Appleton, who had studied in Germany and who, by the early seventies, had won the support of a number of men in different disciplines. Thus, of the eight contributors to the volume *Endowment of Research* (1876), only two were scientists (and of these, one, H. C. Sorby, was a distinguished amateur). This movement was therefore inspired by the achievements of German learning and scholarship in general, rather than by the practical achievements of German science. In fact, in 1873 the Devonshire Commission reported that on no subject were academic witnesses so united as on the desirability of research training in scientific education. In America, this same belief led to the foundation of Johns Hopkins in 1876.

How did German and English universities compare in the matter of training in scientific research at this time? At Owens College, Manchester, H. E. Roscoe had built up what was certainly the biggest and very probably the best school of chemistry in this country. Roscoe, trained under Bunsen, believed that research should have a place in scientific education. He was a member of the Appleton group, and a paper he published in 1874[16] contained the observation that German chemical manufacturers were by then insisting that the chemists they employed should have had a research training. Thus the ideals of pedagogy and utility concur! In 1869 new chemical laboratories for Owens College were designed following the best German practice, and by 1871 they accommodated some sixty students. In 1874 the—for England—unprecedented step of appointing a second professor of chemistry was taken, when Carl Schorlemmer was made professor of organic chemistry. In the twenty years the College had existed, 1851-71, some 84 original papers in chemistry had been published.

But Owens College, in chemistry the best that England could achieve, was the only university institution in the great textile areas of the north. In 1865 A. W. Hofmann had returned to Germany, and for nine critical years there was no higher teaching of organic chemistry in England. So when Kekule's theory came out in 1865 there were few in this country competent to understand it, and very few of these indeed would be in industrial occupations. In comparison with the achievements of Owens College, in the six years 1866-71, some 80 papers were published from the chemistry department at the University of Leipzig (or 89 in the seven years 1866-72). Indeed, some six times as many papers on chemistry were published in Germany in 1866 as in Britain, and this highly unfavourable ratio was maintained until the end of the century: in 1899 some two-thirds of the world's original chemical research came from Germany.[17] As early as 1872 Kolbe had claimed: "Why is German chemical industry now at a higher level than that of England and France? Because in England and France government policy does little for chemical education and Germany has outstripped both countries in that respect. Indeed it is through the scientific laboratories that the chemical industry of Saxony has contributed so much to the wealth and prosperity of the state."[18] A point of view in interesting contrast to Liebig's observation of 1844 that England was not the land of science, for "only works which have a practical tendency command respect, while the purely scientific works, which possess far greater merits, are almost unknown . . . in Germany it is quite the contrary."[19]

Statistics can be produced to show not only that Germany endowed its universities much more generously than did England, but also that there were disproportionately more students in Germany than in England. But of course the social arrangements in the two countries were not the same, and great caution would be necessary in interpreting such statistics. Here I will merely observe that in 1890 there were twice as many German academic chemists as there were British (101:51), while by 1900 the proportion had dropped slightly in favour of Britain: 16 Germans to 10 British, or, taking account of populations, 12 Germans to 10 British—a rough indication which would suggest that the founding of new universities in Britain was tending to redress the balance. But, as the German technological universities have been excluded from this comparison, the case in favour of Britain has been overstated.

III. THE PRACTICAL CONSEQUENCES

There is, then, a general relationship between the numbers and quality of higher education in chemistry on the one hand and the development of advanced industrial techniques, such as systematic applied science, on the other.

The adoption of the simplest laboratory techniques by British industry began, it seems, only after the Exhibition of 1851. The first step appears to have been taken by a somewhat obscure ironmaster of Dudley, Samuel Blackwell, who was responsible for the introduction of analytical, or control, chemists into iron works. It is true that the simple control, or standards, laboratory may develop in the course of time into a research laboratory, as for example the National Physical Laboratory has done. But this process is by no means inevitable and we must ask in which industry and for what reason the systematic research laboratory originated.

Systematic research means the employment of salaried research scientists engaged solely or substantially in research, on a permanent basis and to be replaced by others when they leave. Scientific research in the form of an attack on one or two definite problems, to be wound up when the problems are solved, has been carried on in industry for a long time; there was, indeed, much of it during the eighteenth century. But this *ad hoc* procedure is not the same as institutionalized research which is a permanent feature, a continuing activity.

On this definition it seems certain that systematic industrial research began in the synthetic dyestuffs industry. The exact date is uncertain, but it must have been about 1870, and the location must have been one or more of the German firms which were, by 1900, to dominate the industry. At the end of the century one of these firms employed scores of graduate chemists on research, while many others worked on production, development, sales and services. There was no such scope for the employment of research scientists in the mechanical engineering industries, such as power generation, machine tools, textiles, etc. It is not evident, on the face of it, why this should have been so. The industries related to the biological sciences, such as agriculture, fishing, etc., were similar to the mechanical industries in this respect.[20]

When Perkin retired, in 1874, from the industry he had founded, it was prospering. He later gave as his reason for retirement his desire to return to chemical research. But, "a much more weighty consideration than this" was, said his son W. H. Perkin, the recognition that the firm could only be carried on successfully if a number of trained chemists could be recruited and employed in the all-important work of making new discoveries. "I remember," he says, "that enquiries were made at many of the British universities in the hope of discovering young men trained in the methods of organic chemistry, but in vain. There cannot be any doubt that the manufacturer of organic colouring matter during the critical years 1870-1880 was, owing to the neglect of organic chemistry by our universities, placed in a very difficult and practically impossible position."[21]

According to F. M. Perkin, another of Perkin's sons, the original firm was in the early 1870's faced with the problem of rapidly expanding production in order to meet a sharp increase in demand. This meant more research chemists: "research chemists however could not be obtained."[22] True, German chemists were available, but they tended, after a while, to return to Germany, where their English experiences naturally made them much sought after. In the new dyestuffs industry the pressure of rapid change was very great: it was a matter of new colours, of new processes and the improvement of old ones.

This rings only partly true: the inevitable problems of growth in size and complexity are certainly familiar to students of management; but, as regards the absence of suitable scientists, it may be wondered why Perkin & Son were apparently unable to train their own research workers.[23] Surely, the opportunity of training for research under the leading English organic chemist of the time, who was also the founder of a unique industry, should have brought in a number of able young men. In this respect a comparison between Perkin and James Watt is not inapt: both were brilliant scientists and technologists, both founded revolutionary new industries, both had brilliant sons. Nearly a century before, Watt, with his partner, Matthew Boulton, had faced the same problem—shortage of talent—in an even more acute form. But whereas Watt and Boulton succeeded in attracting able men from all quarters— Southern, Clegg, Peter Ewart, Murdoch and others, so that the firm was called the "science school of Soho"[24]—Perkin apparently sold his firm when the universities were unable to supply trained research chemists. Watt's son, James Watt, junior, later managed his father's firm with great success; Perkin's two elder sons were academics all their lives, and his youngest became a consultant. I conclude that Perkin and his sons lost, or never acquired, a taste for chemical industry, and so they sold the firm. I infer that the shortage of trained chemists was not so critical.

Brooke, Simpson and Spiller, who bought Perkin's firm, were in fact able to recruit two leading colour chemists: Raphael Meldola, who was with them from 1877 to 1885, and A. G. Green, who stayed from 1885 to 1894. But both men considered that their talents were being wasted and resigned disappointed.[25] Brooke, Simpson and Spiller eventually went bankrupt early in the twentieth century.

Levinstein's, the Manchester firm, were more successful and achieved a reputation for scientific innovation. It would have been remarkable if no firm in the area of Roscoe's large and active school of chemistry had been able to achieve a competitive position in this field! Ivan Levinstein himself was a life governor of Owens College, wherein he endowed a bursary and made donations to the physics and chemistry laboratories. This support for scientific education was manifested only in the late 'nineties, and it has been reported that during much of this period this firm lacked a research atmosphere.[26]

On the Yorkshire side of the Pennines the firm of Read Holliday also achieved modest success together with a somewhat better record in scientific innovation. This firm recruited such scientists as they could find—the majority appear to have been Germans—and by 1890 they had a research laboratory in which about four or five scientists were continuously employed upon research. They may well, therefore, have instituted the first systematic research laboratory in Britain, some years before those at Widnes (1892) and Ardeer (1896).[27]

"It is characteristic of the synthetic colour industry that it is subject to continuous change through the introduction of new products: this was true at the start and remains true today."[28] In an industry like this the advantage must be with the organization which has a good supply of trained research chemists to pioneer new products and chemical engineers to ensure their efficient production and sale. The balance of advantage then would lie with Germany. And the sound Baconian point was made, by Sir James Dewar in 1902, that: "It is in the abundance of men of ordinary plodding ability thoroughly trained and methodically directed, that Germany has at present so commanding an advantage."[29]

Many reasons can be found to account for the very slow development of the industry in England: patent laws, alcohol duties, trade discrimination, the tendency of investors to favour such sound securities as Consols, railways, coalmines, etc. But to produce reasons like this is merely to explain things away. In America and France, too, where patent laws, duties, etc., were presumably very different, the industry failed to develop. The only point clearly established is that Germany had more research scientists than England, America or France and certainly more chemical engineers than England. Only those firms which could recruit, without much difficulty, research chemists and chemical engineers could hope to keep up with what had become, by 1914, a veritable flood of new dyestuffs.

Systematic research was one of the necessary factors that gave Germany the lead in this industry; and this in turn was made possible by a supply of research scientists; the final product of an admirably developed education system.[30] But, and this is very important, the English manufacturers *apparently* made no attempt to overcome this handicap by training their own men. And even when men like Meldola and Green were recruited they were not given a proper chance to employ their talents. Towards the end of the century the numbers of graduate and research scientists in England increased rapidly. However, and here I must use parochial evidence, during the first twenty years of the University of Leeds Dyeing Department each one of the four largest German firms made more donations of dyestuffs than the two largest English firms combined. The English firms did not contribute to the funds of the University, nor did their members serve on advisory committees or similar bodies. It should be added that the Department of Dyeing had a high standard from the beginning; A. G. Perkin joined the staff in 1882 and later became Professor. In view of the unenterprising attitude of the industry, it is not surprising that at the end of the century Perkin found it difficult to get students to do research in this subject. Very briefly, the English dyestuffs firms do not seem to have appreciated the value of scientific education and research experience, from the time that W. H. Perkin, senior, left the industry up to the outbreak of the 1914 war.

A short quotation from a report made by Meldola in 1902 indicates the way in which the industry had been penetrated by German firms:

> The manufacturers of colours are themselves—especially the Germans—not only keeping the dyers supplied with a constant succession of new colouring matters, but it is the custom of the German firms to send round their own experts to teach the dyers how to use these new products or to issue such detailed instructions that practically nothing is required of the dyer in the way of scientific knowledge.[31]

The tentative conclusion is that while systematic research was a necessary ingredient of German success, short-sighted management sealed English failure. Generally speaking the difficulties encountered by inefficient organizations are the consequences of their inefficiency and not the cause of it. Of the two best English firms in the industry, Levinstein's took eighteen years to acknowledge the fact that there was a scientific department of dyeing at Leeds University; Read Holliday, being nearer at hand, took only twelve years; the four biggest German firms took at most two years.

My conclusions here must be very tentative, and it seems to me that there is considerable scope for detailed studies of the internal development and external relations of English firms during this time. Such studies should be comparative, taking full account of foreign experience as well as the development of the relevant sciences.

The importance of studies of this sort for the histories of both science and technology (and one cannot separate them) is that here, in the coal tar colour industry, for the first time systematic applied science made its impact: the "invention of the method of inventions," a Baconian ambition, was made when, in Germany, industry borrowed from the universities the distinctive institution of research laboratories and recruited from the same source the men to staff them. A developed educational system ensured a supply of research scientists and other men technically qualified to staff the industry. This first successful attempt at systematic applied science constituted an enormous object lesson, and no really progressive industry could thereafter afford to neglect applied science. In fact, one could not write the history of the application of science in the electrical industry without some reference to the triumphs of the dyestuffs industry. At some unknown date the majority of students under training in chemistry laboratories must have intended to seek an industrial career and not one in the teaching professions. In Germany this probably occurred before 1914; in England certainly after 1918. I do not know when it happened in America or France.

The specialization of higher scientific education in England arose as an administrative consequence of the educational system, and I have tried to show elsewhere[32] that it had nothing to do with industrial demand. I have no reason to believe that the specialization of scientific education in Germany was any more a product of industrial demand. If any practical applications were expected of German academic chemistry in the earlier days, those applications would be in medicine or agriculture; the scientific development of the coal-tar colour industry was an unlooked-for bonus. I have tried to show in this paper that at no time before 1914 did the industry in which science was most readily applicable significantly influence English scientific education.

IV. CONCLUSION

It seems that whether our emphasis is on research or on liberal education, the end is specialization. How does such specialization affect the development of science? No doubt large numbers of scientists in institutions can, sufficiently endowed and well led, produce original papers indefinitely.[33] But is there no danger that an essential component of science, the endeavour to achieve unity, may be lost in a welter of detailed papers?

The older specialization did not at any rate impel the student to make a career for life of his undergraduate discipline. No one thought it unusual if a graduate mathematician subsequently became a lawyer, a clergyman, a naturalist, and so on. But the rise of the modern research institution has in effect consolidated specialization and made it almost a life-long discipline. It is very hard for the individual, especially in his formative years, to escape from such social sanctions, from what is explicitly expected of him, yet demonstrably much of science has been built up by men who, following their scientific intuitions, have abandoned one discipline for another. How then, in the face of these institutional difficulties, to secure the welfare of science without prejudicing the benefits of applied science is a problem which is worthy of the attentions of a modern Francis Bacon![34]

NOTES

1. There appears to be no generally accepted expression for this mode of invention.

2. Early supporters of technical education in England favoured the teaching of the "sciences underlying the arts." Apparently this meant teaching the "public," or carefully edited science of learned journals, textbooks, etc. This is science at its most general and with little or no reference to the particular instances which may be the main concern of technicians. This possibly impeded the development of scientific technical education.

3. A proposed "Davy School of Practical Chemistry" did not materialize.

4. S. d'Irsay, *Histoire des universités* (Paris, 1935) II, 290 ff.

5. R. Vallery-Radot, *The Life of Pasteur* (London, 1906) 196.

6. For the chemical laboratories at Berlin and Bonn, see: *Report of the Department of Science and Art* (1866). For practical chemistry in the early days of the École Polytechnique, see: G. Pinet, *Histoire de L'École Polytechnique* (Paris, 1887) 366.

7. *Annalen der Chemie und Pharmacie*, XXXIV (1844) 97, 339.

8. C. Newman, *Evolution of Medical Education in the Nineteenth Century* (London, 1957).

9. Lord Rayleigh, *John William Strutt, Third Baron Rayleigh* (London, 1924).

10. British chemists who made important contributions to the studies of heat and/or electricity include Black, Cavendish, Dalton and Davy. Through the work of men like Joule, Kelvin and Maxwell these subjects became part of natural philosophy. But public recognition of "physics" did not come immediately, and a curious example of what we should regard as a misuse of the word is provided by Walter Bagehot's *Physics and Politics* (1873). Here by "physics" was meant natural selection. The word *physics* was imported either from France (cf. Biot, *Traité de physique*) or from Germany.

11. Sir Robert Robinson, in *Endeavour*, XV (April, 1956) 94.

12. S. Miall, *A History of the Chemical Industry* (London, 1931) 66 ff.; L. F. Haber, *The Chemical Industry during the Nineteenth Century* (London, 1957) 80, 128, 162.

13. Leopold Cassella & Co.

14. South Lancashire and the West Riding of Yorkshire strongly supported the Mechanics Institute movement between 1825 and 1851.

15. Among the sciences thought to be of the greatest national importance were surveying, meteorology and solar physics. See *Royal Commission on Scientific Instruction and the Advancement of Science* (Devonshire Commission).

16. H. E. Roscoe, *Original Research as a Means of Education* (1874).

17. F. Rose, *Report on Chemical Instruction in Germany and the Growth and Present Condition of the German Chemical Industries, Miscellaneous Series, Diplomatic and Consular Reports* (Cd. 430-16) (1901).

18. H. Kolbe, *Das Chemische Laboratorium der Universität Leipzig* (Braunschweig, 1872) xlv.

19. Letter to Faraday.

20. Edison's famous laboratory at Menlo Park, set up in the 1870's, and the well-known agricultural researches carried out at Rothamsted by Lawes and Gilbert were both highly individualist enterprises, and can therefore hardly be taken as typical.

21. W. H. Perkin (Jr.) "The position of the organic chemical industry," *Journal of the Chemical Society,* CVII (1915) 557 ff.

22. F. M. Perkin, *Journal of the Society of Dyers and Colourists,* XXX (10 November 1914) 339 ff.

23. There was a good precedent: R. Calvert Clapham told the Parliamentary Committee on Scientific Instruction (London, 1868) that the alkali works on Tyneside commonly took apprentices into their laboratories for training as (presumably) control chemists. The facilities for scientific training at Durham University were, he said, very inadequate.

24. Conrad Gill and Asa Briggs, *History of Birmingham* (London, 1952) I, 109.

25. James Marchant (ed.), *Raphael Meldola* (London, 1916) 4, 65. A. G. Green later said that his employers told him that his discovery of Primuline was a "pretty experiment" and he should show it to the Chemical Society!

26. For much of my information about the industry in the north of England I am indebted to Mr. C. M. Mellor, a student at Leeds University.

27. D. W. F. Hardie, *A History of the Chemical Industry in Widnes* (Birmingham, 1952); F. D. Miles, *A History of Research in the Nobel Division of I.C.I.* (Birmingham, 1955).

28. Sir Robert Robinson, op. cit.

29. Presidential Address to the 1902 British Association.

30. An important feature of the German system was the number of schools of chemistry in mutual competition. In England there were, with the possible exception of Roscoe's at Manchester, no schools of this sort between 1865 and 1901. And, if we are to believe contemporary reports, German scientists were very successful in imparting a love of learning and a zeal for research.

31. A private report on the Dyeing Department at the Yorkshire College (Leeds University).

32. D. S. L. Cardwell, *The Organisation of Science in England* (London, 1957) 116-18.

33. F. Paulsen was referring to much the same thing when he used the expression "pseudo-productivity": *The German Universities* (London, 1906) 173.

34. There remains the more general question of the relationship between scientific and non-scientific education. It should be remembered that the mutual exclusiveness of scientific and literary education was explicitly condemned by H. G. Wells and Sir J. J. Thomson, and before them by a long line of educational thinkers. It is satisfactory that the existence of this gap (as Wells termed it) should now be under criticism, but it is depressing that the same criticisms have, apparently, to be made in every generation.

JOSEPH BEN-DAVID

The Universities and the Growth of Science in Germany and the United States

THE STRUCTURE OF THE GERMAN UNIVERSITY

In Germany the number of university students doubled between 1871–76 and 1892–93 from 16,124 to 32,834; in 1908–19 it was 46,632. In the institutes of technology, which were given university status in 1899, the numbers rose from about 4,000 in 1891 to 10,500 in 1899. The growth of the academic staff was somewhat slower, but it started earlier (1,313 in 1860, 1,521 in 1870, 1,839 in 1880, 2,275 in 1892, 2,667 in 1900 and 3,090 in 1909).[1]

There was a spate of new developments in science which gave rise to new disciplines, and to new fields of specialisation within established fields. Many of these developments, especially in physics, chemistry and bacteriology, had important implications for industry, and were *vice versa* spurred on by the tools placed at their disposal by industry, such

Joseph Ben-David, "The Universities and the Growth of Science in Germany and the United States," *Minerva* 7 (Autumn-Winter, 1968–9): 1-35. Reprinted by permission of the author and the publisher.

as measuring and other precision instruments. The social sciences, such as psychology, sociology and economics, which had by then a venerable history, also assumed their present-day disciplinary form at about that time. And just as the natural sciences started to be linked up with industry, the social sciences influenced and were in their turn influenced by practical problems, such as urbanisation, industrial relations, crime and suicide, etc.

The centres of most of these intellectual developments (with the exception of economics) were the German-language universities in Germany, Austria, Switzerland and a few other places. A disproportionate part of the work took place there, and even work done in other countries was to a large extent inspired by and related to the work done in the important German centres. Probably a majority of American, Russian and Japanese scientists received their actual training in science at German universities and even the English and French workers were often decisively influenced by their postgraduate studies or visits at German universities.[2]

Yet in Germany itself there were serious problems in the universities. While the numbers of students and staff increased, and while there was an even greater increase in the expenditure of the universities because of the steeply growing expense of research, no modifications were made in the organisation of the university.[3] Officially it remained a corporation of professors, even though the ratio of the latter to other academic ranks—extraordinary professors, *Privatdozenten* and to institute assistants—who had no formal academic standing at all—changed drastically. This was particularly visible in the experimental natural sciences and in the social sciences which had the greatest potentiality for growth. In the case of the former, the development was probably due to the growth of research institutes which encouraged professors in experimental science to regard their respective fields as feudal domains. The growth of the social sciences was prevented mainly by the difficulty of keeping political controversy apart from empirical inquiry in these ideologically sensitive fields. All this led to a sense of frustration and hopelessness in the academic career manifested by the rise of trade union-like organisations in the lower ranks. The *Vereinigung ausserordentlicher Professoren* was founded in 1909, the *Verband deutscher Privatdozenten* in 1910, and two years later the two organisations were fused into the *Kartell deutscher Nichtordinarier.*[4]

The difficulty experienced by the aspiring scientists and scholars was largely a consequence of the conservatism of university organisation and the professorial oligarchy which dominated it. The professors, who as a corporate body were the university, prevented any important modification of the structure which separated the "institute," where research took place, from the "chair," the incumbent of which was a member of the university corporation. The former was like a feudal fief of the latter.[5] The result of this was that while the increase in research activity fostered an unbroken progression from the mere beginner to the most experienced and successful leader in a field, the organisation of the university obstructed this progression as a result of the gap in power and status between the professor who had a chair and all the others who did not.

A closely related manifestation of this conservatism of the highly privileged university faculties was their resistance to any innovation of a practical or applied character. Not only did they not admit engineering studies to the university, but they resisted the granting of academic degree-awarding powers to institutes of technology (the right was nonetheless conferred by the emperor in 1899); the recognition of the *Realgymnasium* as qualifying preparation for the universities and many other significant proposals for reform were steadily opposed by the professors.[6] Their resistance to the study of bacteriology and psychoanalysis has been described elsewhere.[7] The social sciences never became really independent of other—largely unrelated—fields until after the Second World War.[8]

Thus, while a kind of class tension was built up within the established fields, there was increasing resistance to the institutional provision for the cultivation of new fields. If the latter were innovations of an intradisciplinary or pure scientific character, they were usually accommodated within the university, but often in a way which made them subordinate to older disciplines. There were obstinate and long drawn out debates about the theoretical importance of new fields to justify the establishment of any new chair.[9] These debates were often conducted in terms of personal qualifications of particular candidates, obscured the real issues and also introduced a great deal of personal bitterness into academic matters. Had the universities been organised as departments, these matters might have been treated in a quite impersonal fashion.

Where the innovation was of an applied kind, the attitude was, as has been said, usually negative. Scientific doubts about innovations were often exaggerated, and where there were no doubts, such as in the case of engineering—it was hard to assert that engineering was less scientific than medicine or theology—there were arguments about the dilution of the values of disinterested science and scholarship and of impersonal service. Or, it was simply argued, certain lines of research should be pursued by industry rather than the universities.[10]

Many of the arguments had a great deal of validity, and many of them, too, were probably made in good faith. It was always possible to adduce very good arguments for the maintenance of the existing state of affairs. Furthermore, the existing structure worked; its improvement required at most only minor modifications. It was also usually easy to argue that changes in institutional structure would have all kinds of undesirable effects. Such arguments are often useful, provided that those who have the greatest stake in the existing state of affairs do not have the power to make the final decisions.

In Germany they had this power. Even though on occasions the faculties were overruled by divisional chiefs of the ministries of culture, these occasions were exceptional and were often interpreted as interference with academic freedom.[11] At a time when there was a need to make frequent new decisions in academic and scientific policy, there was in Germany neither a central body nor university administration relatively independent of the faculties to make those decisions. Instead, there was endless argument among those concerned and no decision.

As a result, the great opportunities for an expansion and a more productive adaptation of university studies and research to the occupational and political system, and cultural life in general, were not fully taken advantage of in Germany. Apart from the academic recognition of the technological institutes, the expansion of the range of studies at the universities took place in a selective manner. Among the existing and well-established fields only mathematics and physics expanded rapidly, perhaps as a result of the competition of the new technological institutes in these fields. In the other well-established fields, there was little expansion.[12] Intellectually important innovations, such as physical chemistry, physiological chemistry, earlier bacteriology and other fields too were only grudgingly granted academic recognition.[13] Specialists in these fields received titles of *Extraordinarius,* or institute head, but they were only very sparingly given the rank of *Ordinarius*—the only "real" professors—and then not through the foundation of new chairs, but through the appointment of individuals to existing chairs with loosely defined terms of reference. Sociology, political science and economics were only rudimentarily developed as independent fields. The main developments took place through the establishment of new chairs in clinical medicine and in the increasing number of languages, literatures and histories taught in the humanistic faculties.[14] Such innovations as there were disclosed the very conservative approach to the disciplinary innovation: fields in clinical medicine were considered as substantive specialisations within the established basic medical disciplines, and the new humanistic studies were similarly considered as new substantive applications of the philological method as developed in classical studies.[15] Thus conservatism in the structure of the universities was accompanied by conservatism in substantive intellectual matters as well.

As a result, although German leadership in science and scholarship continued on the whole until the 1930s, in many of the newer fields the centres moved to Britain and the United States. Both of these latter countries attained their new eminence because they managed to make those changes in their universities which were prevented in Germany. The most important of these changes consisted of the establishment of a departmental structure, instead of the system of "chairs"; in large part the more even and undisturbed development of new fields and the much smoother working relationship between academic research and its applications attained in Britain and the United States can be attributed to the departmental system. The beginnings of all this were made in England. The full-scale development, however, took place in the United States. In the following I shall attempt to show how the departmental structure came into existence in the course of attempts to transplant the German type of university to the United States or to graft it on to the then existing American college structure.[16]

THE GRADUATE SCHOOL IN THE UNITED STATES

The changes which occurred in the United States consisted in some cases of simply carrying to their logical conclusions developments which had started in Germany. This was the case in the development of the graduate school and the organisation of university research. On the other hand, in training for the professions and to an even greater extent in undergraduate education the German influence was adapted to more indigenous American or rather to a common American-British tradition.

The crucial step in the importation of the European model was the establishment of the graduate school. Although there was properly speaking no graduate school in Germany—and there is still practically none—those who initiated the graduate school in the United States believed that they were closely following the German model.[17]

German and other European universities trained their students for a single-level degree. Early in the nineteenth century when the system was established it was in fact possible to give a complete and well-rounded training in any branch of science and scholarship at this level. After all, many outstanding scientists were still amateurs, and a single professor could invariably master a whole field. The philosophical faculty of the German university (which included all the humanistic and scientific subjects) provided a scientific or scholarly education up to the highest standards. But not all who obtained the degree were qualified to do research. The conception of a professionally qualified research worker did not exist early in the nineteenth century anywhere, since research was considered as a charismatic activity which could be successfully pursued only by an inspired few. The university could, however, and the German university did in fact, make a serious attempt to teach at the highest level everything that could be taught in the major academic disciplines.

By the end of the century, however, the single degree programme became an anachronism. The university still pretended that its degree course was at the highest scientific level, and some of the teaching more or less comformed to this ideal. But even in these cases it was impossible to obtain a training for independent research within the confines of such a programme. Those who were to become research workers acquired their specialised knowledge and skills informally as assistants working with professors in the research institutes, usually attached to the "chairs," where they had the benefits of doing serious research and of contact with a number of more advanced assistants. The uncompromising level of the degree course was more than the student who did not intend to enter research could usefully assimilate, yet it was not enough for those who wished to enter a professional research career. The training of the latter remained informal. Its main shortcoming was that it was difficult for the student to acquire an all-round training in his field, because he worked with a single teacher. This also created a situation of dependence on a teacher who often behaved in a highly arbitrary and authoritarian manner and it gave rise to feelings of insecurity among those who aspired to a research career. As long as one was not appointed to a university chair, one remained an assistant in a bureaucratic framework with little independent professional standing, even if one was an advanced research worker performing important tasks in research as well as in the training of beginners.[18]

For the American and British (and perhaps other foreign) students who went to Germany, all those shortcomings were not obvious. They were an already highly selected group who possessed a first degree and occasionally even some research experience. The problems of the academic career in Germany did not disturb them, since their own careers were not dependent on the German professor. The fact that the training was not adapted to the needs of the German student who had to acquire well-rounded skills made it all the more appropriate to the needs of visiting graduate students, who often had clear-cut ideas about what to study and with whom. Nor were they, apparently, acutely aware of the problems which arose from the bureaucratic subordination of the assistant to the heads of the institutes. As welcomed visitors they did not have any difficulty in being admitted to institutes or in moving from one institute to another. From their point of view the institutes were no more than parts of the university where research and training for research were performed.[19]

One of the results of this was that when the American or British scholars returned to their countries advocating the adoption of the German pattern, they did not make any distinction between the chair and the institute. Although they knew that German professors personally acted in a very hieratic manner, they were unaware of the structural counterpart. They did not see how different the departmental structure was from the combination of chair and institute which they admired and thought they were establishing in their own universities. Nonetheless, the departmental structure eliminated the anomaly whereby a single professor represented a whole field, while all the specialisations within that field were practised only by members of research institutes who were merely assistants to the professor.

What the American pioneers in the establishment of graduate schools had in mind were students like they themselves had been in Germany, possessing a first degree and committed to a professional career in research. In Germany, research was not recognised as a profession; it was a sacred calling or vocation for very few persons, who needed no formal training beyond what was offered for the standard degree courses; there was no conception of a career leading to the top by gradual steps. The highest positions were rewards for exceptional accomplishment rather than the orderly culmination of a career. In the United States there was from the very beginning an important innovation in the way the idea of the university as a research-based teaching institution was conceived. The idea that research and teaching at the graduate university must not be determined by anything except the state of science and the creativity of the professors was put into effect much more radically than in Germany. The uncompromising "idealism" implicit in this view had as a consequence a far better organised system for training professional research workers. In Germany all students of science or humanistic subjects were required to study their subjects in a highly specialised way, not in order to use them later in life (except for a small minority who followed academic and scientific careers), but because this was considered good for them by those who were in authority. In the United States only the graduate student was invited to pursue science or scholarship for its own sake and for him this was a preparation for a career of research. If he did not want to become a research worker, he could limit his education to the traditional undergraduate college or else he could go to a professional school. The graduate school could therefore concentrate on the training of research workers.

THE PROFESSIONAL SCHOOL

The professional school was another structure which enabled the American university to avoid the intellectually constricting influence of the Germanic professorial system. In its undergraduate form the professional school in America started as a pragmatic experiment in the land grant colleges during the 1860s.[20] But at the postgraduate level its development was parallel to that of the graduate school in the arts and sciences; it was to some extent also an outgrowth of trends inherent in the state of science in about 1900.

According to the conception prevailing in the German universities during the first half of the nineteenth century, the basic scientific and humanistic disciplines had a monopoly of higher education, including that of physicians, lawyers and clergymen. This monopoly was based on the assumption that teaching at university level had to be creative, and based on original research, and that serious research only existed in the basic scientific and humanistic disciplines. This approach was usually less than optimal for the training of students for the practical professions. Even admirers of the German system admitted that the clinical training of the British doctor was superior to that of his German counterpart. But the emphasis on the basic medical fields was justified by the not unreasonable argument that the practical side of medicine could be acquired in apprenticeship outside the university.[21]

During the second half of the nineteenth century, however, there arose a new kind of research which invalidated the assumption that creative research existed only in the basic fields. The discovery of the bacterial genesis of illness, the growing amount of engineering research (especially in electricity), psychoanalysis and in a way all social science research, were not "basic" in the accepted sense of the word. The questions asked by inquirers in these fields did not derive from the state of any given discipline. For instance, for the professional physiologist and pathologist seeking to understand bodily functions in physical and chemical terms, the statistical inquiry of Ignaz Semmelweiss into the etiology of puerperal fever made no theoretical sense. And the same applied to the discovery by Pasteur and others of the bacterial causation of illness.[22] From the point of view of "normal, puzzle-solving" science, these investigators asked the wrong questions and got meaningless answers. The fact that some of these answers had dramatic practical uses made things even more disturbing.

This kind of research grew into a regular activity. It assumed the characteristics of a discipline. There was a permanent exchange of information between groups of research workers, they agreed on what constituted a problem and what were the proper models of research to solve them, and they trained entrants into the field, as basic scientists did, even though the relationship of this inquiry to basic scientific theory was often obscure. What has been called applied or problem-oriented science had begun and certain aspects of it acquired the social structure of academic disciplines. I shall use the word "quasi-discipline" to distinguish them from those fields which originated from attempts to solve problems defined by the internal traditions of a given science.[23]

With the rise of this quasi-disciplinary research, however, the whole question of the relationship between higher education and professional training was reopened. There was great pressure to make engineering an academic field, and there were similar pressures from other "quasi-disciplines."

The attitude towards these developments at the German universities was, with a few exceptions, negative. As it has been pointed out, they preferred to define the task of the university conservatively, and to leave this type of research to other institutions.[24]

This might have been a satisfactory solution had these other institutions been able to compete with the universities on equal terms, as was the case in physics, mathematics and perhaps to some extent in chemistry. In these fields institutes of technology, the Kaiser Wilhelm Gesellschaft, and to some extent industry, provided alternative opportunities for research. There were, however, problems here too. In chemistry, for instance, there were applied research laboratories in industry, but training took place in the universities. This retarded the development of chemical engineering as a profession.

Furthermore, even in the institutes of technology, the acquisition of advanced research skills continued to be dependent on personal apprenticeship. Finally, even in the new non-academic research institutions, research was not professionalised as a career, so that the research worker who was neither a professor nor a head of an institute had to work in a stiffly hierarchical structure which curtailed his scientific freedom and initiative. But since the opportunities were expanding rapidly, in consequence of the recognition of the technological institutes and the foundation of the Kaiser Wilhelm Gesellschaft, these limitations probably did not seriously impair development until the First World War.

In the life sciences, which were an exclusive preserve of the universities, the situation was more difficult. As has been pointed out, the universities opposed the development of bacteriology. This too was left to special institutions. They also did little for physiological chemistry. Clinical research was developed up to a point by the numerous *Privatdozenten* and *Extraordinarii* who had very strong incentive to stay in the university, although it offered them no professional career, because to do so furthered their professional medical practice, both intellectually and financially. The official structure of the universities however took relatively little cognisance of these developments. New chairs were established, but research was dominated by the basic disciplines, and the training of practitioners was little influenced by the new developments. The German universities did not accept the idea that the university had an active role to play in medical practice and in making practitioners more effective users of research, by encouraging research relevant to practice, and by actually training the student in the detailed skills of medical practice in an environment where research constantly tested and modified these skills.[25] Students were still largely taught what was considered the intellectual basis of their profession, and were expected to acquire by their own efforts the skills needed for either research or practice after graduation. But the relationship between the intellectual basis and the practice which had existed in the first half of the nineteenth century was completely changed by the end of it, and this change was not sufficiently reflected in the medical faculties.

This attitude was reversed in the United States. There the principle was accepted that universities were training students for the intellectual-practical professions and that this was justified because of their important scientific basis. As a result, even the most research-oriented schools interpreted their task of enriching the scientific element of the professions as an obligation to encourage quasi-disciplinary research relevant to professional work, and to train practitioners capable of benefiting from research. The most conspicuous and successful instance was the development of clinical research in medicine at Johns Hopkins University and a few other places. Instead of emphasising the invidious difference between basic and clinical research, given the theoretical and experimental deficiencies of the latter, attempts were made to create university hospitals with conditions that approximated as nearly as possible to the conditions of an experimental laboratory, and to use these facilities for the improved training of physicians.

Similar policies were pursued in engineering, agriculture and education. The relevant departments of the universities considered it their task to create as far as possible and as rapidly as possible a research basis for these various professions, and to develop that basis into quasi-disciplines with training programmes, higher degrees, learned associations, journals and textbooks. Scientists in the established disciplines had many misgivings about the danger of blurring the borderlines between disciplinary science and problem-oriented research which—as has been shown—often lacked theoretical significance. In many cases this criticism was justified; the determination to conduct research relevant to the training functions of the university resulted, at times, in research which was irrelevant theoretically as well as practically.[26] What needs emphasis at this point, however, is that here, too, a function which was implicit in the state of science in Europe but could not be properly fitted in the existing conceptions and organisation of scientific work so that it developed through a variety of exceptions and improvisations, became defined, organised and standardised in the universities of the United States.

This transformation occurred in the same way as the transformation —or rather the emergence—of the graduate school in the basic scientific and humanistic disciplines. The visiting Americans in Germany were not very sensitive to the invidious distinctions which existed in that country towards academically unrecognised or not fully recognised fields. For them a *Privatdozent* or an *Extraordinarius* doing interesting research in an institute or university hospital often seemed (as he often was) a pioneer rather than someone specialising in a field which made him *nicht ordinierbar,* i.e., unsuitable for promotion to a chair.

The reason for these differences probably lay in the fact that, unlike their German counterparts, the American academics interested in the creation of more scientific professional schools did not in the beginning possess a monopoly of professional education such as existed in Germany.[27] Rather they had to contend with a powerful British-American tradition of thorough practical training. This was not only the tradition defended by the survivors of the pre-scientific age in university faculties, but also the freedom of the students to choose between types of university. They insisted on being trained thoroughly in practice and did not want to start learning how to practise their professions after leaving the university.

As a result the reform of professional education under the impact of modern science did not lead in the United States to the abandonment of the earlier tradition of learning how to do things through practical experience. The conception of scientific research which liberally included problem-oriented research was fully compatible with this practical orientation. Both the professional school and the graduate school of arts and sciences were conceived as places where students were trained for a particular professional practice and both endeavoured to bring the student up to a point where he would be capable of working on his own.

ORGANISED RESEARCH IN THE UNIVERSITIES

The introduction of graduate training in the basic scientific and humanistic subjects and the active support of problem-oriented research related to professional training lowered the barriers against organised research in the universities. As the function of universities was to train people to perform and apply research up to the highest standards, there had to be within the universities up-to-date research laboratories and other facilities. These facilities were not only necessary in order to make it possible for the professors to pursue their own research, but also for the training of graduate students. Furthermore, since the universities abandoned their qualms about training and research for practical purposes, there were few limitations on the kind of research functions that universities could engage in. Finally, the existence of a departmental structure in teaching probably made it easier to assimilate into the university the administrative arrangements for research.

In agriculture, education, sociology, nuclear research (and probably also in other fields), the universities pioneered research on a scale which far exceeded the needs of training students, and was from the very outset an operation distinct from teaching.[28] The research organisations developed in some of the schools of agriculture, medicine, and even in basic scientific departments became by 1900 a challenge to European science and served as an incentive for the establishment of new research organisations, such as the Kaiser Wilhelm Gesellschaft and the British research councils. This, then, was another function which had started in the German universities where professors had their small research institutes. But their growth within the European universities was curtailed by the rigidity of the university structure. When transferred to the United States further growth took place which was then partially imitated in Europe. But this imitation did not lead to comparable growth, nor did it take place within the universities; it led only to specialised non-university research institutes.[29]

THE GROWTH OF NEW DISCIPLINES: STATISTICS AS A CASE IN POINT

The differentiation of higher education into three sections—undergraduate college, graduate school and professional school—and the provision for research which was at times only loosely tied to teaching, opened virtually unlimited possibilities for the establishment of new fields. The growth of disciplines, such as the social sciences, comparative literature, musicology and other fields, was indirectly stimulated by their popularity as undergraduate subjects. Undergraduate demand led to a demand for teachers trained in these subjects, and hence departments, and even in some cases to Ph.D. programmes in these subjects. There was little risk, therefore, in initiating a disciplinary or quasi-disciplinary organisation in a field which had some promise of either intellectual or practical value. Given the immense variety of interests to which the university catered, there was a demand for teachers of an equally wide range of subjects. This in turn created a demand for post-graduate training, which in turn had a corresponding impact on the diversity and the creation of departments.

The departmental structure reduced even further the risk of enter-prise. New specialities could easily be accommodated and nurtured within existing departments—which always had a considerable degree of heterogeneity—until they were strong enough to operate indepen-dently.

A good example of this is the development of statistics. Both as a field of mathematics, and as a tool which can be applied to a very great variety of problems, statistics has a venerable history in Europe, going back to the seventeenth century. In the nineteenth century there was an important professional movement initiated and led by Quetelet for the improvement and propagation of statistics.[30] Yet, as an academic field, statistics had remained a very marginal affair, and it did not develop a scientifically based professional tradition. The basic work done by mathematicians was usually unknown to the practitioners, and there was little continuity and coherence either in the theoretical or in the practical work.[31]

The reason for this state of affairs was that those who were most creative in statistics were usually mathematicians or physicists who were uninterested in changing their disciplinary affiliation. Or else they were amateurs interested in solving practical problems rather than in initiating fundamental research.

In order to make statistics into an academic discipline there would have had to be a group of persons within the universities interested in identifying themselves as statisticians. These could have come only from those interested in the uses of statistics to the extent that they were capable of communicating with and learning from mathematicians interested in probability. Potential sources for the emergence of such a group were the geneticists, economists, social scientists, and psycholo-gists aware of the statistical nature of their problems. But only an occasional few among these took a serious interest in statistics, since most important contributions in these fields consisted of experimental and observational studies where statistical methods played a relatively limited role. The advocates of quantitative methods were often among the relatively less creative persons in their respective professions and the whole approach still had to prove its utility. Even where the utility was obvious, the statistical techniques involved were simple and there was no unequivocal evidence that more intensive statistical work would be the best way to improve the field. In the German academic system, therefore, where one person had to represent an entire already estab-lished field, it was unlikely that he would be chosen with much consid-eration given to his competence in the marginal subject of statistics.[32]

To the extent that chairs in statistics were established in continental Europe, these were still-births. Rather than reflecting the converging interests of a number of sciences in the statistical method, these chairs came into being as a result of non-academic pressure on the universities. Usually the universities resisted such pressure, but they were willing to make compromises in cases which were deemed academically unimportant, where there was a legitimate state interest involved and where the subject could be kept at a distance from more important academic concerns. Since the law faculties were training grounds for prospective civil servants, there was a long tradition of providing courses in political science and administration in the law faculties. These were narrow courses of study with little academic standing and little practical utility. Statistics were added to these studies. Within the law faculty, statistics had little or no relationship to either mathematics or the biological and social sciences which had a potential interest in it. Those appointed to chairs in the subject were usually persons who had their basic training in law.[33] Thus, whatever statistical work went on in Europe in and outside the universities, the chairs in statistics had little share in it and could not serve as centres for the emergence of a discipline.

The development in the United States contrasts sharply with this. The existence of flexible and expanding departments, with many more or less independent posts, made it possible for all the increasing variety of academic users of statistics, in biology, education, psychology, economics, sociology, etc., each to develop its own specialists in the field.[34] In the beginning the large majority of workers were too poor mathematically and had too narrow a view of their field to do important work. By the 1920s there arose a growing awareness of the shortcomings and a demand for a sounder mathematical basis. There emerged certain centres for serious statistical work, as at Iowa State University, stimulated by the needs of the agricultural research station connected with that university.[35] Still, for advanced training in the theoretical aspects of their subject, the resources in the United States were still insufficient.

Young American statisticians went, therefore, to Britain, which in the twenties and thirties was the centre of statistical research.[36] Having benefited from the British training, important centres for statistics arose in the United States during the late thirties, especially around Hotelling at Columbia and Wilks at Princeton.[37] They were later joined by several young Europeans who obtained their mathematical training in central and eastern Europe and in Britain.[38] During the Second World War, an additional impetus to the development of statistics was given by the creation and operation of the Statistical Research Group.

This wartime cooperation had probably reinforced the sense of practising a common and distinct discipline. (It did not, however, create this consciousness which may be dated, at the latest, from 1935, when the

Institute of Mathematical Statistics was founded, and probably even earlier than that.)[39] Demands for the establishment of separate university departments of statistics were voiced at the meetings of the American Statistical Association. The first establishment of a separate department occurred at the University of North Carolina in cooperation with the state university of the same state, where, as in Iowa, there was important agricultural research interest in the subject. The establishment of this department was rapidly followed by other universities, including the most prestigious ones. This led to the enlargement of the number of practitioners of the statistical discipline and, with it, the development of more theoretical work in the field, which has helped in its definition as an academic discipline.[40]

In addition to the departmental structure, which made possible the extension of statistical work in an increasing number of scientific fields, a crucial role in this development was played by the involvement of the university in training and research in applied fields. In the early decades of this century, statistics was primarily considered as a tool of applied research. Whereas, on the continent of Europe this kind of research was never considered to be suitable for the universities, in America universities undertook to provide for this type of work too.

This explanation is supported by the single significant parallel to the United States in the development of statistics, namely, that of Britain. In fact, as far as contributions to the theory go, those of the British were much more important than the contributions of the American statisticians. Britain also preceded the United States in the establishment of the first chair in statistics, at University College, London, in 1933.[41] The theoretical superiority of the British work does not need a great deal of explanation. That country had a far more developed scientific tradition at that time, and a less abstract school of mathematics than the United States.[42] Hence it was easier in Britain for a few first-rate minds to acquire the necessary mathematical background for statistics than it was in the United States.

THE EXTERNAL CONDITIONS: DECENTRALISATION AND COMPETITION

The diffusion of the innovations and the eventual assumption of its present multiplicity of functions by the American university has not occurred as a result of a preconceived plan. In the formative years of the system between the 1850s and about 1920 there was a wide range of ideas about the functions proper and improper to a university, and the arguments used in the debates were in many cases the same as in Europe. But the effects of these ideas were very different, in consequence of the difference between the ecology of American academic institutions and their European counterparts.

In Europe the procedure for university innovation was to convey the ideas to the government which then rendered a decision between the conflicting viewpoints based on a more or less public debate of the issue.[43] In the United States, however, there was no central authority, or even informal "establishment," to lay down policy for the whole country. Therefore there was no concerting of opinion on a national scale or organised action to press the government to put certain schemes into action, or at least support them. Rather, the protagonists of an idea tried to realise their schemes in institutions in which they worked.[44] There were of course state-supported universities, as in Europe. But they were not the only ones, and they were far from enjoying monopolistic advantages. The most prestigious and the wealthiest universities were private corporations. Thus the system was far more decentralised than in Germany. There different states competed with each other. In the United States the state universities not only competed among themselves; they also had to compete with the private universities.

Decentralisation was not, however, the only condition which made the American system more receptive to innovations. A more important condition was the absence of important monopolies conferred on the system as a whole. Early in this century, lawyers, doctors, teachers and civil servants—to the extent that the latter were trained at all, except "on the job"—were often trained outside the universities, and the most important middle-class career was business, which did not at that time require either formal or certified training. The universities had to prove that they were useful and worthy of support. This is reflected in the development of the courses and subjects offered in the various universities.

THE INTERNAL CONDITIONS: THE STRUCTURE OF THE AMERICAN UNIVERSITY

Since the universities had constantly to adapt to innovations to maintain standing, and to compete for personnel and resources to do so, it was impossible that they should be run either in the civil service manner according to fixed personnel establishments and regulations, or in the manner of wholly autonomous corporations of teachers, scholars and scientists. Hence the imitation of the German model did not involve the adoption of the German system of university government. The changes which occurred in this respect paralleled the changes which occurred in the organisation of business. Until the 1860s college presidents were the managers of their institutions, acting on behalf of the trustees who formed the corporation which enjoyed legal ownership of the physical assets of the college. The rise of the new universities was accomplished by a new type of president, who combined the qualities of autocrat, statesman and entrepreneur. He was still very much the dominant figure, but increasing size, increasing complexity of task and increasing self-esteem on the part of academic staffs of increasing eminence, required that he be capable of delegating authority and acknowledging claims to academic freedom. This group of presidents nurtured the growth of the many universities of the present day. They laid the foundation of the present-day structure of government by much less powerful presidents responsible to a board and assisted by a number of full-time academic administrators, such as vice-presidents, deans, etc. The president had to be an entrepreneur modifying his policies and university organisation in an ever-changing situation, and trying to push his university ahead in its class through careful forward planning and rapid exploitation of new ideas.[45]

In order to operate effectively under these conditions, the subunits of the university had to be (a) flexible enough to carry out all the diverse functions of the university as well as to adjust to new ones; (b) autonomous, so as to be able to make changes in courses of study, teaching arrangements and staff recruitment without undue delay; and (c) of sufficient size to perform effectively training and research functions in fields requiring many kinds of specialisation.[46]

The most important of the units which emerged was and is the department in the basic arts and sciences and the larger professional schools (the smaller professional schools are themselves departments). It is the American substitute for a European "chair" plus institute. Instead, however, of a single person fictitiously representing a broad area of research, this task is given to a group which may actually represent the subject in its entirety.

This development had also occurred in Britain. There, however, the structure of the department was steeply hierarchical, with, usually, one professor directing the work of several juniors. In the United States the departments were from the beginning much more equalitarian, since they comprised several teachers of the same rank. In Britain the authority of the head of the department extended even to scientific matters, e.g., in making decisions as to the kind of research which should be done in the department; remnants of such authority still persist. In the United States the departmental chairman came increasingly to deal primarily, and then only, with administrative matters. With regard to research his task came to be to obtain provision for it from the central authorities of the universities and from outside patrons, rather than to direct it intellectually.

The size of the American department and the presence of a number of professors within it made possible the growth of the department, and within the department the formation of independent research units composed of one or several teachers and graduate students, the introduction of relatively independent sub-specialities without raising the question of what was within the discipline and what was outside it, and the increasing tolerance for interdisciplinary interests on the part of at least a few members of the department without seriously affecting the work within the discipline.

Institutes in the United States—unlike those in Germany which were established to facilitate the work of a single professor—are seldom attached to particular departments and practically never to particular professorships. They are often interdisciplinary ventures.[47] Their purpose has been either "mission-oriented" research, to bring to bear the contribution of several disciplines on the exploration of a single problem, e.g., human development, urban studies, etc., or to share a single piece of equipment, e.g., an accelerator, among different groups of research workers. The departments were well established as the basic unit of the university by the beginning of the century; institutes began after the First World War.[48]

THE RESULTS OF THE SYSTEM: THE PROFESSIONALISATION OF RESEARCH

These developments have rapidly transformed the role of the scientist. By the first decade of the century there emerged the conception of the professionally qualified research worker. A Ph.D. in the humanistic or scientific subjects assumed the same dignity as the M.D. in medicine. Those who possessed the title were considered qualified for research just as an M.D. was qualified to practise medicine.

The requirement of a Ph.D. made suitable candidates scarcer, and raised, thereby, the market value of those who possessed the degree. But its principal effect was to create a professional role which implied a certain ethos on the part of the scientist as well as his employer. The ethos demanded that those who received the Ph.D. must keep abreast of scientific developments, do research and contribute to the advancement of science; while the employer, by employing a Ph.D., accepted an implicit obligation to provide him with the facilities, the time and the freedom for continuous further study and research which were appropriate to his status.

This was a new departure from the particular status of the college teacher in the United States in the nineteenth century; they were then the employees of presidents or trustees, both of whom were accustomed to treat the teachers in a very authoritarian fashion, as if they were no more than the assistants of the president, helping him to do the job for which he was responsible. It also entailed an important departure from European usages. The role and career of the research worker was not one of the central elements of German science organisation—which was the only one that mattered as late as 1900. There research was not considered as a profession. In spite of all the growth of research within and outside the universities, the official acknowledgement and provision for the scientist's role had not changed throughout the nineteenth century. Scientific achievements were considered as sacred things, as expressions of the deepest and most essential qualities of a specially gifted person, which had nothing to do with institutional provision. Research was, so the fiction had it, a voluntary, non-paid activity. A certain number of posts, mainly professorial, had something like an official charisma (*Amtscharisma*). Those who had this status enjoyed also very great freedom, few and relatively circumscribed duties, great honour, quite a good income and complete security of tenure. These positions were not stages of an occupational career, and the freedoms and privileges which attached to them did not carry over to scientists who did not hold such elevated positions. The professor was not in principle paid for research, but occupied a role with a stipend which made it possible for him to do research as he wished. The *Privatdozent* could also, if he could arrange it, do research, but no provision was made for him to do it;[49] he not only received no salary, but he had no officially provided funds for research. If he worked in the laboratory, he did so on the professor's sufferance.

According to this view, research which was directly paid for was not considered as research; it had none of the metaphysical pathos of the deepest expression of a creative spirit. It was simple and bureaucratic work, which could be (and often was) as narrowly and specifically prescribed as the employer (such as the professor heading the institute) wished it to be.[50] Academic freedom in this scheme was the freedom of a privileged estate. This might have fitted the state of science early in the nineteenth century, when scientists were few and when amateurs still played an important role in science. But at the end of the century, when scientific research ceased to be an amateur activity, it was a poor and invidious way of ensuring the growth of science. At that stage only an arrangement which combined regular employment with individual autonomy and scientific responsibility of the research worker could provide a satisfactory solution.

The new conception of the scientist's role as a professional role on the one hand and the flexible structure of the university with its openness towards innovation on the other introduced a great many changes in the hitherto prevailing relationship between academic organisation and science. Although American professors might spend as much time on academic administration as their European counterparts, most of it concerns departmental affairs which are directly related to teaching, research and personnel matters in the field of most immediate interest to them.[51] There is a much more selective involvement in administrative work because they are inclined to do administrative work and allow themselves to gravitate towards becoming academic administrators. They may participate in these affairs as experts advising the dean or the president, in whom large powers still reside, in which case their tasks are parallel to those of the "staff" in other large organisations; or finally they act as watchdogs of the autonomy of the academic staff to prevent the administration from doing anything that interferes with this autonomy. Here they appear as representatives of a professional body within a polycentric, pluralistic system of the allocation of power.

Such institutions as the senate and faculty assembly do not have much importance in the United States. Presidents are appointed by the trustees, although staff representations and consultants play an important part in determining who is appointed, and the deans are part of the administration and not elected heads of the faculties. The American professor is not legally a member of the university corporation. He has been from the very outset a professional, employed by an organisation to perform certain loosely defined services. His loyalty to the organisation often becomes very pervasive and deep but it is also often limited by economic and professional considerations. He has particularly in the period since 1945 regarded it as right to insist that the university he serves provide optimal conditions for the exercise of his scientific capacities, and that it provide him with the freedom and the backing to establish those conditions for himself from funds which he seeks to obtain from outside the university.

What is called academic freedom in the United States is not the autonomous self-government of the senior teachers as a corporate body directing the affairs of the university as a whole, but the guarantee of freedom from interference by an administration representing a lay board with the direction of his work and the expression of his views, and from interference originating from outside the university and mediated through the lay board and the administrators of the institution.[52]

The constitutional history of the American university is the history of the devolution of authority in intellectual and academic matters from the board of trustees and the president to the department and its individual members. This, coupled with the vigour of strong presidents, is the source of the unequalled adaptiveness and innovativeness of the American university and the social structure of scientific research in America.

SOME RESULTS OF THE SYSTEM

The emergence of the scientific role in the American university is intimately connected with mobility of American scientists, which is in its turn the most important element in the adaptiveness of American universities to new possibilities in research and training. There used to be (and still is) great mobility in the German system also. But there mobility was strictly circumscribed by the structure of the academic career and the hierarchy of universities. People went from one place to another either to obtain higher rank or to be at a more famous university (which usually implied better facilities and a more attractive intellectual setting).[53] In the United States, there is, in addition, a great deal of mobility motivated by an individual's assessment of what he wants intellectually at a particular stage of his intellectual career or what he desires in income. Scientists may go from a high position at a first-rate university to a less prestigious university in order to get an institute or a department, or improved facilities and conditions of work. Retired members of the most famous universities do not consider it beneath their dignity to go to teach at a small college. And similar considerations influence academics to move out of the academic system altogether. In connection with this, scientists have become less identified with their universities than with their discipline, although usually they very much prefer to work in the atmosphere of a university.[54] There exists a *professional community* of scientists or scholars in each field, and one's standing in this community is a more important matter than in other countries.

One of the tangible manifestations of the importance of the professional community is the relatively greater importance of professional-scientific associations in the United States than in continental Europe. They play a more important role in publications, their conventions are most important affairs, and there is a closer relationship between the scientific and professional aspects of their activities than in Europe (the British situation is closer to the American).[55]

Only in the United States has there been a general and early recognition that there is no necessary contradiction between creative accomplishment in research and the organisation of research. This absence of prejudice against organised research and its effectiveness through standardisation made it much easier to devise increasingly complex and sophisticated types of organised research. Thus departments, research institutes, and laboratories soon outgrew their European counterparts in complexity as well as in size. By the thirties and perhaps even before, the difference reached a stage where in some fields European scientists were no longer able to compete effectively with American workers in their respective field.[56]

RESEARCH IN INDUSTRY AND GOVERNMENT

The rise of scientific entrepreneurs and administrators, the professionalisation of research careers and the rise of standardised procedures for staffing, equipping and costing different types of research made scientific research into a transferable operation. Administrators would move from university administration to the administration of large industrial or governmental research laboratories and establish research units of the same kind that existed in the universities. And research workers could work in any of these settings without having to change markedly their professional identities or give up their expectations or standards.

Of course the practice of scientific research in organisations which have non-scientific goals presents the possibility of conflict. Instead of pursuing intellectually promising leads, the researcher may be required to engage in scientifically less interesting problems. He may, furthermore, be limited in his freedom to communicate and cooperate with his colleagues who work elsewhere so as to safeguard industrial or military secrets.

The attitudes developed in universities could not provide a ready-made answer to these problems, but they created a basis for a pragmatic approach to them. First of all they helped to make up a culture partly shared by industry and government which defined what could legitimately be expected of scientists. In this way, the culture of university science helped to create a congenial environment in non-academic institutions for university-trained scientists.

In consequence, industrial research was given considerable autonomy and a long time-span to show its creativity. The industrial research worker was not considered just as any employee to be assigned at will to all kinds of "trouble-shooting" tasks. In these favourable circumstances, there arose a type of research worker continuously and fully engaged in product development. This role first appeared outside the university perhaps in the laboratory of Thomas A. Edison, where it was performed by self-educated inventors. Gradually it was assumed by trained scientists and engineers and became more integrated with the complex of activities regarded as falling within the jurisdiction of professional scientists.[57]

A very large variety of modes of supporting training and research by government and industry without direct involvement on their part in activities for which they are unqualified also emerged from extension of the research activity beyond the limits of the university. The most common are research or training grants, contracts and donations. The advantages are that they are (a) given to persons and organisations of proven competence; (b) give the recipients sufficient freedom to devise their own plans, and at times, even change their original scheme as soon as they find out that it is not the most fruitful one; (c) encourage constant re-evaluation, criticism and comparison of programmes and changes in policy without the necessity of abolishing or drastically changing whole organisations.

The existence of professional research workers and standardised procedures for the organisation of research have been necessary preconditions for this proliferation and flexibility of research activities. The close relationship between universities on the one hand, and government, business and agriculture and the community in general on the other, had been initiated and managed by administrators specialising in academic and scientific affairs (university presidents, officers of foundations, governmental research directors). The emergence of the specialist in university and scientific administration with traditions of initiative and a considerable body of "know-how" has been a *sine qua non* of the recent growth of science in the United States.

A COMPARISON OF SCIENTIFIC ORGANISATION IN THE UNITED STATES AND WESTERN EUROPE

In western Europe, the new functions of science which emerged since the middle of the nineteenth century were grafted on the national systems of higher education which had emerged in the first half of the century. In this system, the universities—and in France, some of the *grandes écoles* also—were the centres of pure science. From the last decade of the nineteenth century they were increasingly supported from the budgets of governmentally financed research organisations and laboratories established from time to time in an *ad hoc* manner. Research aiming at the solution of practical problems, or more generally in fields where the likelihood of practical application was great, took place in segregated and specialised research institutions, usually financed by government and directly accountable to it, and in a few cases in industry. Finally, development research was done in industry, but only in a few instances was it effective and systematic. To make up, therefore, for this deficiency, the government has since the First World War, and more especially since the Second World War, stepped into this field, too, either by establishing applied research institutions of its own, or encouraging trade associations by direct or indirect subsidy to establish and operate such institutions.[58]

In the United States, the trend has been from specialised institutions of higher education to universities performing an increasingly greater variety of functions. And there has been a parallel development from relatively small-scale specialised research institutions to large-scale, multi-purpose ones; such developments occurred both in industrial and in governmental research institutions. In no case was this development foreseen or planned in advance. It was the result of trial and error within a pluralistic and competitive system. Nonetheless, the superiority of large multipurpose organisations seems to have been demonstrated, and with it the hypothesis that research being a cooperative enterprise where ideas and skills can be indefinitely shared, and where the sources of stimulation are probably quite variable, small and segmented institutions cannot compete successfully with large and varied ones. In a large university there will always be some innovating fields, and some generational change to ensure stimulation, whereas in a small specialised and segregated institution the atmosphere may easily become extremely homogeneous. European experience supports this view. The liveliest places scientifically have been the capital cities, such as London, Paris, and at one time Berlin and Vienna, which by virtue of the spatial proximity of many relatively small institutions provided the atmosphere that only very large organisations could provide otherwise.[59]

Large, multipurpose institutions are particularly important in applied or "mission-oriented" research. Such investigations, with goals which are not derived from the normal internal processes of scientific research, are very likely to be interdisciplinary; not only does the mission require it but the attitude of indifference of the administrators towards the dignities of academic disciplines is likely also to favour it. Small specialised research institutes are likely to be more resistant to multipurpose projects; where the director and the senior staff are of the same disciplinary background, they are unlikely to seek out new problems other than those which arise within the framework of their own disciplinary tradition. In a larger, more heterogeneous organisation, the director is less likely to be committed to a particular discipline. Administrators who are interested in results but not in particular disciplines can greatly facilitate the process of bringing in new types of personnel and taking on new problems. Such changes will create crises in a small, specialised research institution. Some persons may have to lose authority or even their jobs in the process. Decisions will, therefore, be delayed.

Since the frontiers between basic and applied work are continually shifting, the establishment of specialised institutions in a field which is promising today may immobilise resources at a future date when other fields have become more interesting. Here too the multipurpose research institution is more effective than one with specialised concerns.

American academic and scientific institutions have thrived because they have learned from experience. They had to learn from experience since their mere existence was no guarantee of their eminence. They had to compete for fame through accomplishment and they had to compete for funds and for persons. They were helped to do this because they had administrators who were not bound by the results and reputations of particular persons and whose concern for the whole institution made them more open to the lessons of experience.

This innovating function was, and to a large extent still is, absent in Europe. Truly self-governing university corporations have rarely been able to exercise much initiative, because of their tendency to represent first of all the vested interests of their members. In effect much of their efforts have always been directed to prevent change and innovation.

Thus scientific policy-making usually devolved on the government. As a result policy has been made at a great remove from its execution; and since it has been made always for the system as a whole there has been little opportunity to evaluate its success, except by comparisons with other countries. Paradoxically, therefore, the nationalisation of the university and scientific research system which was supposed to lead to more objective and better coordinated planning of higher education and research has, as a matter of fact, debilitated the capacity of the systems to learn from experience, because the centralised systems had no con- stitutive feedback mechanisms such as are given by situations where universities and research institutes are free to make innovations and compete with each other, and because there was no room in these systems for the development of executive and entrepreneurial roles specialised in academic and research affairs, not too remote from its day-to-day activities and yet also not completely absorbed into them.

THE BALANCE OF THE SYSTEM

The most obvious results of the system have been the transformation of the relationship between higher education and research on the one hand and the economy on the other. This enterprising system of univer- sities working within pluralistic higher educational, scientific and eco- nomic systems has created an unprecedentedly widespread demand for knowledge and research and has turned science into an important eco- nomic resource.

One decisive question which we have not confronted so far is whether the system has also encouraged scientific creativity for its own sake. After all, even the most effective diffusion and use of science are not necessarily scientifically creative. New knowledge is created by very few people who are interested in it and capable of creating it. And it has been believed by many that making the practice of scientific research into a professional career might inhibit scientists from following up freely the paths opened to them by curiosity and imagination.

As a matter of fact, however, the widespread uses of science have created a very wide foundation for "pure" research, the aim of which is to increase knowledge without consideration for its potential uses. How the practical uses support science for its own sake can be seen from a comparison of the statistics of research expenditure in different countries. The support of research of all kinds per capita of the population or as a percentage of the GNP is greater in the United States than in Europe. Expenditure on basic research is a smaller fraction of the total national expenditure on research than in Europe, but the absolute sum spent on basic research in the United States exceeds by a very great margin the amount spent on it in other western countries, and the same is true of per capita expenditure. (Table 1.) This shows that entrepreneurial applied science which extended research and training to new and often relatively risky fields did not ultimately diminish the share of basic research relative to the society's total resources, as has been feared in Europe, but rather is associated with an increase of this share.

Furthermore, the widespread cultivation of applied research has not led to a loss in the autonomy of science, as was originally feared. Even though the public outlook which prevailed in the United States when the changes under discussion were initiated was of the kind which did not hesitate to judge research by the criterion of short-term utility, it was not forced on the scientific community by a central power or a single source of support. Rather the job of creating the new kinds of institutions was left to academic and research administrators and policy makers, such as university presidents, heads and advisers of foundations, private industrialists and some heads of government departments. Some of these were genuine believers in the value of pure science, others might have been true utilitarians believing only in the value of science applied to something else. But all of them had to face the two highly practical tasks of either earning money by research or obtaining it for research and higher education. In both cases they could succeed only if the research they sponsored or promoted was at a very high level and they had to recruit and retain good scientists for this purpose. If they failed the costs in terms of money and eminence were very high. They could never rest on their laurels; if they did they were forced into a condition of decline by their industrial and academic competitors.

It was learnt that the best way to utilise science for non-scientific purposes was not through subjecting research or teaching to non-scientific criteria, but to aid it in its own immanent course and then to see what uses could be made of the results for productive purposes, for education and for the improvement of the quality of life. The link

TABLE 1.* Gross National Expenditure on Research and Development in the United States and Western Europe Related to National Resources and Analysed by Sector of Performance and Type of Research

	Absolute amount Million US $	Per capita US $	% as of GNP	SECTOR OF PERFORMANCE (% OF TOTAL)				TYPE OF RESEARCH (% OF TOTAL)		
				Business enterprise %	Government %	Other non-profit organisation %	Higher education %	Basic research %	Applied research %	Development %
United States 1963-64	21,075	110.5	3.4	67	18	3	12	12.4	22.1	65.5
France 1963	1,299	27.1	1.6	51	38	—	11	17.3	33.9	48.8
Germany 1964	1,436	24.6	1.4	66	3	11	20	—	—	—
Italy 1963	291	5.7	0.6	63	23	—	14	18.6	39.9	41.5
United Kingdom 1964-65	2,160	39.8	2.3	67	25	1	7	12.5	26.1	61.4
Austria 1963	23	3.2	0.3	64	9	1	26	22.6	31.9	45.5
Belgium 1963	137	14.7	1.0	69	10	1	20	20.9	41.2	37.9
Netherlands 1964	330	27.2	1.9	56	3	21	20	27.1	36.4	36.5
Norway 1963	42	11.5	0.7	52	21	2	25	22.2	34.6	43.2
Sweden 1964	257	33.5	1.5	67	15	—	18	—	—	—

* Compiled from Organisation for Economic Cooperation and Development, *The Overall Level and Structure of Research and Development Efforts in OECD Member Countries* (Paris: OECD, 1967), p. 57 and p. 59.

between science on the one hand and industry and government on the other was not established by the industrialists or the civil servants giving instructions to scientists. Rather there has been a constant and subtle give and take between professional scientists who had a fair idea of what they wanted and could do, and the potential users of science in the professions, industry and government. This mutually advantageous interchange was established and has been kept alive by academic and research entrepreneurs acting as organisers and interpreters between the interlocutors.

The economy has benefited from science, but a large enough proportion of the benefits has been ploughed back into research to ensure systematically organised pure research in an increasing number of fields. What began to appear around the middle of the nineteenth century in Germany, namely, a group of workers, usually the students of a great innovator, concertedly working on a coherent set of ideas until they had exploited all its potentialities, has become the normal state of affairs in the United States. Due to their secure economic base—which was never established in Europe—these activities are now pursued in the United States regularly and in a constantly widening range of fields. Scientific growth, to the extent that it can be measured by manpower figures, resources invested in science, or publications, has been accelerating, with the United States setting the pace and forcing it on other countries which have found it increasingly difficult to stay in the race. This is the positive side of the balance of the system.

THREATS TO THE SYSTEM

There are also, however, negative aspects. One of these is the delicacy of the balance between the internal structures and traditions of scientific and scholarly creativity and the demands of the economic and political powers. This balance is more delicate in the United States than elsewhere because the pluralistic entrepreneurial structure and expansive system of science and higher education requires a much greater involvement of the universities in the affairs of society. This is the price paid for the greater support of science and scholarship.

Until the 1940s this involvement had typically taken two forms. Universities and colleges were at times pressed or impelled to institute degree courses in occupations which had practically no actual or prospective scientific content, and to accredit courses of study with little intellectual content. A similar but more legitimate external influence led to the great extension of professional training at the universities in fields which had had a genuine but still only potential and undeveloped scientific or scholarly content. The early efforts of land grant colleges in agricultural and engineering education and the establishment of schools of education, business, social welfare and several other fields belong to this category.

Judging from the vantage point of the present, these attempts have not caused serious long-term damage to the system. They were considered by the group of creative and devoted academic scientists, scholars and administrators either as evils to be contained, or as challenges which spurred them on to extend serious research and study to these new fields. As a result, some of the worst anomalies have been either eliminated or contained without seriously diluting the quality of the system as a whole, and the professional schools have managed to raise the intellectual content of their courses, and have constantly grappled with this problem.

Again, it was the professional university administrators who played an important part in neutralising the noxious consequences of such "service functions." The pressure for the institution of these intellectually problematic courses was exerted on (and in some cases initiated by) the university administrations. The scientists and scholars in the faculties of arts and sciences usually had little incentive or opportunity to become involved in these practical matters. Usually they regarded such involvement as a threat to science and scholarship. In this situation university presidents had to act as mediators between the demands of the external environment, which drew the university towards greater involvement in the service of the community, and the requirements of the academic community, which demanded the greatest possible freedom for concentrating on pure science and scholarship. If we disregard small colleges serving particular local, religious or ethnic groups, the greatest pressures were probably exerted by unenlightened state governments. These had the power, through their control of financial support, to force state universities into performing various non-academic services. In principle, boards of trustees of private universities had similar powers, but in practice, at the most important private universities, they tended to share the academic rather than the non-academic outlook, at least in their capacity as trustees. Other groups, such as professional and voluntary associations, could only try to influence the universities by offering them support in exchange for the establishment of professional schools and the performance of similar services.

As a result, the leading private universities which were the centre of the system had to contend only with a relatively limited amount of pressure to compromise their standards.[61] They were sustained by the prestige of their scholars and scientists, by the fame of their institutions —not always exclusively intellectual—and despite the equalitarian deference system in the United States, intellectual eminence, relative economic independence and faithful and well-placed trustees succeeded in protecting the autonomy of intellectual activity.

Since the Second World War the outside pressure has not only been exerted on presidents, but also on the members of the teaching and research staff. Since then there has been a great acceleration of what Weinberg has called "the force-feeding" of scientific growth.[62] As has been shown, this process had started earlier, and it began probably with the setting up of graduate schools in different professions. But since the Second World War the trend has penetrated the basic humanistic and scientific disciplines as well. Pure scientists and scholars have increasingly changed their habits of working merely on what they think intellectually the most promising thing to work on, and watching for intellectual challenge and stimulation from others. Instead they very often try to "attack" a broad and vaguely promising field through massive research effort, thinking that although much of the total effort will be wasted, a few important findings hit upon in the course of it will justify the whole effort. Indeed, a certain amount of present-day "big science" e.g., bubble chamber research, is of this sort. Much of this force-feeding does not arise from external pressure but from the ambitions of academic scientists whose imagination is aroused by the awareness of the large resources available. It is not possible yet to assert whether the powerful self-renewing traditions of concentrated pure science will be able to withstand the attractions of "big science."

Another problem which arises from the present system is the virtual monopoly of the federal government in the financing of research.[63] Thus far there is no evidence that this has hurt science, and there is much evidence that it has helped. Perhaps the grant system and other measures of decentralisation which were deliberately incorporated into the system have been sufficient to stay the threat of political pressure on science, and especially on the universities. Yet the negligibility of deleterious effects so far might be a function of the very great expansion of financial resources which made it almost unnecessary to make choices. When expansion has slowed down, a situation might arise where political considerations interfere seriously with scientific choice. It has indeed been argued that such influence has actually been exerted imperceptibly but on a large scale through shifting the focus of research towards militarily relevant fields of science.[64]

This extraordinarily proliferating scientific and scholarly activity released by the diversification of the internal structure of university departments, by institutional pluralism and incessant enterprise has now grown so prominent in American society that it calls forth the danger of the potential politicisation of the university. This problem has become quite acute in the United States twice during the last 25 years. Thorough and indiscriminate anti-communism of the decade following the war disturbed the calm of the universities; more recently, left-wing groups within the universities have sought to politicise them. McCarthy and his ilk wished to suppress "un-American activities" in the universities to guarantee the security of American society; the "new left" wishes to use the universities as bases for the disruption of American society. Both have wished for conflicting purposes to politicise the universities for extra-academic purposes.

Never before has there been such a large proportion of the 18-25 age-group at the universities for such long periods as there are at present. A very large proportion of young people in the years when they are potentially most sensitive to public affairs and potentially most available for politics are now at universities. The universities are now in a position in which they can become very important centres of political activity.

There is finally another new element in the situation. One of the distinctive features of the United States system has always been the willingness of graduates, especially of those who possessed only a first degree, to enter all kinds of occupations. This prevented the emergence of a significant group of university graduates who, either because of the specificity of their training or the level and content of their social aspiration, were unwilling to enter any but a few prestigious and well-remunerated occupations. The resulting existence of a large number of "unemployed intellectuals" had much to do with the alienation and radicalisation of intellectual politics in Europe in the first third of the present century. This phenomenon has been virtually absent in the United States.

But this situation might not continue. The sudden rise of graduate education in fields for which there is no specific demand, and where the criteria of competence are not quite unequivocal, might create a problem of excess supply, alienation and radicalisation of students and intellectuals in the United States. If that happens it will certainly be an unforeseen consequence of the dethronement of the haughty professor of the Hohenzollern epoch.

NOTES

1. Lexis, W. (ed.), *Die deutschen Universitäten: für die Universitätsausstellung in Chicago* (Berlin: A. A. Ascher, 1893). See Vol. 1, p. 119 and p. 146, and for other countries, p. 116; Lexis, W. (ed.), *Das Unterrichtswesen im deutschen Reich,* Vol. I, *Die Universitäten* (Berlin:

A. A. Ascher, 1904), pp. 652-653; and Paulsen, Friedrich, *Geschichte des gelehrten Unterrichts an den deutschen Schulen und Universitäten vom Ausgang des Mittelalters bis zur Gegenwart* (Berlin and Leipzig: Vereinigung Wissenschaftlicher Verleger, 3rd ed., 1921), Vol. II, pp. 696-697. For Scientific developments see Lexis, W., *op. cit.*, 1904, pp. 250-252.

2. The following table shows the numbers of foreign students in Germany from the countries which later became the most important centres of science.

Foreign Students from Selected Countries at German Universities
1835-36—1891-92

	NUMBER OF STUDENTS AND YEAR			
COUNTRY	1835-36	1860-61	1880-81	1891-92
Britain	26	42	71	137
Russia	64	156	204	407
United States	4	77	173	440[a]

[a]The total of 446 is given for the whole American continent, but judging from previous years this included only a few individuals from countries other than the United States. SOURCE: Lexis, W., *op. cit.*, 1893, p. 128. These figures might not include advanced students who did not work for a particular degree. For more qualitative descriptions, see the sources cited in footnote 19 and Cardwell, D.S.L., *The Organisation of Science in England* (London: Heinemann, 1957), pp. 50-51, p. 106 and p. 135.

3. The following table shows the development of the budget of the Prussian universities during this period (in Reichsmarks):

Year	Recurrent Expenditure	Capital Expenditure	Special Funds For Personnel
1871	4,150,254	831,572	—
1881-82	5,573,775	1,565,188	294,094
1897-98	11,662,343	2,171,554	655,952
1907-08	16,647,269	3,983,272	995,674

SOURCE: Paulsen, Friedrich, *op. cit.*, pp. 704-05. For the virtual absence of organisational change, see *ibid.*, p. 700 and p. 706.

4. For the ratio of full professors to other ranks, see Lexis, W., *op. cit.*, 1893, p. 146, and *op. cit.*, 1904, p. 653, and for the differences between fields see Ferber, Christian von, "Die Entwicklung des Lehrkörpers der deutschen Universitäten und Hochschulen, 1864–1954," in Plessner, H. (ed.), *Untersuchungen zur Lage der deutschen Hochschullehrer* (Göttingen: Vandenhoeck und Ruprecht, 1956), pp. 54-61 and p. 81. For the foundation of the various unions, see Paulsen, Friedrich, *op. cit.*, p. 708. The whole problem has been surveyed in Busch, Alexander, *Die Geschichte des Privatdozenten* (Stuttgard: F. Enke, 1959), and "The Vicissitudes of the *Privatdozent:* Breakdown and Adaptation in the Recruitment of the German University Teacher," *Minerva*, I, 3 (Spring, 1963), pp. 319-341. The difficulties in the social sciences are described in Oberschall, Anthony, *Empirical Social Research in Germany* 1868-1914 (Paris and The Hague: Mouton, 1965), pp. 1-15 and pp. 137-145.

5. See Ferber, Christian von, *op. cit.*, pp. 62-74.

6. Cf. Paulsen, Friedrich, *op. cit,* pp. 725-726 and pp. 737-738.

7. Ben-David, Joseph, "Roles and Innovation in Medicine", *American Journal of Sociology*, LXV, 6 (May, 1960), pp. 557-568.

8. The single exception in respect of social science was the University of Frankfurt am Main. Paulsen, Friedrich, *op. cit.*, p. 710.

9. See Zloczower, A., *Career Opportunities and the Growth of Scientific Discovery in Nineteenth Century Germany, with Special Reference to Physiology* (Jerusalem: Hebrew University, 1966), pp. 38-44.

10. Such attitudes were prevalent even in physics, which was the least conservative discipline in about 1900. See Lexis, W., *op. cit.*, 1904, pp. 250-252. In other fields the opposition was more vocal and dogmatic. For the opposition to the training of high school teachers in educational methods, see Paulsen, Friedrich, *op. cit.*, pp. 275-276 and pp. 711-712, and for the similar refusal to incorporate training for the new medical specialties in the medical schools, see Lexis, W., *op. cit.*, 1904, pp. 129-130 and pp. 151-152.

11. The best known and most controversial figure of these interfering civil servants was F. Althoff, who was head of the universities section of the Prussian Ministry of Education during 1897-1907; see Paulsen, Friedrich, *op. cit.*, p. 706.

12. See Ferber, Christian von, *op. cit.*, pp. 71-72, and Zloczower, A., *op. cit.*, pp. 101-125.

13. *Ibid.*, pp. 114-115 (about physiological chemistry). Even in such a theoretically important field as physical chemistry, there were only five institutes—Leipzig, Berlin, Giessen, Göttingen and Freiburg—and five subordinate positions *(Extraordinarii* who had special departments in institutes headed by someone else) in Breslau, Bonn, Heidelberg, Kiel and Marburg in 1903, almost 20 years after Ostwald founded the first chair in the field in Leipzig (1887) and more than 20 years after the publication of his famous textbook and the launching of a journal in the field; see Lexis, W., *op. cit.*, 1904, pp. 271-273.

14. See Ferber, Christian von, *op. cit.*, pp. 54-61.

15. For a survey of the many beginnings of the social sciences, none of which produced a real academic response in Germany, see Oberschall, Anthony, *op. cit.*, pp. 1-5.

16. This is the same sort of question as that dealt with by ecologists who study the effects of the transplantation of plants or animal species from one environment to another. Appropriately the approach was introduced to the study of the diffusion of university institutions by Sir Eric Ashby (in association with Mary Anderson). See Ashby, Eric, *Universities: British, Indian, African. A Study in the Ecology of Higher Education* (London: Weidenfeld and Nicolson, 1967). See also for a discussion of the sociological issues involved, Shils, Edward, "The Implantation of the Universities: Reflections on a Theme of Ashby," *Universities Quarterly*, XXII, 2 (March, 1968), pp. 142-166.

17. See Veysey, Lawrence R., *The Emergence of the American University* (University of Chicago Press, 1965), pp. 160-161 and p. 166. The desire to follow the German model as closely as possible was especially marked in the faculties of the graduate schools. University presidents tended to be more pragmatic.

18. See Zloczower, A., *op. cit.*, pp. 64-66.

19. Interesting accounts of the experiences of American students in Germany can be found in Perry, Ralph Barton, *The Thought and Character of William James* (Boston: Little, Brown and Company, 1935), Vol. I, pp. 249-283, and Fleming, Donald, *William H. Welch and the Rise of Modern Medicine* (Boston: Little, Brown and Company, 1954), pp. 32-54, and pp. 100-105. For the situation in the twenties and the thirties with an interesting attempt to compare the perceptions of the visiting outsiders with those of the German students, see Weiner, Charles, "A New Site for the Seminar: The Refugees and American Physics in the Thirties," to be published in *Perspectives in American History*, Vol. 2 (Charles Warren Center for Studies in American History, Harvard University, 1968).

20. For an exposition of the tradition of the land grant colleges, see Morrill, James Lewis, *The Ongoing State University* (Minneapolis: The University of Minnesota Press, 1960); and for an evaluation, see Bowman, Mary Jean, "The Land Grant Colleges and Universities in Human Resource Development," *Journal of Economic History*, XII, 4 (December, 1962), pp. 547-554.

21. The best exposition and the most convincing attempt at the justification of the German university is to be found in the different writings of Abraham Flexner, namely, *Universities: American, English, German* (Oxford University Press, 1930), and *I Remember* (New York; Simon and Schuster, 1940). For an exposition of some of the shortcomings see Paulsen, Friedrich, *op. cit.,* pp. 710-738.

22. See Ben-David, Joseph, *op. cit.*

23. This term may be useful to distinguish that kind of applied research which assumes the form of an academic discipline, from applied research which does not. It is impossible to say what this difference is due to, but it is probably related to the intellectual quality of the innovation and the usefulness of training people in the field.

24. See footnotes 12, 13 and 14.

25. See Flexner, Abraham, *Medical Education: A Comparative Study* (New York: Macmillan, 1925), pp. 221-225.

26. *Ibid.* See also Fleming, Donald, *op. cit.,* p. 110 (about the superiority of training at Johns Hopkins). For a criticism of attempts at misconceived efforts to develop certain fields into quasi-disciplines, see Flexner, Abraham, *op. cit.,* 1930, pp. 152-177.

27. How this monopoly led to an erosion of the function of training for professional practice is amply illustrated by Paulsen, Friedrich, *op. cit.,* pp. 225, 261, 262-264, 269, 274-275, and pp. 711-714.

28. For a partly very critical description of the growth of institutes not related to teaching see Flexner, Abraham, *op. cit.,* 1930, pp. 110-124; for the opposite view, see Morrill, James Lewis, *op. cit.,* pp. 24-37.

29. In Germany the most important organisations belong to the Max Planck (previously Kaiser Wilhelm) Gesellschaft; in Britain to the various research councils; and in France to the Centre National de Recherche Scientifique (CNRS). There are governmental and private research institutions in the United States but most research in the basic sciences is performed in universities.

30. See Clark, Terry, "Institutionalisation of Innovation in Higher Education: Empirical Social Research in France, 1850–1914" (unpublished doctoral thesis, faculty of political science, Columbia University), pp. 19-21.

31. See Clark, Terry, "Discontinuities in Social Research: The Case of the *Cours Elémentaire de Statistique Administrative'',* *Journal of the History of the Behavioral Sciences,* III, 1 (January, 1967), pp. 3-16.

32. The best-known case to illustrate this point was the attitude to Mendel's work by one of the most outstanding botanists of his time and the ensuing fate of his discovery. See Barber, Bernard, "Resistance by Scientists to Scientific Discovery", *Science,* 1 September, 1961, pp. 596-602.

33. See Clark, Terry, *op. cit.,* and for the situation in Germany, Lexis, W., *op. cit.,* 1893, pp. 598-603.

34. See Fitzpatrick, Paul J., "The Early Teaching of Statistics in American Colleges and Universities", *The American Statistician,* X, 5 (December, 1955), pp. 12-18; Glover, James W., "Requirements for Staticians and Their Training", *Journal of the American Statistical Association,* XXI (1926), pp. 419-424, which includes detailed information on the teaching of statistics in departments of mathematics, economics, and social science, schools of business, education, public health, psychology and agriculture.

35. The leading figure at Iowa was Henry L. Rietz, A Cornell-trained mathematician, who prior to his appointment as professor of mathematics at Iowa had been for more than 10 years professor of mathematics at the University of Illinois and statistician of the University of Illinois College of Agriculture. His first publication was a 32-page appendix to a treatise on breeding (1907); see *Annals of Mathematical Statistics,* XV (1944), pp. 102-104; Weida, F. M., "Henry Lewis Rietz 1875–1943," *Journal of the American Statistical Association,* XXXIX (1944), pp. 249-251. For a detailed description of the history of this centre, see Dodson, J. C., "The Statistical Program of Iowa State College," *The American Statistician,* II, 3 (June, 1948), pp. 13-14.

36. Hotelling went to Rothamsted in 1929 and those working in London included Samuel S. Wilks (1932-33) and Samuel A. Stouffer. For the beginnings of the movement towards mathematical statistics in the 1920's, see Craig, A. T., "Our Silver Anniversary," *Annals of Mathematical Statistics,* XXXI (1960), pp. 835-837.

37. 95 fellows of the Institute of Mathematical Statistics in 1967 received their Ph.D.s in the United States. The largest numbers received their doctorate from Columbia and Princeton—17 from each—followed by North Carolina and the University of California at Berkeley with nine doctorates from each. But many of those who did not receive their degrees from these universities were influenced in one way or another by these centres, especially by that at Columbia. This information is based on an analysis of data from *Statisticians and Others in Allied Professions* (American Statistical Association, 1967), and from *American Men of Science* (Tempe, Arizona: J. Cattell Press, 1962).

38. Of the foreigners who came to the United States during the 1930's were A. Wald (Columbia), J. Neyman (University of California, Berkeley) and several others.

39. See Craig, A. T., *op. cit.*

40. The department of statistics at the University of North Carolina was founded in 1946-1947.

41. The incumbent of the chair was Egon Pearson.

42. For this information I am indebted to Professor Leo Goodman of the University of Chicago.

43. The oldest tradition of this kind of scientific policy-making exists in France. See Gilpin, Robert, *France in the Age of the Scientific State* (Princeton: Princeton University Press, 1968), pp. 77-123. In Germany central scientific policy-making became a matter of importance only in the 1890's; see Schreiber, Georg, *Deutsche Wissenschaftspolitik von Bismarck bis zum Atomwissenschaftler Otto Hahn* (Cologne and Opladen: Westdeutscher Verlag, 1954), pp. 25-36. In both of these cases the assumption of this function by the central government marked the beginning of the end of the scientific supremacy of these countries. The British tradition has been somewhat different, since central government in Britain has usually preferred to leave a great deal of discretion to independent bodies, like the University Grants Committee, the research councils and the universities. This has also ensured a greater degree of pluralism in Britain than in continental Europe. See for the nineteenth century, Cardwell, E. T. S., *op. cit.,* Armytage, W. H. G., *Civic Universities* (London: Ernest Benn, 1955). For the present, see Kneller, George F., *Higher Learning in Britain* (London: Cambridge University Press, 1955); and the OECD *Reviews of National Science Policy* which are available for the majority of the OECD member countries.

44. See Veysey, Lawrence R., *op. cit.,* pp. 10-18, pp. 81-88, and pp. 158-159.

45. See Veysey, Lawrence R., *op. cit.,* pp. 302-311.

46. *Ibid.,* pp. 321-332 on the development of departments.

47. Committees have also been interdisciplinary organisations, but they have been mainly for teaching and training and not so much for research.

48. See Flexner, Abraham, *op. cit.,* 1930, pp. 110-111. Committees had apparently not yet existed in the twenties, otherwise they would probably have been mentioned by Flexner. A search in the catalogues of the University of Chicago showed that at that university they began to appear in the 1930's.

49. See Busch, Alexander, *op. cit.,* pp. 109-117.

50. *Ibid.,* pp. 70-71.

51. Professors of chemistry do not sit on committees to appoint professors of sociology!

52. See Hofstadter, R., and Metzger, Walter P., *The Development of Academic Freedom in the United States* (New York: Columbia University Press, 1955), pp. 396-412, for the development of the specifically American conception of academic freedom.

53. Cf. Zloczower, A., *op. cit.,* pp. 29-38.

54. See Kornhauser, William, *Scientists in Industry: Conflict and Accommodation* (Berkeley: University of California Press, 1962), p. 71 ff.; Mareson, Simon, *The Scientist in*

Industry (New York: Harper, 1961), pp. 52-57, and Kornhauser, William, "Strains and Accommodations in Industrial Research Organisations in the United States", *Minerva*, I, 1 (Autumn 1962), pp. 30-42.

55. The statement concerning journals is based on counts of physics journals at the University of Chicago Library and information from experts on other fields.

56. For the superiority of physics laboratories and other arrangements for physical research, see Weiner, Charles, *op. cit.;* for medical research, see Flexner, Abraham, *op. cit.,* 1925, pp. 221-226.

57. With the exception of a few large industrial research laboratories, there is little research performed in European industry. For the differences in total investment in general, and development work in particular, see Table 1 below.

58. See OECD, *Reviews of National Science Policy: France* (Paris: OECD, 1966), pp. 41-43; *United Kingdom, Germany* (Paris: OECD, 1967), pp. 60-66.

59. See Ben-David, Joseph, *Fundamental Research and the Universities* (Paris, OECD, 1968), pp. 67-75.

60. Compiled from Organisation for Economic Cooperation and Development, *The Overall Level and Structure of Research and Development Efforts in OECD Member Countries* (Paris: OECD, 1967), p. 14, p. 57 and p. 59.

61. "Centre" is used here to refer to an integral part of a system which serves as a model for the rest. See Shils, Edward, "Centre and Periphery," in *The Logic of Personal Knowledge: Essays presented to Michael Polanyi* (London: Routledge and Kegan Paul, 1961), pp. 116-130, and "Observations on the American University," *Universities Quarterly*, XVII, 2 (March, 1963), pp. 182-193.

62. See Weinberg, Alvin M., *Reflections on Big Science* (Cambridge, Mass.: The MIT Press, 1967), p. 106.

63. See Organisation for Economic Cooperation and Development, *Reviews of National Science Policy: United States* (Paris: OECD, 1968), p. 191, for the sources of funds for research by United States universities and colleges. Of a total of $2,510 million spent on research and development, $1,740 million was provided by the federal government.

64. Weinberg, Alvin M., *op. cit.,* pp. 65-84.

ALEXANDER VUCINICH

Mathematics in Russian Culture

The great mathematical tradition in the Soviet Union is a legacy of the *ancien régime*. Its development has been comparatively short: at the time when the West was working on the complex machinery of the calculus, Russia did not have even an elementary textbook in mathematics. During the XVIIth century only one mathematical book was published in Russian providing elementary knowledge useful to the "sellers and buyers."[1] The first textbook, prepared by L. F. Magnitskii, appeared in 1703 and for several decades was the most widely used source of mathematical knowledge in Russia. Not an original work but an encyclopedia of translated articles on various topics in arithmetic, geometry, astronomy, and navigation, this book marked the beginning of mathematics in Russia as a specific scientific discipline and as a subject taught on elementary and secondary levels. The cumulative effects of the most

Alexander Vucinich, "Mathematics in Russian Culture," *Journal of the History of Ideas* 21 (April–June 1960): 161-79. Reprinted by permission of the publisher.

striking historical and cultural developments which have molded the mathematical tradition in Russia are the subject of our discussion.

During his Western trips Peter I was given ample opportunity to learn that the development of those practical sciences in which he was primarily interested as gunnery, cartography, land surveying, and navigation depended on mathematics. He fully realized that his ambitious modernization program hinged to a great extent on an extensive dissemination of mathematical knowledge in Russia. He had no foolproof designs and was compelled to resort to experiments. His initial plan of sending young Russians to Western schools brought no desirable results: the Russians were unaccustomed to travelling abroad in search of education and had difficulty in adjusting themselves to the Western way of life and intellectual atmosphere. The traditional, deeply rooted antagonism toward foreigners, which was beginning to thaw after the 1650's, was still a force of great importance. The few Russians who were educated in the West served mostly as translators of foreign books, particularly in mathematics and navigation, and none of them became known through original work.[2]

Peter's next program included the opening of secular schools. In 1698 he invited Fergharson, a young British mathematician from Aberdeen University, to come to Russia and organize the teaching of mathematics and navigation. As a result, in 1701 the School of Mathematical and Navigational Sciences was opened in Moscow. The curriculum included arithmetic, algebra, and plane and spherical geometry, all taught primarily by Fergharson, who for this purpose wrote several manuals.[3] In 1715 this school was transferred to St. Petersburg where it became known as the Naval Academy. One of the tasks assigned to Fergharson's school was to train teachers for elementary schools which were opened in all provincial capitals and were known as ciphering schools, a name indicative of the emphasis placed on the dissemination of elementary mathematical knowledge: their avowed aim was to teach "numbers and certain parts of geometry."[4] In 1715 the first teachers were sent to the provinces. The country was not yet ready for formal education and the ciphering schools brought but meager initial results. The Church persuaded Peter to allow the children of the clergy to attend the parish schools where little if any emphasis was placed on the sciences. The tradespeople succeeded in obtaining permission which freed their children from the obligation of attending these schools. In 1722 there were forty-two ciphering schools; by 1744 only eight were left.[5] Gradually they went out of existence and were partially replaced by garrison schools, opened at the headquarters of each regiment and taught by army officers; these schools for a long time were the major secular training centers of literate citizenry and a source of teachers. The basic contribution of these schools was that their emphasis on the teaching of mathematics became a tradition never to be pushed to a secondary position.

Peter's contribution to the development of mathematical thought in Russia went even further. Through his encouragement, and often under his direct guidance, a number of scientific works were translated into Russian and through them the names and contributions of Euclid, Descartes, Leibniz, and Newton became known in Russia. The influence of the translated works was not very strong because Russia had virtually no readers who possessed the requisite elementary knowledge in mathematics.[6]

The last step planned by Peter was the creation of a scientific forum. After personal observation of the work of the Royal Society in London and the Academy of Sciences in Paris, and conversation with Leibniz and with Christian Wolff, he decided to found an academy of sciences which would differ from its Western European models in that its members would not only pursue investigations in their respective disciplines but would also serve as professors of the country's first university. Peter did not live to see his last project carried out. He died in 1725, a half year before the arrival of the first members of the St. Petersburg Academy of Sciences.

As a result of recommendations and direct assistance by Wolff, the newly founded Academy boasted of four distinguished mathematicians: Jacob Hermann, who previously taught at the universities of Padua and Frankfurt on the Oder, two promising sons of the great Swiss mathematician Johann Bernoulli (Nicholas and Daniel), and Christian Goldbach. It was in 1727 that the Academy admitted to its ranks the youthful Leonhard Euler, who was to emerge as the brightest star in the entire history of this scientific body and as the world's leading mathematician of the XVIIIth century.

Euler stayed in Russia from 1727 to 1741 and from 1766 to 1783 (the year of his death). In 1741, because of "unfavorable conditions" for scientific work and a more attractive offer by Frederick II, he joined the Berlin Academy of Sciences. He returned to Russia twenty-five years later on the invitation of Catherine II. Although in 1741 he chose to leave St. Petersburg, Euler's sentimental attachment to Russia was unwavering. He came to Russia as a young man of nineteen years of age, he received his apprenticeship in the St. Petersburg Academy of Sciences, and it was here that he produced papers which were eagerly awaited in the intellectual centers of the West. Although the reasons why he left Russia in 1741 will probably never be fully explained, it is quite conceivable that he was motivated by a search for more advantageous working conditions. During the 1730's the Academy was mismanaged and torn by continuous strife between the Russian and German factions. The government authorities, unheeding Peter's legacy, showed little understanding and appreciation for scientific theory. To make Euler "earn" his government stipend he was given many practical assignments unworthy of his genius. He even had difficulty in getting adequate living quarters.

Fergharson and Magnitskii introduced the rudiments of modern mathematics to Russia; Euler covered the whole mathematical spectrum. He wrote introductory texts of algebra and popular accounts of mathematical achievements, but he also scaled the heights of mathematical wisdom. He systematized the entire contemporary knowledge of the calculus and indicated the new paths of theoretical development and practical application which were more fully explored by his numerous successors. He freed Descartes' geometry of the last ancient strangleholds and gave it a full algebraic basis. The texts and manuals of Magnitskii and Fergharson were too elementary to prepare the reader to understand the great mathematical achievements of the XVIIth century; Euler continued where Fermat, Descartes, Leibniz, and Newton left off. Fergharson and Magnitskii introduced mathematics as a school subject; Euler introduced it as a field of scientific inquiry. "We learned everything we know from Euler," exclaimed the French mathematician Laplace at the end of the XVIIIth century, thus summing up the place of Euler in the mathematics of his century. "All the noted mathematicians of the present-day [stated Condorcet] are his pupils: there is no one of them who has not found himself by the study of his works, who has not received from him the formulas, the method which he employs; who is not directed and supported by the genius of Euler in his discoveries."[7]

Euler, more than any other individual, made the St. Petersburg Academy a known and respected institution in the educated world of the XVIIIth century. Through his rich personal correspondence with scholars of the West he brought the Academy into direct contact with many leading scholars and learned societies. He raised the standards of its scientific output to enviable heights, a lofty challenge to his colleagues and disciples. It is because of Euler's contributions to the Academy's *Commentarii* that Daniel Bernoulli (who by that time had left Russia) wrote to him: "I have no words with which to explain how eagerly the people everywhere inquire about the St. Petersburg memoirs. . . ."[8] Euler, more than any other person, made the intellectual traffic between Russia and Western Europe a two-way street.

Euler belonged to the group of imported scholars who developed a genuine affection for their adopted homeland. He was not a part of the powerful German faction in the Academy which placed various obstacles on the road of budding Russian scholarship, nor, for that matter, did he side with Russian firebrands who deliberately magnified the dangers inherent in the domination of foreign scholars in the Academy. He encouraged the authorities to assist promising Russian youths in their search for specialized scientific education. During the twenty-five years of his membership in the Berlin Academy of Sciences his ties with the St. Petersburg Academy were as strong as ever. He continued to send his contributions to the *Commentarii* and *Novi Commentarii;* in fact, during the Berlin period Euler published 109 works in Russia, as against 119 in Berlin.[9] In 1750–55 he received from St. Petersburg mathematical papers written by Russian students for criticism. He was asked to appeal to certain foreign scholars who had left Russia to return to St. Petersburg and rejoin the Academy. When the internal conflict within the Academy led to the loss of its prestige in the West Euler was requested to use his great influence to help restore the earlier reputation of the learned society. His Berlin residence became the home of a number of Russian students. In 1752–56 he provided quarters and tutoring to S. Ia. Rumovskii and S. K. Kotel'nikov, who later made noted mathematical contributions.[10] Euler's *Elements of Algebra* initiated a whole generation of young Russians into the mysteries of mathematics and became a model followed by the textbook writers at the end of the XVIIIth century. Thanks to the encyclopedic nature of his knowledge and his powerful and adaptable style of writing, Euler addressed himself to three distinct audiences: through his papers in the *Commentarii* and the periodicals which succeeded it he communicated with the scientific élite made up primarily of foreigners; through books like the *Elements of Algebra* he reached the students of the academic university, and, finally, through his popular scientific articles in the *St. Petersburg Journal* and his *Letters to a German Princess* he became known to the general reading public in Russia.[11]

Although Euler's work in mathematics established a highly honored place for this science in Russia, it was not before the twilight of his long career that he could point to individual mathematicians as his disciples. When he came to Russia the country had no young men adequately trained to pursue higher studies in mathematics. The academic university could not develop into a serious institution of higher learning because the academic gymnasium, the only school of its kind in the country, made no appeal to the Russian youth and could not become a going concern. The first students were brought from abroad and later the university was forced to recruit its students from theological schools whose curricula provided no room for mathematics and generated no scientific aspirations or even tolerance toward the realm of scientific thought.[12]

There was another reason why Euler's influence in Russia did not become paramount before the last third of the century. In the late 1740's —while Euler was in Berlin—Mikhail Lomonosov began to assert himself in many fields of scientific endeavor. This versatile and talented scientist—the first Russian scientist *par excellence*—began to exert an influence on the ideas and aspirations of young Russian scholars. By his temperament, philosophy, and scientific attitude he was an antidote to Euler. While both were possessors of knowledge of encyclopedic magnitude, Lomonosov was an experimentalist with Cartesian leanings and with little need for mathematics, while Euler was a great adherent of the Newtonian tradition which led to the elevation of mathematical analysis as the principal instrument for the understanding of natural-scientific problems. Lomonosov made serious efforts to expand the experimental basis of natural science. Euler covered a wide range of interests but his mathematical bent dominated all his scientific thinking. "It cannot be denied that, in general, his attention is more occupied with the analysis itself, than with the subject to which he is applying it; and that he seems more taken up with the instruments, than with the works, which they are to assist him in executing."[13] Euler as a mathematician travelled a scientific road unhampered by ideological pressures from established political and church authorities; he himself engaged in a search for demonstrations of the existence of God without bringing his Calvinist devotion into conflict with the scientific and philosophical principles of mathematical analysis.

Lomonosov never expressed any direct challenges to the religious creed of Eastern Orthodoxy but he did ask questions on the structure of the basic material elements of which the world is made, questions which were incompatible with the established religious beliefs; he also appealed to the quasi-literate clergy to be more tolerant of the work done by scientists.[14] Euler worked in a field which had a tradition and eager and ready protagonists; Lomonosov asked more basic philosophical questions which found little support in tradition and generated little enthusiasm in the contemporary world of scholarship, primarily because of its theoretical and mathematical unreadiness to cope with them. Lomonosov had no central interest: he covered the wide gamut from rhetoric and Russian history to physical chemistry and geology. He composed odes and tragedies and translated librettos, and at the same time wrote profound papers in the various natural sciences. In the history of Russian science he proved to be more an inspiration to young scholars than the founder of a serious school of scientific thought. He showed Russia's readiness to cope with scientific problems on the highest level.[15] However, this readiness found its most formidable expression not in Lomonosov's Cartesianism but in the Newtonian tradition of Euler. In the Academy there was not one person who could be identified as Lomonosov's follower, although there were several scholars who were initially trained, encouraged, and inspired by him. When Euler died (1783), eight academicians—exactly one-half of the total number—were his disciples, including four Russians.

Outside the Academy there was no important scientific center. The University of Moscow, founded in 1755, suffered from a chronic shortage of both students and professors. During the first five years of its existence the university did not offer a course in mathematics, and when it was introduced it was taught for a while by A. A. Barsov, who soon realized that his talents lay elsewhere and shifted to the teaching of literature and philology. Subsequently mathematics was taught by D. S. Anichkov, who also lectured on logic and metaphysics and who after several years of teaching defended a dissertation on natural theology.[16] He did, however, translate several manuals of arithmetic, algebra and plane trigonometry from Latin.[17] The calculus was not taught in Moscow University before the beginning of the XIXth century.[18] During the late 1820's and the early 1830's mathematics was taught at the University of Moscow by Shchepkin who was known as an excellent lecturer but poor scholar and I. A. Davydov who, according to one of his contemporaries, was at the same time a "philosopher, philologist, historian, critic, Latinist, Hellenist and mathematician" but who actually was a dilettante.[19] Until well into the XIXth century it was the Academy which dominated the scientific thought in Russia, and it was mathematics which ruled supreme in the Academy.

The contributions of Euler's disciples to Russian culture were considerable, despite the fact that none of them was an outstanding mathematician of his time. During the last years of Euler's life, some of them helped the old master to carry on his scientific work without interruption. Nicholas Fuss proved to be Euler's Eckermann; his sympathetic attitude, mathematical skill and enviable working capacity helped the blind Euler in the preparation of some three hundred papers, most of which were published posthumously.[20] W. L. Krafft, who arrived in Russia in 1767, helped Euler in the writing of the *Dioptrica* (3 vols.), an integrated presentation of his earlier efforts to build optical instruments. P. B. Inokhodtsev (together with I. Iudin) translated into Russian Euler's *Elements of Algebra* (published in 1768/69) from a German manuscript. S. Ia. Rumovskii translated the *Letters to a German Princess* and published several papers on the integration of functions and differential equations.[21] S. K. Kotel'nikov, one of Euler's favorites among Russian mathematicians, was the head of the team which translated Christian Wolff's book on the "foundations of mathematics" to which he added an original survey of the principles of the differential and integral calculus (the first study in Russian on mathematical analysis).[22] M. E. Golovin translated Euler's *Théorie complette de la construction et de la manoeuvre des vaisseaux* (1778). In 1789 Golovin published *Plane and Spheric Trigonometry with Algebraic Proofs* which was based on Euler's ideas and was designed to serve as a textbook.

We need not go into details on the productivity of Euler's students to indicate those basic areas of their work through which they had enhanced the rôle of mathematics in Russian culture. In the first place, in their scientific work they pursued the avenues of mathematical inquiry mapped by Euler. Some of them confined their scientific work to pure mathematics, while the others channelled their energy toward mechanics, optics, meteorology, and astronomy. However, the unity of these scholars was expressed in their unequivocal adherence to mathematical analysis as the principal instrument in the hands of the natural scientist. In the second place, most of them distinguished themselves as teachers in various schools. Golovin brought high standards in the teaching of mathematics to the newly founded St. Petersburg Pedagogical Seminarium, Inokhodtsev and Rumovskii to the academic gymnasium, Kotel'nikov to the naval school, and Fuss to the infantry school. The teaching activity of these scholarly professors was matched by their equally important participation in the publication of mathematical textbooks. Almost every one of Euler's eight disciples had been active in the preparation of textbooks. Golovin's textbook in trigonometry (1789) stood out for decades because of its excellence and clarity. In 1783 N. Fuss published his *Leçons d' algèbre* which in 1789 appeared in its first Russian edition. Many other books were printed during the closing decades of the XVIIIth century, including several translations of works popular in Western Europe. At the time when the academic university had come to a virtual standstill (and eventual demise) and when Moscow University (the country's only other school of higher education) was at first without mathematics and then with hardly more than its rudiments, this textbook productivity was a remarkable phenomenon. It was a testimony to a high level of mathematical instruction in various military and pedagogical schools, and to the appreciation of mathematics in the general segments of a numerically small educated public. All the textbooks written by Euler's students reflected the mathematical interests of the great master. Although they did not add new substance to mathematical analysis, without them the Eulerian tradition would not have carved for itself such a prominent place in Russian culture.

At the end of the XVIIIth and the beginning of the XIXth century, several students of Euler participated in various curriculum committees and were noted for setting high standards of mathematical instruction. It is interesting that the general plan for secondary schools emphasized that the teaching of mathematics should begin early because "it cultivates the mind and sharpens the power of reasoning."[23] The charter of

5 August, 1789 of the so-called main public schools stipulated that most of the curriculum be devoted to arithmetic, algebra, geometry, mechanics, physics, natural history, geography, and civil architecture.[24] The same was repeated by the charter of gymnasiums and elementary schools of 5 November, 1804.[25] At the beginning of the XIXth century Fuss and Rumovskii were two of the three members of the committee on textbooks established by the Main Administration of Schools and played an important rôle in raising the standards of gymnasium mathematics textbooks.[26]

The outstanding contribution of Euler's disciples was that they took mathematics outside the comparatively isolated walls of the Academy and planted its roots firmly in the Russian educational system. As scientists they were a strong link between Euler and Russia's great mathematicians of the XIXth century, headed by N. I. Lobachevskii, M. V. Ostrogradskii, and P. L. Chebyshev and his formidable school. As educators they introduced high standards in general and professional secondary schools and paved the way for the high level of mathematical instruction in the new universities. In 1796 S. E. Gur'ev, the author of a book on "transcendental geometry and differential calculus," was the first Russian mathematician to become a member of the Academy who did not receive his education in the Academy's gymnasium and university. He was educated in the Artillery and Engineering Cadet School and spent some time in England where he studied hydraulic works.[27] T. F. Osipovskii, a graduate of the St. Petersburg Pedagogical School, acquired sufficient knowledge to write a university textbook in mathematics. He also translated Laplace's *Mécanique céleste* into Russian and wrote essays in which he attacked Kant's interpretation of the concept of space and time as subjective categories.[28] N. I. Lobachevskii, who received his formal education from Kazan gymnasium and Kazan University, presented the first logically coherent challenge to the fifth postulate of Euclid's geometry. When his teacher Johann M. Bartels came to Kazan University in 1808 to teach pure mathematics, he was impressed with the quality of knowledge which his students had brought from the gymnasium and found in them an alert and appreciative audience in his efforts to interpret the capital works of Euler, Lagrange, and Gauss.[29]

It is no wonder then that by the end of the second decade of the XIXth century, D. M. Perevoshchikov found enough readers in Russia to publish a six-volume mathematical encyclopedia. Although the teaching of mathematics at Moscow University continued to be poorly developed, it had achieved high standards in the universities of Kazan and Kharkov. In St. Petersburg University, founded in 1819, "there was never a dearth of competent teachers of mathematics and they always had receptive and well prepared students."[30] The output of mathematical literature continued to dominate the scientific field. From 1801 to 1806, for example, nineteen original works were published in Russian on various mathematical subjects as compared with fourteen original works in all physical and biological sciences.[31] Foreign visitors and scholars began to take note of the Russian interest in and "proclivity" toward mathematics. Bartels wrote in his reminiscences about Kazan University that at the first meeting with his mathematics class in Kazan he talked extemporaneously on various general themes but the students' questions compelled him to confess to them that the high level of their knowledge came as a surprise to him and he promised to deliver prepared lectures in the future.[32] Rommel, who taught at Kazan University from 1806 to 1815, also noted the emphasis given to mathematics and the various natural sciences in the gymnasium curriculum at the expense of classical literature and languages.[33] A. Erman, the author of *Travels in Siberia,* noted that the Russians "have a decided preference for mathematical studies in which they often succeed."[34]

During the first quarter of the XIXth century the general increase in the publication of mathematical studies, textbooks, and manuals came in the wake of an intensified Western influence. To the loyal and industrious custodians of the Eulerian tradition were added the champions of new intellectual currents. The German scholars who initiated the teaching of mathematics at Kazan University brought to Russia the orientation of youthful K. F. Gauss, ·whose intellectual daring led him to conclude that there may be other geometries as valid as Euclid's, a contention which was destined to find its proof and first logically consistent illustration in the work of Lobachevskii and Bolyai. Another influence emanated from France where such giants as Lagrange, Legendre, Fourier, Ampère, Laplace, Poisson, and Cauchy outlined whole new disciplines based on general mathematical formulas, but not before they had made substantial contributions to mathematical analysis and had thus essentially completed the work of Euler. Russia's most illustrious disciples of the "French school" were V. Ia. Buniakovskii and M. V. Ostrogradskii. Buniakovskii received the degree of doctor of philosophy from the University of Paris in 1825. He wrote the first Russian book on mathematical foundations of the theory of probability which he tried to apply to the study of various demographic problems. Ostrogradskii, a Ukrainian by birth, opened a new page in the history of mathematics in Russia. Until his time the Russian mathematicians, preoccupied with writing textbooks and surveys of general or specialized mathematical fields, did not produce anything to attract the attention of Western scholars. The turning point came in 1825 when Augustin-Louis Cauchy gave Ostrogradskii credit for having produced a new demonstration for one of his previously published formulas.[35] Subsequently, Ostrogradskii was elected a member of the American Academy of Arts and Sciences (1834). In 1826 the Paris Academy of Sciences accepted, and subsequently published, his paper on wave-motions of liquids in cylindrical containers. He wrote many papers in mechanics (including the voluminous *Cours de mécanique céleste*, St. Petersburg, 1831) and made a noted contribution in the field of the calculus of variations.[36] He became one of the most influential members of the St. Petersburg Academy of Sciences.

During the first three decades of the XIXth century three distinct mathematical orientations had their Russian exponents. Euler's disciples, after they had performed a meritorious scientific, educational, and civic function, finished their course by 1820 with N. Fuss' ironical rebellion against Lobachevskii's acceptance of the metric system (a product of the French Revolution). The "French school"—with Buniakovskii and Ostrogradskii as its most eminent protagonists—was in a sense a logical continuation of the time-honored Eulerian school. It had roots and tradition and also powerful spokesmen in academic councils. The third orientation had Lobachevskii as its only champion. Neither Buniakovskii nor Ostrogradskii were willing to accept Lobachevskii's revolutionary ideas in geometry; in 1832 and 1842 Ostrogradskii wrote adverse reviews of Lobachevskii's pioneering work. It was said that Ostrogradskii, who enjoyed the eminence of the first internationally recognized Russian mathematician, nurtured a supercilious attitude toward Lobachevskii, who, in addition to having been a "provincial scholar," wrote in a language devoid of lucidity and precision.[37] Lobachevskii's scientific unorthodoxy and thoughts of revolutionary proportions were the main reasons why neither the West nor Russia was ready before the 1860's to pay much heed to his work.

As the disciples of Euler were disappearing from the historical scene and Lobachevskii waited for recognition (which came posthumously), the French mathematical orientation became paramount in Russia. In the studies of Pafnutii Chebyshev, who worked in many mathematical fields, the tradition of the French mathematicians (the application of mathematical analysis to the natural sciences and technology) and the Eulerian school (particularly the theory of numbers) were fruitfully blended and opened new vistas for original work. He was the dominant figure in Russian mathematics during the second half of the XIXth century and his disciples held chairs in most of the country's universities. Several of them made contributions of general acclaim (A. M. Liapunov, A. A. Markov, A. N. Korkin, E. I. Zolotarev, G. F. Voronoi). During his long teaching career he offered courses in higher algebra, theory of numbers, theory of elliptic functions, probability theory, and the calculus of finite differences. In 1874 he was elected a *membre associé* of the Paris Academy of Sciences, having been the first Russian to receive this high honor. The versatility and productivity of his talents, as well as his unusual ability to communicate with his peers and students, enabled him to found what came to be known as the St. Petersburg school of mathematics. It was mathematics which provided the background for the creation of the first Russian scientific school.

Mathematics, together with astronomy, was the scientific area in which the most original work was done during the first six decades of the XIXth century. Yet, at this time the number of mathematicians was still exceedingly small. The university charter provided for only two positions to be filled by mathematicians (one in pure mathematics and one in applied mathematics). For this reason, for example, the announcement of courses of St. Petersburg University for 1836/37 listed only the calculus, algebra, analytical geometry, descriptive geometry, and theoretical mechanics.[38] However, mathematics developed more rapidly than any other science. It established itself as the leading science which attracted promising young Russian scholars. The roster of the Academy of Sciences from 1800 to 1860 shows that of four academicians in physics only one was Russian, of six astronomers, again only one was Russian, of seven chemists two were Russians, of eight botanists one was Russian, but of seven mathematicians six were Russians.

In the 1850's the West began to take serious note of Chebyshev's mathematical contributions. During the subsequent decades the growth of mathematics in Russia proceeded at an accelerated pace and reached a particularly high point with the formation of the Moscow school of mathematics in the 1920's which concerned itself with the most abstract mathematical areas and abandoned the tradition of the Petersburg school that each problem should be formulated with a clear view as to its practical applicability. It was not before the beginning of the XXth century, when the publications of Chebyshev's disciples began to appear in increasing numbers, that the West began to recognize fully the existence of a strong mathematical tradition in Russia.[39]

There is no single event or development which holds the key for the understanding of the rôle of mathematics in Russian culture. The ability of Peter I to grasp the fundamental place occupied by this science in the system of knowledge, the Euler accident (he was originally employed by the Academy to fill the chair of physiology), the versatile activity of Euler's students, the recognition of Ostrogradskii's mathematical work in the West, and Chebyshev's very essential contributions in the theory of numbers and the theory of probability are but a few of the major developments in the history of mathematics in Russia. The conditions which influenced the creation of the great mathematical tradition in Russia are to be sought in a combination of events and developments and not in any peculiarities of the Russian "psyche." Mathematics was introduced in Russia at the time when there was a concrete need for it but a need which was felt by a small group of noblemen who surrounded Peter I and not by the society at large. As a matter of fact, mathematics, as well as the other natural sciences, was introduced—or imposed on Russian culture—by the same coercive measures as were most other reforms of Peter I. It was forced upon Russian society at the time when all its classes and estates shared a common enmity toward formal education and scientific inquiry. We have seen what happened to Peter's ciphering schools when the scorching blows of his whips were removed. We have also mentioned the fact that Peter imported not only the first scientists-professors but also the first students. Animosity or apathy toward education and science remained unmitigated during the entire XVIIIth century. In 1780, in twenty-three private boarding schools in St. Petersburg, of 500 students only 200 were Russians, and of seventy-two teachers twenty were Russians.[40] In 1809, the government, in order to fill the classrooms of the newly founded gymnasiums and universities, made the level of completed education a criterion for promotions in the all-dominating government service.[41] Even with this policy the government was forced to depend on stipends as the essential weapon for luring young men to the universities and to allow all "free classes" access to higher education. The government monopoly in the sponsorship of education and science took another turn after 1828; when formal education had become an important part of Russian culture and young men from all walks of life were seeking formal training, the government began to make both gymnasium and university education a monopoly of the gentry. In 1833, according to official data, 78.9 percent of all gymnasium students were children of gentry and government officials. By 1863 this percentage was slightly reduced (72.3 percent).[42] In 1864/65, 67.2 percent of the university students belonged to these two groups.

Since from the very beginning mathematics in Russia was a "government science" one would expect that it would be dominated by a narrow practical orientation. Such, however, was not the case: the practical emphasis did not limit the scope of mathematical inquiry and did not close the doors to the most abstract branches of this science. It was the scholar himself whose authority counted in interpreting the adjective "practical." Euler set up the model for this interpretation which has become a strong tradition of its own; he wrote on many "practical" subjects in astronomy, navigation, optics, and mechanics, but most of these writings were actually highly theoretical and general studies in mathematical theory. Euler's treatise on mechanics, published in 1736, was the first great scientific work in which mathematical analysis was made the principal instrument for the study of motion.[43] In Russian universities for a long time "theoretical mechanics" was the chief representative of "applied mathematics," a tradition which produced such great scholars as N. A. Zhukovskii, A. N. Krylov, and S. A. Chaplygin. The St. Petersburg school of mathematics has consistently emphasized its practical orientation; yet its main contributions have been in the most abstract areas of the theory of numbers and probability theory. It seems paradoxical that during the first decade of Soviet rule when the government made the initial efforts to mobilize the country's scientists to participate in various government-sponsored practical projects, the most significant scientific achievements were made by such scholars as D. F. Egorov, N. N. Luzin, and A. Ia. Khinchin (theory of sets and theory of functions of real variables), P. S. Uryson and P. S. Aleksandrov (topology), A. N. Kolmogorov (axiomatics of the theory of probability), and many others who worked in the most abstract areas in mathematics; they subsequently have exemplified A. N. Whitehead's contention that "as mathematics grows more abstract, it becomes more effective as a tool for dealing with the concrete."

There was an important reason why mathematics developed in Russia more rapidly than the other sciences: it was less sensitive to ideological pressures and currents which dominated the thinking of government authorities. It was also less susceptible to influences emanating from various metaphysical schools. During the late 1810's and early 1820's, for example, philosophy, natural law, political economy, and ethics came under blistering attacks by M. L. Magnitskii, D. P. Runich, and other champions of conservative mysticism ushered in by the intellectual climate of the Holy Alliance; the subjects attacked were eventually scratched from the list of university courses. History was tightly controlled and the work of historians was subjected to ruthless ideological scrutiny. The physical and biological sciences were carefully watched by church authorities. This control discouraged any serious theoretical work in these sciences, particularly in biology, physiology, and geology. The scholar confined his activities to the mere accumulation of empirical data. These sciences were easy prey to such metaphysical currents as Schelling's *Naturphilosophie* which for two decades blocked the development of the scientific method and which by the middle of the 1830's had begun to retreat before the rising tide of Hegelianism.[44]

Mathematics, of course, could not but feel the negative impact of the anti-intellectualism of the high officials of the Ministry of Religious Affairs and National Education and of the pseudo-scientific orientation of the proponents of the *Naturphilosophie* who substituted deductions from ontological principles for calculation and experiment. Inspired by the then current mysticism, Nikolskii, professor of mathematics at Kazan University, transformed his science into a system of mystical symbols of the "sacred truths" of Christianity.[45] T. F. Osipovskii was expelled from Kharkov University for alleged adherence to ideas incompatible with the new philosophy of mysticism.[46] When Ostrogradskii returned from Paris in 1828, the St. Petersburg police was alerted to keep their vigilant eyes on his movements and associations.[47] It was not before Schelling's influence had begun to lose momentum that academician E. A. Collins could read his paper on "The Influence of Mathematics on the Development and Progress of Natural Philosophy" in which he attacked the dependence of the proponents of the *Naturphilosophie* on sheer intuition and ridiculed their production of "chimerical laws" and their diatribes against Newton. During this entire period, mathematics, with a deeply rooted tradition behind it, suffered very little from external obstacles to which most other sciences were exposed. It carried no strong ideological overtones which would bring it into open conflict with vested political interests and which would force it to make unhealthy adjustments to cultural forces inimical to scientific development. It is because of this "aloofness" of mathematics that in St. Petersburg University the teaching of this subject became an immediate success.[48] In Moscow University the largest number of students who were enrolled in the Physico-Mathematical Faculty (covering all the physical and biological sciences) began to study mathematics exclusively.[49] While the students preferred mathematics because of its philosophical-ideological detachment, the government regarded it as an excellent educational medium for the sharpening of thought and as a discipline least susceptible to uncontrollable ideological influences. It came as no surprise when Graf D. Tolstoi, while engaged in the work on a new high school charter during the late 1860's, demanded that the emphasis be placed on classical languages and mathematics as the subjects allowing teachers a minimum opportunity to express their personal ideas.[50]

P. Pekarskii, the author of the *History of the St. Petersburg Academy,* noted that the flourishing of higher and applied mathematics in Russian universities owed much to the Academy's consistent emphasis on this science and to the teaching and scholarly activities of Euler's students. However, he warned that it would be an error to try to explain "the progress of mathematics in Russia exclusively by the accident that Wolff succeeded in persuading the Bernoulli brothers to come to St. Petersburg and that these attracted Euler. . . . " He added that the comparatively intensive development of mathematics owed much to the detachment of this science from political and religious ideas and values.[51]

The attitude of the government toward mathematics as a school subject was clearly expressed in the emphasis given to the teaching of this subject in various secondary schools. A study of the evolution of the secondary school (gymnasium) curriculum throws a particularly interesting light on this important subject since this school was by its very definition college-preparatory and, therefore, the breeding ground of future scientists. During the XIXth century the gymnasium passed through four distinct phases. From 1803 to 1828 its curriculum, prepared by N. Fuss on the model of French lycées, placed emphasis on the natural sciences and mathematics, including algebra, geometry, plane geometry, and applied mathematics. From 1828 to 1864 the accent was on the study of classical history, literature, and languages. From 1864 to 1871 the natural sciences, along with the classical languages and literature, were stressed, while from 1871 to the end of the century the study of the Latin and Greek languages (without literature) took almost one-third of the entire secondary school curriculum. The fluctuating evolution of this curriculum reflected the political moods of different periods and was an index of the growth of political reaction. It is significant, however, that throughout the century mathematics was the only science whose teaching continued to be emphasized and grew in scope. It is also interesting that the government continued to emphasize mathematics as "particularly important for the refinement of thoughts and their formation, and for the development of penetrating minds and the power of reflexion."[52]

During long intervals in the XIXth century very little emphasis was placed on the teaching of the natural sciences in high schools and on natural-scientific research in universities. Prior to 1860, according to I. M. Sechenov, "research by the staff members of Russian universities was generally poor, and university scholars with Russian names were but few, worked in isolation, and exercised little cultural influence."[53] The gymnasium charter of 1826 eliminated from the curriculum all physical and biological sciences with the exception of physics. Only the children whose parents could afford to send them to private boarding schools were in a position to acquire a solid grounding in these disciplines. This explains why the first outstanding Russian scholars in these fields, who began to emerge during the early 1860's, were predominantly of aristocratic origin. The case with mathematics was quite different. Because of the comparatively solid mathematical curriculum in high schools the universities had a wider source from which to select promising mathematicians. While Chebyshev came from an impoverished aristocratic family, most of his disciples were of middle-class origin. After the 1860's the democratization of mathematics proceeded more rapidly than that of the physical and biological sciences. The crowning point was reached by the rise to eminence of Sophia Kovalevskaia, the first Russian woman-scholar whose works in mathematics earned her such distinctions as corresponding member of the St. Petersburg Academy, the *Prix Bordin* of the Paris Academy of Sciences, praise from such scholars as the great French mathematician Hermite, and a professorship at Stockholm University. The rapid widening of the social base from which mathematicians were recruited was accompanied by an accelerated growth of mathematical studies and the appearance of mathematical societies, "circles," and journals. The *Mathematical Symposium* published by the Moscow Mathematical Society (founded in 1867), was the first Russian journal devoted to a single science which printed articles in the Russian language. It provided an opportunity for young mathematicians to maintain a fruitful contact with the scientific world and to publish their own contributions. As a result, the ranks of producing mathematicians increased rapidly and met the demands for additional teachers which grew with the opening of new universities and technical schools and the modernization of military academies.[54] Toward the end of the century mathematics received a strong impetus from the general efflorescence of the biological and physical sciences. By this time such expressions as "mathematical physics" and "mathematical chemistry" had ceased to sound paradoxical.

Mathematics became a challenge to all social classes and a scientific pursuit fraught with great professional enthusiasm. It was not a mere accident that, when in 1905 the students' unrests led to a temporary closing of St. Petersburg University, a group of six eminent mathematics professors found enough genuine demand to justify the continuation of their lectures on the premises of a high school.[55]

In 1866, in an appeal to the government for postal and other privileges for the *Mathematical Symposium,* Chebyshev stated: "In addition to the great importance of mathematics for general education and to its practical usefulness, this science is of special interest to us in Russia: the declarations of foreign scholars and the history of education in our country show that this is the science toward which we have a special inclination and in which we can, to a smaller or greater degree, compete with foreigners."[56] Chebyshev did not say it, but he had implied that mathematics had become Russia's "national science."

NOTES

1. P. Pekarskii, *Nauka i literatura v Rossii pri Petre Velikom* [*Science and Literature Under Peter the Great*] (St. Petersburg, 1862), 263.

2. *Ibid.,* 5.

3. John Perry, *The State of Russia Under the Present Tsar* (London, 1716), 168, 211-212. See also Pekarskii, 122.

4. Pekarskii, 16.

5. P. Miliukov, *Ocherki po istorii russkoi kultury* [*Essays on the History of Russian Culture*], vol. II, Part II (Paris, 1931), 734-736.

6. P. Pekarskii, *Istoriia Imperatorskoi Akademii nauk* [*History of the Imperial Academy of Sciences*], I (St. Petersburg, 1870), xvi.

7. *Letters of Euler on Different Subjects in Physics and Philosophy. Addressed to a German Princess.* Transl. from French by Henry Hunter, I (2d ed., London, 1802), lxiv.

8. Pekarskii, *Istoriia*, lx-lxi.

9. A. Satkevich, "Leonhard Euler (v dvukhsotniu godovshchinu dnia ego rozhdeniia)" ["Leonhard Euler (On the Occasion of the Two-Hundredth Anniversary of his Birth)"], *Russkaia starina* [*Russian Antiquity*], CXXXII (1907), 483.

10. B. G. Kuznetsov, *Ocherki istorii russkoi nauki* [*Essays on the History of Russian Science*] (Moscow-Leningrad, 1940), 19.

11. A. M. Deborin, ed., *Leonhard Euler, 1707–1783: Sbornik statei i materialov k 150-letiiu so dnia smerti* [*Leonhard Euler, 1707–1783: A Collection of Articles and Materials on the Occasion of the One Hundred and Fiftieth Anniversary of his Death*] (Moscow-Leningrad, 1935), 181.

12. D. A. Tolstoi, "Akademicheskii universitet v XVIII stoletii po rukopisnym dokumentam Akademii nauk" ["The Academic University in the Eighteenth Century on the Basis of Unpublished Documents of the Academy of Sciences"], *Zapiskii Imperatorskoi Akademii nauk* [*Memoirs of the Imperial Academy of Sciences*], vol. LI, Suppl. no. 3 (St. Petersburg, 1885), 3, 12.

13. Leonard Euler, *Elements of Algebra*, transl. from French, with the notes of M. [Jean] Bernoulli, &c, and the additions of M. De la Grange [Joseph Louis] (5th ed., London, 1840), xvi.

14. P. Pekarskii, *Istoriia Imperatorskoi Akademii nauk v Peterburge* [*History of the Imperial Academy of Sciences in St. Petersburg*], II (St. Petersburg, 1873), 1.

15. August L. Schlözer, one of his critical contemporaries, called him a "true genius," who, had he chosen to confine his work to a limited number of disciplines, would have made substantial contributions to science. A. G. Fomin, *et al.*, eds., *Vystavka "Lomonosov i Elizavetinskoe vremia"* [*Exposition "Lomonosov and the Elizabethan Era"*], VII (Petrograd, 1915), 135.

16. Stepan Shevyrev, *Istoriia Imperatorskogo moskovskogo universiteta, 1755–1855* [*History of the Imperial Moscow University*] (Moscow, 1855), 142.

17. *Ibid.*, 150.

18. B. V. Gnedenko, *Ocherki po istorii matematiki v Rossii* [*Essays on the History of Mathematics in Russia*] (Moscow-Leningrad, 1946), 85.

19. T. P. Passek, *Iz dal'nikh let. Vospominaniia* [*From the Distant Past, Recollections*], I (St. Petersburg, 1878), 319-320.

20. Otto Spiess, *Leonhard Euler, Ein Beitrag zur Geistesgeschichte des XVIII Jahrhunderts* (Leipzig, 1929), 193-194.

21. K. V. Ostrovitianov, ed., *Istoriia Akademii nauk SSSR* [*History of the Academy of Sciences of the U.S.S.R.*], vol. I (Moscow-Leningrad, 1958), 348.

22. *Ibid.*

23. S. V. Rozhdestvenskii, *Materialy dlia istorii uchebnykh reform v Rossii v XVIII-XIX vekakh* [*Materials on the History of Educational Reforms in Russia in the Eighteenth and Nineteenth Centuries*] (St. Petersburg, 1910), 106.

24. A. Shchapov, *Sotsialno-pedagogicheskiia usloviia umstvennogo razvitiia russkogo naroda* [*The Social and Educational Conditions of the Intellectual Development of the Russian People*] (St. Petersburg, 1870), 226.

25. *Ibid.*, 226-227.

26. V. Rudakov, "Gimnaziia" ["The Gymnasium"], in: F. A. Brockhaus and I. A. Afron, eds., *Entsiklopedicheskii slovar* [*Encyclopedic Dictionary*] VIII (St. Petersburg, 1893), 699.

27. K. S. Veselovskii, "Otnosheniia imp. Pavla k Akademii nauk" ["The Relations of Emperor Paul with the Academy of Sciences"], *Russkaia starina* [*Russian Antiquity*], XCIV (1898), 5.

28. Timofei Fedorovich Osipovskii, "O prostranstve i vremeni" ["On Space and Time"], in G. S. Vasetskii *et al.*, eds., *Izbrannye proizvedeniia russkikh estestvoispytatelei pervoi poloviny XIX veka* [*Selected Writings of Russian Natural Scientists of the First Half of the Nineteenth Century*] (Moscow, 1959), 45-49.

29. S. T. Aksakov, *Polnoe sobranie sochinenii* [*Complete Works*] (St. Petersburg, 1914), 250 (on excellent quality of mathematics teaching at Kazan gymnasium which he attended at the very beginning of the XIXth century).

30. V. V. Grigor'ev, *Imperatorskii S. Peterburgskii Universitet v techenie pervykh piatidesiati let ego sushchestvovaniia* [*The Imperial St. Petersburg University During the First Fifty Years of its Existence*] (St. Petersburg, 1860), 67.

31. A. Shtorkh and F. Adelung, *Sistematicheskoe obozrenie literatury v Rossii v techenie piatiletiia, s 1801 po 1806 god.* [*A Systematic Review of Literature in Russia During the Five Years from 1801 to 1806*] I (St. Petersburg, 1810), 340.

32. A. P. Shchapov, "Estestvenno-psikhologicheskaia usloviia umstvennogo i sotsial-nogo razvitiia russkogo naroda" ["The Natural and Psychological Conditions of the Intellectual and Social Development of the Russian People"], *Sochineniia* [*Works*], III (St. Petersburg, 1908), 31-32.

33. Shchapov, *Sotsialno-pedagogicheskiia usloviia umstvennogo razvitiia russkogo naroda* [*The Social and Educational Conditions of the Intellectual Development of the Russian People*], 227.

34. Adolph Erman, *Travels in Siberia,* transl. from German by W. D. Cooley, vol. I (London, 1848), 48.

35. Augustin-Louis Cauchy, *Mémoire sur les intégrales définies, prises entre des limites imaginaires* (Paris, 1825), 2.

36. James Clark Maxwell, in his *A Treatise on Electricity and Magnetism,* I, second edition (Oxford, 1881), 117, gave Ostrogradskii credit for having been the first to formulate the theorem which establishes a relationship between a volume-integral and the corresponding surface-integral. See also I. Todhunter, *A History of the Progress of the Calculus of Variations* (Cambridge, 1851), 111-139.

37. B. V. Gnedenko, *Mikhail Vasil'evich Ostrogradskii: Ocherki zhizni, nauchnogo tvorchestva i pedagogicheskoi deiatel'nosti* [*Mikhail Vasil'evich Ostrogradskii: Essays on His Life, Scientific Work and Educational Activity*] (Moscow, 1952), 164-165.

38. Grigor'ev, *op. cit.,* 127ff.

39. Even in the XXth century the recognition was not universal. For example, F. Klein (*Vorlesungen über die Entwicklung der Mathematik im 19. Jahrhundert*), I-II (Berlin, 1926-27), mentioned only Lobachevskii, S. Kovalevskaia, and E. S. Fedorov.

40. Miliukov, *op. cit.,* 745.

41. S. V. Rozhdestvenskii, *Istoricheskii obzor deiatel'nosti Ministerstva narodnogo prosveshcheniia, 1802-1902* [*A Historical Review of the Activities of the Ministry of National Education, 1802-1902*] (St. Petersburg, 1902), 61.

42. "Offitsial'nyia staty i izvestiia: Materialy dlia istorii i statistiki nashikh gimnazii" ["Official Articles and Reports: Materials on the History and Statistics of our Gymnasiums"], *Zhurnal Ministerstva Narodnogo Prosveshcheniia* [*Journal of the Ministry of National Education*], part cxxi, book 2 (St. Petersburg, 1864), 378.

43. *Letters of Euler,* xliii.

44. For the relationship of Schelling's philosophy to the other intellectual currents of the 1820's, see N. P. Pavlov-Silvanskii, *Ocherki po russkoi istorii XVIII-XIX vv.* [*Essays on Russian History of the Eighteenth and Nineteenth Centuries*] (St. Petersburg, 1910), 239ff.

45. Shchapov, *op. cit.*, 221.

46. G. S. Chirikov, "Timofei Fedorovich Osipovskii, rektor Khar'kovskogo univer-siteta, 1820 g." ["Timofei Fedorovich Osipovskii, Rector of the Khar'kov University, 1820"], *Russkaia starina* [*Russian Antiquity*], XVII (1876), 463-490.

47. "Mikhail Vasil'evich Ostrogradskii pod nadzorom pol'itsii" [Mikhail Vasil'evich Ostrogradskii Under Police Surveillance"], *Russkaia starina* [*Russian Antiquity*], CVIII (1901) 341-342.

48. Grigor'ev, *op. cit.*, 70.

49. Passek, *op. cit.*, I, 319.

50. M. N. Kovalenskii, "Srednaia shkola" ["The Secondary School"], in *Istoriia Rossii v XIX veke* [*History of Russia in the Nineteenth Century*] VII (St. Petersburg, 1910), 171-172.

51. Pekarskii, *Istoriia Imperatorskoi Akademii nauk* [*History of the Imperial Academy of Sciences*], I, lxiii-lxiv.

52. Rudakov, *op. cit.*, 700.

53. I. Sechenov, "Nauchnaia deiatel'nost russkikh universitetov po estestvoznaniiu za poslednee dvadtsatipiatiletie" ["Scientific Work in Natural Sciences in the Russian Uni-versities During the Last Twenty-Five Years"], *Vestnik Europi* [*The Messenger of Europe*], VI, (1883), 331.

54. Similar societies were soon organized in Khar'kov (1879), St. Petersburg, Kazan, and Kiev (all in 1890), each of which had its own publications.

55. A. N. Krylov, *Moi vospominaniia* [*My Recollections*] (Moscow-Leningrad, 1942), 143-144.

56. P. L. Chebyshev, *Polnoe sobranie sochinenii* [*Complete Works*], V (Moscow-Lenin-grad, 1951), 401.

JOSEPH NEEDHAM

Science and Society in East and West

When I first formed the idea, about 1938, of writing a systematic, objective, and authoritative treatise on the history of science, scientific thought, and technology in the Chinese culture-area,[1] I regarded the essential problem as that of why *modern* science (as we know it since the +17th century, the time of Galileo) had not developed in Chinese civilization (or in Indian) but only in Europe? As the years went by, and as I began to find out something at last about Chinese science and society,[2] I came to realize that there is a second question at least equally important, namely, why, between the —1st century and the +15th century, Chinese civilization was much *more* efficient than occidental in applying human natural knowledge to practical human needs?

The answer to all such questions lies, I now believe, primarily in the social, intellectual, and economic structures of the different civilizations.

Joseph Needham, "Science and Society in East and West," in Maurice Goldsmith and Alan Mackay, eds., *Society and Science* (New York: Simon and Schuster, 1964) pp. 127-49. Copyright © 1964 by Maurice Goldsmith. Reprinted by permission of The Sterling Lord Agency.

The comparison between China and Europe is particularly instructive, almost a test-bench experiment one might say, because the complicating factor of climatic conditions does not enter in—broadly speaking, the climate of the Chinese culture-area is similar to that of the European. It is not possible for anyone to say (as has been maintained in the Indian case) that the environment of an exceptionally hot climate inhibited the rise of modern natural science.[3] Although the natural, geographical, and climatic settings of the different civilizations undoubtedly played a great part in the development of their specific characteristics, I am not inclined to regard this suggestion as valid for Indian culture. The point is that it cannot even be asserted of China.

From the beginning I was deeply sceptical of the validity of any of those "physical-anthropological" or "racial-spiritual" factors which have satisfied a good many people. Everything I have experienced during the past thirty years, since I first came into close personal contact with Chinese friends and colleagues, has only confirmed me in this scepticism. They proved to be entirely, as John of Monte Corvino said so many centuries ago, "di nostra qualità." I believe that the vast historical differences between the cultures can be explained by sociological studies, and that some day they will be. The further I penetrate into the detailed history of the achievements of Chinese science and technology before the time when like all other ethnic cultural rivers they flowed into the sea of modern science, the more convinced I become that the cause for the breakthrough occurring only in Europe was connected with the special social and economic conditions prevailing there at the Renaissance, and can never be explained by any deficiencies either of the Chinese mind or of the Chinese intellectual and philosophical tradition. In many ways this was much more congruent with modern science than was the world-outlook of Christendom. Such a point of view may or may not be a marxist one—for me it is based on personal experience of life and study.

For the purposes of the historian of science therefore we have to be on the watch for some essential differences between the aristocratic military feudalism of Europe, out of the womb of which mercantile and then industrial capitalism, together with the Renaissance and the Reformation, could be born; and those other kinds of feudalism (if that was really what it was) which were characteristic of medieval Asia. From the point of view of the history of science we must have something at any rate sufficiently different from what existed in Europe to help us solve our problem. This is why I have never been sympathetic to that other trend in marxist thinking which has sought for a rigid and unitary formula of the stages of social development which all civilizations "must have passed through."

Primitive communalism, the earliest of these, is a concept which has evoked much debate. Though commonly rejected by the majority of Western anthropologists and archaeologists (with, of course, some notable exceptions, such as Gordon Childe), it has always seemed to me eminently sensible to believe in a state of society before the differentiation of social classes, and in my studies of ancient Chinese society I have found it appearing through the mists clearly enough time after time. Nor at the other end of the story is there any essential difficulty in the transition from feudalism to capitalism, though of course this was enormously complex in detail, and much has still to be worked out. In particular, the exact connections between the social and economic changes and the rise of modern science, that is to say, the successful application of mathematized hypotheses to the systematic experimental investigation of natural phenomena, remain elusive. All historians, no matter what their theoretical inclinations and prejudices, are necessarily constrained to admit that the rise of modern science occurred *pari passu* with the Renaissance, the Reformation, and the rise of capitalism[4]—it is the intimate connections between the social and economic changes, on the one hand, and the success of the "new, or experimental" science, on the other, which are the most difficult to pin down. A great deal can be said about this, for example, the vitally important role of the "higher artisanate" and its acceptance into the company of educated scholars at this time,[5] but the present essay is not the place for it because we are in pursuit of something else. For us the essential point is that the development of modern science occurred in Europe and nowhere else.

In comparing the position of Europe with China the greatest and most obscure problems are: (*a*) how far and in what way did Chinese medieval feudalism (if that is the proper term for it) differ from European feudalism, and (*b*) did China (or indeed India) ever pass through a "slave-society" analogous to that of classical Greece and Rome? The question is, of course, not merely whether the institution of slavery existed, that is quite a different matter, but whether the society was ever based on it.

In my early days, when I was still a working biochemist, I was greatly influenced by Karl A. Wittfogel's book *Wirtschaft und Gesellschaft Chinas,* written when he was a more or less orthodox marxist in pre-Hitler Germany.[6] He was particularly interested in developing the conception of "Asiatic bureaucratism," or "bureaucratic feudalism," as I found that some Chinese historians called it later on. This arose from the works of Marx and Engels themselves, who had based it partly on, or derived it from, the observations of the seventeenth-century Frenchman François Bernier, physician to the Mogul emperor Aurungzeb in India.[7] Marx and Engels had spoken about the "Asiatic mode of production." How exactly they defined this at different times and how exactly it ought to be defined now is today once again the subject of animated discussions in nearly every country. Broadly speaking, it was the growth of a State apparatus fundamentally bureaucratic in character, and operated by a non-hereditary élite, upon the basis of a large number of relatively self-governing peasant communities, still retaining much tribal character and with little or no division of labour as between agriculture and industry. The form of exploitation here consisted essentially in the collection of taxes for the centralized State, i.e. the royal or imperial court and its regiments of bureaucratic officials. The justification of the State apparatus was, of course, twofold, on the one hand it organized the defence of the whole area (whether an ancient "feudal" state or later the entire Chinese empire), on the other hand it organized the construction and maintenance of public works. It is possible to say without fear of contradiction that throughout Chinese history the latter function was more important than the first, and this was one of the things that Wittfogel saw. The necessities of the country's topography and agriculture imposed from the beginning a vast series of waterworks directed to: (*a*) the conservation of the great rivers, in flood protection and the like; (*b*) the use of water for irrigation, especially for wet rice cultivation; and (*c*) the development of a far-flung canal system whereby the tax-grain could be brought to granary centres and to the capital. The greatest culture-hero of all time in China was a hydraulic engineer.[8] All this necessitated besides tax exploitation, the organization of corvée labour, and one might say that the only duties of the self-governing peasant communities *vis-à-vis* the State apparatus were the payment of tax and the provision of labour power for public purposes when called upon to give it.[9] Besides this the State bureaucracy assumed the function of the general organization of production, i.e. the direction of broad agricultural policy, and for this reason the State apparatus of such a type of society is now receiving the appellation of "an economic high command." Alone in China do we find among the most ancient high officials the Ssu Khung, the Ssu Thu, and the Ssu Nung (Directors of the Multitudes for Engineering Works and for Agriculture). Nor can we forget that the "nationalization" of salt and iron manufacture (the only commodities which had to travel, because not everywhere producible), suggested first in the —5th century, was thoroughly put into practice in the —2nd century. Also in the Han period there was a governmental Fermented Beverages Authority; and there are many examples of similar bureaucratic industries under subsequent dynasties.[10]

Various other aspects of this situation reveal themselves as one looks farther into it, for example peasant production was not under private control or ownership but public control, and theoretically all the land within the whole empire belonged to the Emperor and the Emperor alone. There was at first a semblance of landed property securely held by individual families, but this institution never developed in Chinese history in a way comparable with the feudal fief tenures of the West, since Chinese society did not retain the system of primogeniture. Hence all landed estates had to be parcelled out at each demise of the head of the family. Again in that society the conception of the city-state was totally absent; the towns were purposefully created as nodes in the administrative network, though very often no doubt they tended to grow up at spontaneous market centres. Every town was a fortified city held for the Prince or the Emperor by his civil governor and his military official. Since the economic function was so much more important in Chinese society than the military, it is not surprising that the governor was usually a more highly respected personality than the garrison commander. Lastly, broadly speaking, slaves were not used in agricultural production, nor indeed very much in industry; slavery was primarily domestic, or as some would say, "patriarchal" in character, throughout the ages.

In its later highly developed forms such as one finds in Thang or Sung China the "Asiatic mode of production" developed into a social system which while fundamentally "feudal" in the limited sense that most of the wealth was based on agricultural exploitation,[11] was essentially bureaucratic and not military-aristocratic. It is quite impossible to overestimate the depth of the civilian *ethos* in Chinese history. Imperial power was exercised not through a hierarchy of enfeoffed barons but through an extremely elaborate civil service, which Westerners know of as the "mandarinate," enjoying no hereditary principle of succession to estates but recruited afresh in every generation. All I can say is that throughout nearly thirty years of study of Chinese culture these conceptions have made more sense in understanding Chinese society than any others. I believe that it will be possible to show in some considerable detail why the Asian "bureaucratic feudalism" at first favoured the growth of natural knowledge and its application to technology for human benefit, while later on it inhibited the rise of modern capitalism and of modern science in contrast with the other form of feudalism in Europe which favoured it—I mean of course by decaying and generating the new mercantile order of society. A predominantly mercantile order of society could never arise in Chinese civilization because the basic conception of the mandarinate was opposed not only to the principles of hereditary aristocratic feudalism but also to the value-systems of the wealthy merchants. Capital accumulation in Chinese society there could indeed be, but the application of it in permanently productive industrial enterprises was constantly inhibited by the scholar-bureaucrats, as indeed was any other social action which might threaten their supremacy. Thus the merchant guilds in China never achieved anything approaching the status and power of the merchant guilds of the city-states of European civilization.

In many ways I should be prepared to say that the social and economic system of medieval China was much more rational than that of medieval Europe. The system of imperial examinations for entry into the bureaucracy, a system which had taken its origin as far back as the —2nd century, together with the age-old practice of the "recommendation of outstanding talent," brought it about that the mandarinate creamed off into itself the best brains of the nation (and the nation was a whole sub-continent) for more than two thousand years.[12] This stands in very great contrast to the European situation, where the best brains were not especially likely to arise in the families of the feudal lords, still less among the more restricted group of eldest sons of feudal lords. There were, of course, certain bureaucratic features of early medieval European society, for example the office of the "Counts," the institutions which gave rise to the position of "Lord-Lieutenant," and the widely customary use of bishops and clergy as administrators under the king, but all this fell far short of the systematic utilization of administrative talent which the Chinese system brought fully into play.

Moreover, not only was administrative talent brought forward and settled thoroughly into the right place, but so strong was the Confucian *ethos* and ideal that the chief representatives of those who were not scholar-gentry remained for the most part conscious of their lesser position in the scheme of things. When I was giving a talk recently to a university society on these subjects someone asked the excellent question, "How was it that the military men could accept their inferiority to the civil officials throughout Chinese history?" After all, "the power of the sword" has been overwhelming in other civilizations. What immediately came to my mind in replying was the imperial *charisma* carried by the bureaucracy,[13] the holiness of the written character (when I first went to China the stoves for giving honourable cremation to any piece of paper with words written on it were still to be seen in every temple), and the Chinese conviction that the sword might win but only the *logos* could keep. There is a famous story about the first Han emperor, who was impatient with the elaborate ceremonies devised for the court by his attendant philosophers, till one of them said to him, "You conquered the empire on horseback, but from horseback you will never succeed in ruling it." After which the rites and ceremonies were allowed to unfold in all their liturgical majesty.[14] In ancient times the Chinese leader was often an important official and a general indiscriminately, but

what is important is that the psychology of purely military men themselves admitted their inferiority. They were very often "failed civilians." Of course, force was the ultimate argument, the final sanction, as
in all societies, but the question was—what force, moral or purely
physical? The Chinese profoundly believed that only the former lasted,
and what the latter could gain only the former could keep.

Furthermore, there may have been technical factors in the primacy of
the spoken and written word in Chinese society.[15] It has been demonstrated that in ancient times the progress of invention in offensive
weapons, especially the efficient crossbow, far outstripped progress in
defensive armour. There are many cases in antiquity of feudal lords
being killed by commoners or peasants well armed with crossbows—a
situation quite unlike the favourable position of the heavily armed
knight in Western medieval society. Hence perhaps arose the Confucian
emphasis on persuasion. The Chinese were Whigs, "For Whigs do use
no force but argument." The Chinese peasant-farmer could not be
driven into battle to defend the boundaries of his State, for instance,
before the unification of the Empire, since he would be quite capable
of shooting his Prince first; but if he was persuaded by the philosophers,
whether patriots or sophists, that it was necessary to fight for that State,
as indeed also later for the Empire, then he would march. Hence the
presence of a certain amount of what one might call "propaganda" (not
necessarily in a pejorative sense) in Chinese classical and historical texts
—a kind of "personal equation" for which the historian has to make
proper allowance. There was nothing peculiar to China in this, it is, of
course, a world-wide phenomenon notable from Josephus to Gibbon,
but the sinologist has always to be on the look-out for it, for it was the
défaut of the civilized civilian *qualité*.

Yet another argument is of interest in this connection, namely the fact
that the Chinese was always primarily a peasant-farmer, and not engaged in either animal husbandry or sea-faring.[16] These two latter occupations encourage excessive command and obedience; the cowboy or
shepherd drives his animals about, the sea-captain gives orders to his
crew which are neglected at the peril of everyone's life on board, but
the peasant-farmer, once he has done all that is necessary for the crops,
must wait for them to come up. A famous parable in Chinese philosophical literature derides a man of Sung State who was discontented with
the growth rate of his plants and started to pull at them to help them
to come up.[17] Force therefore was always the wrong way of doing
things, hence civil persuasion rather than military might was always the
correct way of doing things. And everything that one could say for the
position of the soldier *vis-à-vis* the civil official holds good *mutatis mutan-
dis* for the merchant. Wealth as such was not valued. It had not spiritual

power. It could give comfort but not wisdom, and in China affluence carried comparatively little prestige. The one idea of every merchant's son was to become a scholar, to enter the imperial examinations, and to rise high in the bureaucracy. Thus did the system perpetuate itself through ten thousand generations. I am not sure that it is still not alive, though raised, of course, to a higher plane, for does not the Party official, whose position is quite irrelevant to the accidents of his birth, despise both aristocratic values on the one hand, and acquisitive values on the other? In a word perhaps socialism was the spirit of undominating justice imprisoned within the shell of Chinese medieval bureaucratism.[18] Basic Chinese traditions may perhaps be more congruent with the scientific world co-operative commonwealth than those of Europe.

Between 1920 and 1932 there were great discussions in the Soviet Union about what Marx had meant by the "Asiatic mode of production," but very little is known of these in Western countries, because they were never translated. If any copies of the Russian accounts still exist it would be highly desirable to have them republished in Western languages. Although we have never been able to study the results, it is believed that those who opposed any variation from the standard succession—primitive communalism- slave–society-feudalism-capitalism-socialism—gained the day. No doubt the climate of dogmatism which prevailed in the social sciences during the personality-cult period played some role in this situation.[19] We now have younger writers, as in a recent discussion, expressing the great embarrassment felt by English marxists that "feudalism" has become a meaningless term.[20] "Obviously," they say, "a socio-economic stage which covers both Ruanda-Urundi today and France in +1788, both China in 1900 and Norman England, is in danger of losing any kind of specific character likely to assjst analysis. . . . " Subdivisions are desperately needed. The remarkable thing is that these writers do not seem to know much about the original views of Marx and Engels. "The 'Asiatic mode'," one of them says, "has long since tacitly dropped out of use."[21] However, the same writer goes on to pose the problem of the arrested development of certain Asian and African societies very well, and recommends the "rehabilitation of Marx's 'Asiatic mode' or even several modes, to enable" a differentiation in nomenclature between regional variations. The use of the term "proto-feudal" (which I believe I invented myself) is also recommended for a single basic stage which then developed in different ways.

Whenever the name of Wittfogel appears in marxist writings nowadays it is mentioned with aversion. The reason for this is that during the Hitler period Wittfogel migrated to America, where he still works. For many years he has been a great brandisher of tomahawks in the intellectual "cold war," and those writers who regard his recent book *Oriental Despotism*,[22] as propaganda directed against Russia and China both old and new are only too probably correct. Wittfogel now seeks to attribute all abuses of power, whether in totalitarian or other societies, to the principle of bureaucratism; but the fact that he has become a great opponent of the ideas which I and many others favour does not alter the fact that he once upon a time set them forth quite brilliantly himself, and thus I admire his first book while deprecating his last one. Although Wittfogel has perhaps overdone it, I do not regard his theory of "hydraulic society" as essentially erroneous, for I also believe that the spatial range of public works (river control, irrigation, and the building of transport canals) in Chinese history transcended time after time the barriers between the territories of individual feudal or proto-feudal lords. It thus invariably tended to concentrate power at the centre, i.e. in the bureaucratic apparatus arched above the granular mass of "tribal" clan villages.[23] I think it played an important part therefore in making Chinese feudalism "bureaucratic." Of course it does not matter from the standpoint of the historian of science and technology how different Chinese feudalism was from European feudalism, but it has got to be different enough (and I firmly believe that it was different enough) to account for the total inhibition of capitalism and modern science in China as against the successful development of both these features in the West.

As for bureaucracy, it is sheer nonsense to lay all social evil at its door. On the contrary, it has been through the ages a magnificent instrument of human social organization. Furthermore, it is going to be with us, if humanity endures, for many centuries to come. The fundamental problem before us now is the humanization of bureaucracy, so that under socialism not only shall its organizing power be used for the benefit of the ordinary man and woman; but that it shall be known and palpably felt and seen to be so used. Modern human society is, and will increasingly be, based on modern science and technology, and the more this goes on, the more indispensable a highly organized bureaucracy will be. The fallacy here is to compare such a system after the rise of modern science with *any* precursor systems which existed before it. For modern science has given us a vast wealth of instruments from telephones to computers which now, and only now, could truly implement the will to humanize bureaucracy. That will may rest on what is essentially Confucianism, Taoism, and revolutionary Christianity as well as Marxism.

The term "Oriental despotism" recalls, of course, the speculations of the Physiocrats in eighteenth-century France, who were deeply influenced by what was then known of the Chinese economic and social structure.[24] For them, of course, it was an enlightened despotism, which they much admired, not the grim and wicked system of Wittfogel's later imagination. Sinologists throughout the world were impatient with his later book because it persistently selected from the facts—thus, for example, it is impossible to say that there was no educated public opinion in medieval China. On the contrary, the scholar-gentry and the scholar-bureaucrats constituted a wide and very powerful public opinion, and there were times when the Emperor might command but the bureaucracy would not obey.[25] In theory the Emperor might be an absolute ruler, in practice what happened was regulated by long-established precedent and convention, interpreted age after age by the Confucian exegesis of historical texts. China has always been a "one-Party State," and for over two thousand years the rule was that of the Confucian Party. My opinion is therefore that the term "Oriental despotism" is no more justified in the hands of Wittfogel than it was in those of the Physiocrats, and I never use it myself. On the other hand, there are many marxist terms, some old and some now gaining prominence, which I find great difficulty in adopting. For example, in some texts the "imaginary State construct" is contrasted with the "real substratum" of the independent peasant villages—this does not seem to me justifiable, because in its way the State apparatus was quite as real as the work of the peasant-farmers. Nor do I like to apply the term "autonomic" to the village communities, because I think it was only true within very definite limitations. The truth is that we urgently need the development of some entirely new technical terms. We are dealing here with states of society far removed from anything that the West ever knew, and in coining these new technical terms I would suggest that we might make use of Chinese forms rather than continuing to insist on using Greek and Latin roots to apply to societies which were enormously different.[26] Here the term *kuan-liao* for the bureaucracy might come in useful. If we could get more adequate terminology it would also help us to consider certain other related problems. Here I am thinking of the remarkable fact that Japanese society was more similar to that of Western European society, and for that very reason more capable of developing modern capitalism. This has been recognized by historians for a long time past, but recent writings have pinpointed rather precisely the exact ways in which Japanese military-aristocratic feudalism could generate capitalism as the Chinese bureaucratic society could not.[27]

Next I may say something, though it will not be much, about "slave-society." According to my own experiences with Chinese archaeology and literature, for what they are worth, I am not very inclined to believe that Chinese society, even during the Shang and early Chou periods, was ever a slave-based society in the same sense as the ancient occidental cultures with their slave-manned galleys ploughing the Mediterranean and their *latifundia* spread over the fields of Italy. Here I diverge, with deep humility, from some contemporary Chinese scholars, who have been extremely impressed by the "single-track" system of developmental stages of society prominent in marxist thinking during the past twenty or thirty years. The subject is still under intensive debate, and we cannot yet say that certainty has been achieved in any aspect of it. Some years ago at Cambridge we had a symposium on slavery in the different civilizations, in the course of which the participants all had to agree that the actual forms of slavery were very different in Chinese society from anything known elsewhere. Owing to the dominance of clan and family obligations, it was rather doubtful whether anyone in that civilization could have been called "free" in some of the Western senses, while on the other hand (contrary to what many believe) chattel-slavery was distinctly rare.[28] The fact is that no one really yet knows fully what was the status of servile and semi-servile groups in the different periods in China (and there were many very different kinds of such groups), neither Western Sinologists nor even the Chinese scholars themselves. A great amount of research remains to be done, but I think it seems already clear that neither in the economic nor in the political field was chattel-slavery ever a basis for the whole of society in China in the same way as it was at some times in the West.[29]

Although the question of the slave basis of society has a certain importance in so far as it affects the position of science and technology among the Greeks and Romans, it is, of course, less germane to what was originally my central point of interest, namely the origin and development of modern science in the late Renaissance in the West. It could, however, have a very important bearing on the greater success of Chinese society in the application of the sciences of Nature to human benefit during the earlier period, the first fourteen centuries of the Christian era and four or five centuries prior to that. Is it not very striking and significant that China has nothing whatever to show comparable with the use of slaves at sea in galleys in the Mediterranean? Sail, and a very refined use of it, was the universal method of propulsion of Chinese ships from ancient times. China has no records of the mass use of the human motor comparable with the building methods of ancient Egypt. So also it is remarkable that we have never so far come across any important instance of the refusal of an invention in Chinese society due to fear of technological unemployment. If Chinese labour-power was as vast as most people imagine, it is not easy to see why this factor should not sometimes have come into play. We have numerous examples of labour-saving devices introduced at early times in Chinese culture, very often much earlier than in Europe—a concrete case would be the wheel-barrow, not known in the West before the +13th century, but common in China the +3rd century, and arising there almost certainly a couple of hundred years earlier than that. It may well be that just as the bureaucratic apparatus will explain the failure of modern science to arise spontaneously in Chinese culture, so also the absence of mass chattel-slavery may turn out to have been an important factor in the greater success of Chinese culture in fostering science pure and applied in the earlier centuries.

At the present time there is a great ferment among younger European sociological scholars concerning the reconsideration of the problem of the "Asiatic mode of production."[30] This may have been brought about partly because of the importance of such ideas for the interpretation of African societies now emerging from underdeveloped conditions. It is not clear that the restricted categories which had become conventional will altogether account for them. But the greatest stimulus perhaps has been the publication in Moscow in 1939 of a text of Marx himself written in 1857 and 1858 entitled, *Formen die der kapitalistischen Produktion vorhergehen.* This was a kind of trial run for his *Kapital,* and is included in his *Grundrisse der Kritik der politischen Ökonomie,* a collection of basic papers published again in Germany in 1952.[31] It was singularly unfortunate that Marx's text was not known to the participants in the Russian discussions of the twenties and thirties, for it gave the only deep and systematic exposé of his ideas on the "Asiatic mode of production." One great question is whether Marx and Engels regarded this as something qualitatively different from one or other of the classically distinguished types of society in the rest of the world, or only quantitatively different. It is not yet clear whether they saw it as essentially a "transitory" situation (though in some cases it might be capable of age-long stabilization) or whether they thought of "bureaucratism" as a fourth fundamental type of society. Was the "Asiatic mode of production" simply a variation of classical slavery or classical feudalism? Some Chinese historians have certainly regarded it as a special type of feudalism. But sometimes Marx and Engels seemed to speak as if they did consider it as something qualitatively different from slave production or feudal production. There was also always the question how far the conceptions of "bureaucratic feudalism" might be applicable to pre-Columbian America or other societies such as medieval Ceylon. This is the kind of problem to which Wittfogel has applied himself much in recent times, but without satisfactory conclusions (Ceylon is not even mentioned in his index), and the younger sociologists are now attacking it in quite a different light.[32]

I have no doubt that their thinking will greatly elucidate my own problem of the early advanced and later retarded character of Chinese science and technology. In particular, my French friends and colleagues Jean Chesneaux and André Haudricourt have been active in the matter, and what follows is based on some ideas which they have presented to me. It seems clear that the early superiority of Chinese science and technology through long centuries must be placed in relation with the elaborate, rationalized, and conscious mechanisms of a society having the characters of "Asiatic bureaucracy." It was a society which functioned fundamentally in a "learned" way, the seats of power being filled by scholars, not military commanders. Central authority relied a great deal upon the "automatic" functioning of the village communities, and in general tended to reduce to the minimum its intervention in their life. I have already spoken of the fundamental difference between peasant-farmers, on the one hand, and shepherds or seamen, on the other. This difference is expressed epigrammatically in the Chinese terms *wei* and *wu wei*. *Wei* meant the application of force, of will-power, the determination that things, animals, or even other men should do what they were ordered to do; but *wu wei* was the opposite of this, leaving things alone, letting Nature take her course, profiting by going with the grain of things instead of going against it, and knowing how not to interfere. This was the great Taoist watchword throughout the ages, the untaught doctrine, the wordless edict.[33] It was summarized in that numinous phrase which Bertrand Russell collected from his time in China, "production without possession, action without self-assertion, development without domination.[34] Now *wu wei*, the lack of interference, might very well be applied to a respect for the "automotive" capacity of the individual farmers and their peasant communities. Even when the old "Asiatic" society had given place to "bureaucratic feudalism" such conceptions remained very much alive. Chinese political practice and government administration was continually founded upon that non-intervention inherited from ancient Asian society and from the single pair of opposites "villages—prince." Thus all through Chinese history the best magistrate was he who intervened least in society's affairs, and all through history too the chief aim of clans and families was to settle their affairs internally without having recourse to the courts.[35] It seems probable that a society like this would be favourable to reflection upon the world of Nature. Man should try to penetrate as far as possible into the mechanisms of the natural world and to utilize the sources of power which it contained while intervening directly as little as possible, and utilizing "action at a distance." Conceptions of this kind, highly intelligent, sought always to achieve effects with an economy of means, and naturally encouraged the investigation of Nature for essentially Baconian reasons. Hence such early triumphs as those of the seismograph, the casting of iron, and water-power.

It might thus be said that this non-interventionist conception of human activity was, to begin with, propitious for the development of the natural sciences. For example, the predilection for "action at a distance" had great effects in early wave-theory, the discovery of the nature of the tides, the knowledge of relations between mineral bodies and plants as in geo-botanical prospecting, or again in the science of magnetism. It is often forgotten that one of the fundamental features of the great breakthrough of modern science in the time of Galileo was the knowledge of magnetic polarity, declination, etc.; and unlike Euclidean geometry and Ptolemaic astronomy magnetical science had been a totally non-European contribution.[35] Nothing had been known of it to speak in Europe before the end of the +12th century, and its transmission from the earlier work of the Chinese is not in doubt. If the Chinese were (apart from the Babylonians) the greatest observers among all ancient peoples, was it not perhaps precisely because of the encouragement of non-interventionist principles, enshrined in the numinous poetry of the Taoists on the "water symbol" and the "eternal feminine?"[37]

However, if the non-interventionist character of the "villages—prince" relationship engendered a certain conception of the world which was propitious to the progress of science it had certain natural limitations. It was not congruent with characteristically occidental "interventionism," so natural to a people of shepherds and sea-farers. Since it was not capable of allowing the mercantile mentality a leading place in the civilization, it was not capable of fusing together the techniques of the higher artisanate with the methods of mathematical and logical reasoning which the scholars had worked out, so that the passage from the Vincean to the Galilean stage in the development of modern natural science was not achieved, perhaps not possible. In medieval China there had been more systematic experimentation than the Greeks had ever attempted, or medieval Europe either, but so long as "bureaucratic feudalism" remained unchanged, mathematics could not come together with empirical Nature-observation and experiment to produce something fundamentally new. The suggestion is that experiment demanded too much active intervention, and while this had always been accepted in the arts and trades, indeed more so than in Europe, it was perhaps more difficult in China to make it philosophically respectable.

There was another way also in which medieval Chinese society had been highly favourable to the growth of the natural sciences at the pre-Renaissance level. Traditional Chinese society was highly organic, highly cohesive, the State was responsible for the good functioning of the entire society, even if this responsibility was carried out with the minimum intervention—one remembers that the ancient definition of the Ideal Ruler was that he should sit simply facing the South and exerting his virtue *(tê)* in all directions so that the Ten Thousand Things would automatically be well governed. As we have been able to show over and over again, the State brought powerful aid to scientific research.[38] Astronomical observatories, for example, keeping millennial records, were part of the civil service, vast encyclopaedias not only literary but also medical and agricultural were published at the expense of the State, and scientific expeditions altogether remarkable for their time was successfully accomplished (one thinks of the early +8th century geodetic survey of a meridian arc stretching from Indo-China to Mongolia, and the expedition to chart the constellations of the southern hemisphere to within 20° of the south celestial pole.[39] By contrast science in Europe was generally a private enterprise. Therefore it hung back for many centuries. Yet the State science and medicine of China was not capable of making, when the time came, that qualitative leap which happened in occidental science in the +16th and early +17th centuries.

Some Asian scholars have been suspicious of the idea of the "Asiatic mode of production" or "bureaucratic feudalism" because they have identified it with a certain "stagnation" which they thought they saw in the history of their own societies. In the name of the right of the Asian and African peoples to progress they have projected this feeling into the past and have wished to claim for their ancestors exactly the same stages as those which the West had itself gone through, that Western world which had for a time dominated so hatefully over them. It is, I think, very important to clear up this misunderstanding, for there seems no reason at all why we should assume *a priori* that China and other ancient civilizations passed through exactly the same social stages as the European West. In fact, the word "stagnation" was never applicable to China at all; it was purely a Western misconception. As I show elsewhere,[40] a continuing general and scientific progress manifested itself in traditional Chinese society, but this was violently overtaken by the exponential growth of modern science after the Renaissance in Europe. China was homoeostatic, cybernetic if you like, but never stagnant. In case after case it can be shown with overwhelming probability that fundamental discoveries and inventions made in China were transmitted to Europe, for example, magnetic science, equatorial celestial coordinates and the equatorial mounting of observational astronomical instruments,[41] quantitative cartography, the technology of cast iron,[42] essential components of the reciprocating steam-engine such as the double-acting principle and the standard interconversion of rotary and longitudinal motion,[43] the mechanical clock,[44] the boot stirrup and the efficient equine harnesses, to say nothing of gunpowder and all that followed therefrom.[45] These many diverse discoveries and inventions had earth-shaking effects in Europe, but in China the social order of bureaucratic feudalism was very little disturbed by them. The built-in instability of European society must therefore be contrasted with a homoeostatic equilibrium in China; the product I believe of a society fundamentally more rational. What remains is an analysis of the relationships of social classes in China and Europe. The clashes between them in the West have been charted well enough, but in China the problem is much more difficult because of the non-hereditary nature of the bureaucracy. This is a task for the future.

In recent decades much interest has been aroused in the history of science and technology in the great non-European civilizations, especially China and India, interest, that is, on the part of scientists, engineers, philosophers, and Orientalists, but not, on the whole, among historians. Why, one may ask, has the history of Chinese and Indian science been unpopular among them? Lack of the necessary linguistic and cultural tools for approaching the original sources has naturally been an inhibition, and of course if one is primarily attracted by +18th and +19th century science European developments will monopolize one's interest. But I believe there is a deeper reason.

The study of great civilizations in which *modern* science and technology did not spontaneously develop obviously tends to raise the causal problem of how modern science did come into being at the European end of the Old World, and it does so in acute form. Indeed, the more brilliant the achievements of the ancient and medieval Asian civilizations turn out to have been the more discomforting the problem becomes. During the past thirty years historians of science in Western countries have tended to reject the sociological theories of the origin of modern science which had a considerable innings earlier in this century. The forms in which such hypotheses had then been presented were doubtless relatively crude,[46] but that was surely no reason why they should not have been refined. Perhaps also the hypotheses themselves were felt to be too unsettling for a period during which the history of science was establishing itself as a factual academic discipline. Most historians have been prepared to see science having an influence on society, but not to admit that society influenced science, and they have liked to think of the progress of science solely in terms of the internal or autonomous filiation of ideas, theories, mental or mathematical techniques, and practical discoveries, handed on like torches from one great man to another. They have been essentially "internalists" or "autonomists." In other words, "there was a man sent from God, whose name was . . . " Kepler.[47]

The study of other civilizations therefore places traditional historical thought in a serious intellectual difficulty. For the most obvious and necessary kind of explanation which it demands is one which would

demonstrate the fundamental differences in social and economic structure and mutability between Europe on the one hand and the great
Asian civilisations on the other, differences which would account not
only for the development of modern science in Europe alone, but also
of capitalism in Europe alone, together with its typical accompaniments
of protestantism, nationalism, etc., not paralleled in any other part of
the globe. Such explanations are, I believe, capable of much refinement.
They must in no way neglect the importance of a multitude of factors
in the realm of ideas—language and logic, religion and philosophy,
theology, music, humanitarianism, attitudes to time and change—but
they will be most deeply concerned with the analysis of the society in
question, its patterns, its urges, its needs, its transformations. On the
internalist or autonomist view such explanations are unwelcome. Those
who hold it therefore instinctively dislike the study of the other great
civilizations.

But if you reject the validity or even the relevance of sociological
accounts of the "scientific revolution" of the late Renaissance, which
brought modern science into being, if you renounce them as too revolutionary for that revolution, and if at the same time you wish to explain
why Europeans were able to do what Chinese and Indians were not,
then you are driven back upon an inescapable dilemma. One of its horns
is called pure chance, the other is racialism however disguised. To attribute the origin of modern science entirely to chance is to declare the
bankruptcy of history as a form of enlightenment of the human mind.
To dwell upon geography and harp upon climate as chance factors will
not save the situation, for it brings you straight into the question of
city-states, maritime commerce, agriculture, and the like, concrete factors with which autonomism declines to have anything to do. The
"Greek miracle," like the scientific revolution itself, is then doomed to
remain miraculous. But what is the alternative to chance? Only the
doctrine that one particular group of peoples, in this case the European
"race," possessed some intrinsic superiority to all other groups of peoples. Against the scientific study of human races, physical anthropology,
comparative haematology, and the like, there can, of course, be no
objection, but the doctrine of European superiority is racialism in the
political sense and has nothing in common with science. For the European autonomist, I fear, "we are the people, and wisdom was born
with us." However, since racialism (at least in its explicit forms) is
neither intellectually respectable nor internationally acceptable, the autonomists are in a quandary which may be expected to become more
obvious as time goes on.[48] I confidently anticipate therefore a great
revival of interest in the relations of science and society during the
crucial European centuries, as well as a study ever more intense of the

social structures of all the civilizations, and the delineation of how they differed in glory, one from another.

NOTES

1. In course of publication since 1953 by the Cambridge University Press as *Science and Civilisation in China* (7 vols.), with the collaboration of Wang Ling, Lu Gwei-Djen, Ho Ping-Yü, Kenneth Robinson, and Tshao Thien-Chhin. For all documentation in detail the reader is referred to this work. References to it in what follows are abbreviated as SCC. The present contribution originated as a letter replying to my friend Dr. Mikulas Teich of Prague, who sought information preparatory to an essay-review of SCC which is appearing in *Historica*. The social and economic background of science and technology in Chinese culture will, we hope, be discussed with thoroughness in SCC, Vol. 7. I have already, however, adumbrated the general line of argument from time to time, as in my "Thoughts on the Social Relations of Science and Technology in China," *Centaurus*, 1953, **3**, 40, and earlier in my "Science and Social Change,"*Science & Society*, 1946, **10**, 225, an essay written during enforced leisure on the Burma Road in 1944. See also SCC, vol. 3, pp. 150 ff.

2. In this contribution Chinese references will not be expected, but I should like to say that both for the instruction of their writings and the inspiration of personal contact, no words can express my indebtedness over many years to the scholars and scientists of China—I mention at random only a few—Hou Wai-Lu, Hsü Shih-Lien, Kuo Mo-Jo, Kuo Pên-Tao, Li Shu-Hua, Shih Shêng-Han, Thao Mêng-Hou, Thang Phei-Sung, Wu Su-Hsüan, Wên I-To.

3. Cf. the writings of E. Huntington, for instance, *Mainsprings of Civilisation*, Wiley, New York, 1945, reprinted Mentor, New York, 1959.

4. The great stumbling-block here for the internalist school of historiography of science (see below, p. 117) is the question of historical causation. Scenting economic determinism under every formulation, they insist that the scientific revolution, as primarily a revolution in scientific ideas, cannot have been "derivative from" some other social movement, such as the Reformation or the rise of capitalism. Perhaps for the moment we could settle for some such phrase as "indissolubly associated with. . . ." The internalists always seem to me essentially Manichaean; they do not like to admit that scientists have bodies, eat and drink, and live social lives among their fellow-men, whose practical problems cannot remain unknown to them; nor are the internalists willing to credit their scientific subjects with subconscious minds.

5. This factor was much emphasized and elaborated by the late Edgar Zilsel. Its importance has recently been recognized by a medievalist whom no one could suspect of marxism, A. C. Crombie, in his "The Relevance of the Middle Ages to the Scientific Movement" in *Perspectives in Mediaeval History*, ed. K. F. Drew & F. S. Lear (Rice University Semicentennial Volume, Univ. Chicago Press, Chicago, Ill., 1963), p. 35. See also his "Quantification in Mediaeval Physics" in *Quantification*, ed. H. Woolf, (Bobbs-Merrill, Indianapolis and New York, 1961), p. 13.

6. Hirschfeld, Leipzig, 1931. I also learnt much from a golden little book by H. Wilhelm, the son of the great sinologist Richard Wilhelm, *Gesellschaft und Staat in China*, Vetch, Peiping, 1944. It is most unfortunate that this nonmarxist work has long been quite inaccessible, and that there has never been an English translation of it.

7. *The History of the Late Revolution of the Empire of the Great Mogul*, originally published in French, Paris, 1671; many times republished, as by Dass, Calcutta, 1909. See the famous letter of Marx to Engels, 2 June 1853.

8. Ta Yü, Yü the Great, one of the legendary ancient emperors, believed to have preceded the historic Shang Kingdom of the late −2nd millennium.

9. Today they do not have to give it, but are paid at the ordinary commune rate per labour day, and the work is done by the country people at slack times (A. L. Strong, *Letter from China*, 1964, No. 15). This principle of the rational and maximal utilization of man-power is one which goes back more than two thousand years in Chinese history, and its timing was one of the functions of the "economic high command."

10. Cf. H. F. Schurmann, *The Economic Structure of the Yuan Dynasty*, Harvard University Press, Cambridge, Mass., 1956, pp. 146 ff.

11. This must not be taken to mean that industry and trade were poorly developed in the Middle Ages, on the contrary, especially in the Southern Sung in the +12th and +13th centuries, they were so productive and prosperous that the continuance of the typical bureaucratic forms is what surprises.

12. A remarkable sidelight on this will be found in the paper by Lu Gwei-Djen & J. Needham, "China and the Origin of (Qualifying) Examinations in Medicine," *Proc. Roy. Soc. Med.* 1963, **56**, 63.

13. One should add the high moral standards of Confucianism which exerted great social pressure throughout the ages upon the members of the mandarinate.

14. See SCC, Vol. I, p. 103.

15. The following argument was first put forward by H. G. Creel, and will bear his name in my full exposition of it.

16. This contrast was, I think, first appreciated by André Haudricourt.

17. See SCC, Vol. 2, p. 576.

18. Of course, the medieval mandarinate was part of an exploiting system, like those of Western feudalism and capitalism, but as a non-hereditary élite it did oppose both aristocratic and mercantile ways of life.
Cf. the work by C. Brandt, B. Schwartz, & J. K. Fairbank, *A Documentary History of Chinese Communism*, Harvard Univ. Press, Cambridge, Mass., 1952, and J. Needham, "The Past in China's Present," *Centennial Review*, 1960, **4**, 145, 281; *Pacific Viewpoint*, 1963, **4**, 115; French tr. *Comprendre*, 1962, **21**, 261; **23**, 113.

19. During the subsequent decades there have been many distinguished sociological studies of Asian cultures by Russian sinologists, usually avoiding, however, the concept of the "Asiatic mode of production."

20. J. Simon, in *Marxism Today*, 1962, **6** (No. 6).

21. J. Simon, loc. cit.

22. Yale Univ. Press, New Haven, 1957. Reviewed, *inter alia*, by J. Needham, *Science & Society*, 1959, **23**, 58. Among the many critiques of Wittfogel's ideas may be mentioned an interesting recent study from the juristic point of view by Orlan Lee, "Traditionelle Rechtsgebräuche und der Begriff d. Orientalischen Despotismus," *Zeitschr. f. vergl. Rechtswiss.* 1964, **66**, 157.

23. And competing successfully with the remnants of tribal or military hereditary aristocracy.
On the *ethos* of the villages see some noteworthy remarks by V. Dedijer in *The Times*, 18 November 1963, entitled "The Chinese Theory of Revolution."

24. On this see L. A. Maverick's *China a Model for Europe*, Anderson, San Antonio, Texas, 1946, which includes a translation of F. Quesnay's *Le Despotisme de la Chine*, Paris, 1767.

25. Cf. Liu Tzu-Chien, "An Early Sung Reformer, Fan Chung-Yen," in *Chinese Thought and Institutions*, ed. J. K. Fairbank, Univ. Chicago Press, Chicago, Ill., 1957, p. 105.

26. This idea was recently put to me, though in another context, by a chemical colleague at Hongkong University.

27. See, for example, the recent monograph by N. Jacobs, *The Origin of Modern Capitalism and Eastern Asia*, Univ. Press, Hongkong, 1958, notable also for the excellence of its index. The author is a Weberian sociologist who executes the remarkable feat of making no mention of Marx and Engels. Evidently the Departments of History of Economics and History of Science at Hongkong inhabit separate ivory pagodas (cf. p. 147).

28. See E. G. Pulleyblank, "The Origins and Nature of Chattel-Slavery in China," *Journ. Econ. & Soc. History of the Orient,* 1958, **1**, 185.

29. The subject is under active discussion, as may be seen from T. Pokora, "Existierte in China eine Sklavenhaltergesellschaft?," *Archiv. Orientalni,* 1963, **31**, 353, and E. Welskopf, "Probleme der Periodisierung d. alten Geschichte; die Einordnung des alten Orients und Altamerikas in die Weltgeschichtliche Entwicklung," *Zeitschr. f. Geschichtswiss,* 1957, **5**, 296. Welskopf believes that the veteran sinologist E. Erkes went too far in his denial of the existence of slavery in ancient China, but that the pendulum afterwards swung much too far in the other direction. See Erkes' *Das Problem der Sklaverei in China,* Akad. Verlag, Berlin, 1952, and *Die Entwicklung der chinesischen Gesellschaft von der Urzeit bis zur Gegenwart,* Akad. Verlag, Berlin, 1953. A splendid collection of source material on slavery in the Han period is due to C. M. Wilbur, "Slavery in China during the Former Han Dynasty (—206 to +25)," *Field Museum of Nat. Hist. Pubs.* (Anthropol. Ser.), 1943, **34**, 1-490 (Pub. No. 525).

30. See especially the review by J. Chesneaux, "Le Mode de Production Asiatique; une nouvelle Étape de la Discussion," *Eirene* (in the press, 1964). Among the many scholars who have been contributing to this movement it is indispensable to mention F. Tökei, "Les Conditions de la Propriété foncière à l'Époque des Tcheou," *Acta Antiqua Acad. Scient. Hungar.,* 1958, **6** (No. 3-4), and *Sur le "Mode de Production Asiatique"* (mimeogr.), Centre d'Etudes et de Rech. Marxistes, Paris, 1963, part of a forthcoming book. See also M. Godelier *La Notion de Mode de Production Asiatique* (mimeogr.), Univ. Paris, 1964.

31. Dietz, Berlin. I am told that two English translations of the *Formen* have been made and that an English edition is in preparation. The *Formen* is also expected to appear before long in French.

32. On the Ceylonese situation, where hydraulic works were very remarkable and abundant but generated no mandarinate, see E. R. Leach, "Hydraulic Society in Ceylon," *Past and Present,* 1959 (No. 15), 2.

33. Cf. SCC, Vol. 2, p. 564.

34. SCC, Vol. 2, p. 164; from *The Problem of China,* Allen & Unwin, London, 1922, p. 194.

35. An aspect of the darker side of this is given in the partly autobiographical account of my old friend Kuo Yu-Shou, *La Lune sur le Fleuve Perle,* Bonne, Paris, 1963.

36. See J. Needham, "The Chinese Contribution to the Development of the Mariner's Compass," *Scientia,* 1961, **55**, 1; *Actas do Congresso Internacional de História dos Descobrimentos,* Lisbon, 1961, Vol. 2, p. 311.

37. Cf. SCC, Vol. 2, p. 57.

38. SCC, Vols. 2, 3, 4, 6 *passim.* See also my "Poverties and Triumphs of the Chinese Scientific Tradition" in *Scientific Change,* ed. A. C. Crombie, Heinemann, London, 1963; French tr. *Pensée,* 1963, No. 111.

39. See A. Beer, Ho Ping-Yü, Lu Gwei-Djen, J. Needham, E. G. Pulleyblank, & G. I. Thompson, "An Eighth-century Meridian Line; I-Hsing's Chain of Gnomons and the Prehistory of the Metric System," *Vistas in Astronomy,* 1961, **4**, 3.

40. "China's Scientific Influence on the World," a contribution to *The Legacy of China,* ed. R. Dawson, Oxford University Press, 1964.

41. J. Needham, "The Peking Observatory in +1280 and the Development of the Equatorial Mounting," *Vistas in Astronomy,* 1955, **1**, 67.

42. Cf. J. Needham, *The Development of Iron and Steel Technology in China,* Newcomen Society, London; 1958, reprinted Heffer, Cambridge, 1964; French tr. *Revue d'Hist. de Sidérurgie,* 1961, **2**, 187, 235; 1962, **3**, 1, 62.

43. See my Earl Grey lecture at the Newcastle University, 1961, "Classical Chinese Contributions to Mechanical Engineering," and my Newcomen Centenary Lecture, "The Pre-Natal History of the Steam-engine," *Trans. Newcomen Soc.* (in the press).

44. Cf. J. Needham, Wang Ling, & D. J. de S. Price, *Heavenly Clockwork,* Cambridge Univ. Press, 1960.

45. Some of the multifarious influences of Chinese inventions and discoveries on the pre-Renaissance world have recently been emphasized by Lynn White in his *Mediaeval Technology and Social Change*, Oxford Univ. Press, 1962.

46. Such is the adjective generally applied to B. Hessen's famous paper "on the Social and Economic Roots of Newton's *Principia*," delivered at the International Congress of the History of Science at London in 1931 (reprinted in *Science at the Cross-roads*, Kniga, London, 1932). It was certainly in plain blunt Cromwellian style. But already half a dozen years later R. K. Merton's remarkable monograph, "Science, Technology and Society in Seventeenth-Century England," *Osiris*, 1938, **4**, 360-632, had achieved a considerably more refined and sophisticated presentation. Much is owing also to the works of E. Zilsel, several of which were published in the *Journal of the History of Ideas*, and all of which ought to be collected in a single volume.

47. Though off the rails at various points, J. Agassi is entertaining on this topic in his monograph *Towards a Historiography of Science*, Mouton, the Hague, 1963 (*History and Theory, Studies in the Philosophy of History*, Beiheft 2). The "inductivist" historians of science, he says, are chiefly concerned with questions of whom to worship and for what reasons; but he does not like the "conventionalists" much better. With this particular quarrel I am not here concerned, but it is surprising that Agassi did not make more use of the work of Walter Pagel, which would have supported some of his arguments strongly. On the whole, Agassi takes his own stand for autonomism, regarding marxism as one of the failings of inductivists, and believing that contention between different schools was the main factor in the development of science. As his monograph comes from the University of Hongkong, he seems to have encysted himself with extraordinary success from all contact with Chinese culture—at any rate, so far. Cf. p. 139.

48. D. J. de S. Price, a valued collaborator of our own, knows much of the Asian contributions, but in his *Science since Babylon*, Yale Univ. Press, New Haven, Conn., 1961, follows a "hunch" of Einstein's and favours chance combinations of circumstances as the evocators of Greek and Renaissance science. A. R. Hall, in "Merton Revisited," *History of Science*, 1963, **2**, **1**, attacks anew what he calls the 'externalist' historiography of science, but significantly keeps silence about the problem posed by the Asian contributions. If he had taken a broader comparative point of view his arguments about the European situation might have carried more conviction. A. C. Crombie (op. cit.) alone of the three, shows a real consciousness of the slow social changes which permitted the intellectual movements of the late Middle Ages and the Renaissance to bring modern science into being in the European culture-area, but even he pays little attention to their economic concomitants.

HERBERT H. KARP
SAL P. RESTIVO

Ecological Factors in the Emergence of Modern Science

INTRODUCTION

The "scientific revolution" which occurred in Western Europe beginning in the sixteenth century is the source of a critical and persistent problem in the history of scientific activity: why did modern science emerge in Western Europe and not elsewhere?[1] The two major contributors to the sociology of the scientific revolution are Robert K. Merton and Joseph Ben-David. Their work has followed the conceptual guidelines established by Max Weber for the study of modern science. Weber did not consider the stimulation of science by emerging capitalism a factor in the origin of modern science; but he noted that "the *technical* utilization of scientific knowledge was certainly encouraged by economic considerations, which were extremely favourable to it in the Occident" (1958, p. 24). Merton, following Weber, explored the reciprocal relations between science and other institutional spheres (1970). His conclusion was that science became "accredited and organized" as a consequence of changes in cultural values favorable to scientific activity

(1970, p. 55). Arguments for the utility of science arose in a variety of institutional spheres, including religion, the economy, and the military. These arguments were "mainly prelude" to the institutionalization of science (Merton 1970, p. xxiii). More recently, Ben-David has elaborated the utilitarian hypothesis in conjunction with his analysis of the establishment of the scientific role (1965). The importance of this analysis lies in Ben-David's explicit recognition of the institutionalization of science as a precondition for the scientific revolution.

Ben-David has formulated the problem we are considering in terms of the "scientific role:" "What made certain men in seventeenth-century Europe, and nowhere before, view themselves as scientists and see the scientific role as one with unique and special obligations and possibilities?" (1965, p. 15). Ben-David's adherence to the utilitarian thesis expounded by Weber and Merton is illustrated by the conditions he considers necessary for the scientific view to emerge: "either there had to be some striking scientific discoveries of practical value convincing people that the practice of science was an economically worthwhile occupation, or there had to be a group of persons who believed in science as an intrinsically valuable preoccupation and who had a reasonable prospect of making their belief generally accepted, even before science proved its economic worth" (1965, p. 15). The social structural conditions necessary for the development and establishment of the scientific role, Ben-David argues, were (1) the "established church authority was weak and doctrine hence open to individual interpretation," and (2) there was a "class of people" (a) oriented to economic and technological change, (b) utopian rather than practical in their "intellectual policies," and (c) "expansive in power and influence but not yet in a position of authority and responsibility" (1965, p. 48).

Joseph Needham has arrived at conclusions consonant with Ben-David's basic thesis by studying a negative case, China. China, where the development of modern science "should have been intellectually possible" (Ben-David 1964, p. 457), has been a crucial comparative case for students of the West's scientific revolution. Needham began his work to determine" why *modern* science (as we know it since the +17th century, the time of Galileo) had not developed in Chinese civilization (or in Indian) but only Europe?" (1964, p. 127). Early in his research, a second equally important question occurred to him: "why, between the -1st century and the +15th century, Chinese civilization was much *more* efficient than occidental in applying human natural knowledge to practical human needs?" (1964, p. 127). Needham's thesis is that Chinese bureaucratic feudalism was obviously not an obstacle to scientific and technological growth; but it was an obstacle to a "scientific revolution." The opposition of the mandarinate to any and all social activities which might threaten their position prevented the autonomous development of science (Needham 1964, pp. 132-33). The structural basis for the realization of the mandarinate opposition was the centralization of power in the bureaucratic center; historical China was characterized by a centralized bureaucratic structure that, in Needham's words, "arched above the granular mass of 'tribal' clan villages" (1964, p. 137). There was little opportunity for increasing the mobility, status, and autonomy of individuals or "classes," the most striking examples being "merchants" and "scientists." At this point, Needham introduces a crucial idea; he attributes the centralization of power to "the spatial range of public works (river control, irrigation, and the building of transport canals) transcending time after time the barriers between the territories of individual feudal or proto-feudal lords" (1964, p. 137). This raises questions about the influence of environmental conditions on scientific activity. Human ecology provides a framework for studying the complex interactions implied in such questions.

The salient characteristic of human ecological analysis is the study of population aggregates adjusting to their physical environment by means of a technology and patterns of social organization (Duncan and Schnore 1959; Duncan et al. 1960; Duncan 1964). In considering the application of analysis in human ecology to the study of the scientific revolution we examine (1) the significance of autonomy in science for the emergence of modern science, and (2) the ecological factors in the differentiation and autonomous development of social activities. These ideas are developed in a comparative study of China and Western Europe as "ecological units." Our analysis suggests an alternative to the "utilitarian hypothesis" and the conceptualization of reciprocal institutional influences expounded by Weber, Merton, and Ben-David. The emergence of Protestantism, modern capitalism, and modern science can be viewed as *parallel* institutional responses to a set of underlying ecological conditions. The subjective factors stressed in the utilitarian hypothesis and in Merton's conception of institutional relations (whose formulation is more explicit than either Weber's or Ben-David's) are, on this view, manifestations of these conditions and the process of institutionalization in science.

AUTONOMY IN SCIENCE

In the first chapter of his *Science and the Social Order* (1952), Bernard Barber answers the question "What is science?" by proposing to view science as a social activity. Rather than viewing science as "disembodied items of guaranteed knowledge and . . . a set of logical procedures for achieving such knowledge," he views it as a special kind of thought and behavior realized in different ways and to different degrees in different societies. It is important, he notes "to see the source of science in the generic human attribute of empirical rationality" (1952, pp. 26, 52).

Scientific activity has occurred in all societies from prehistoric and ancient times to the present. But it is not until 1500 A.D. in Western Europe that the foundation is established for the emergence of science as a visible, differentiated part of the social structure, organized as an activity providing a livelihood for members of society occupying the position "scientist."

Following Weber and Ben-David, we define modern science as the pursuit of scientific activity by "a specific group of persons" occupying positions as "scientists" in specialized settings (e. g., scientific societies, research laboratories, universities), all of whom "regard the scientific investigation of nature as a major source of truth about the world," all of whom submit the results of their activity to be "used and judged" by other scientists, and all of whom are rewarded (given sustenance) for their roles in "publicly recognized institutions enjoying far-reaching autonomy" (Weber 1958, pp. 15-16; Ben-David 1964, p. 459). The significance of defining modern science in sociological terms is that the development of what are usually considered the distinctive attributes of modern science—e. g., systematic and continuous interaction between theoretical explanation and experimentation, the mathematization of hypotheses, and universalism and rationalism as basic normative orientations—was historically dependent on the institutionalization of scientific activity. Only an autonomous social activity could have supported the combination of "continuity and rapid innovation" identified with the emergence of modern science (Ben-David 1964, p. 459). Ben-David, for example, has argued that given "several generations of intellectuals conceiving of themselves as scientists—with the motivations and obligations entailed in that—the Greeks could undoubtedly have applied themselves to the discovery of a less cumbersome method of mathematical notation and have made many of the scientific advances accomplished in the sixteenth and seventeenth centuries and subsequently" (1964, p. 459). This argument might be valid too for fourteenth century Arab science (Pines 1963); and "several generations" of "scientists" might have made a "scientific revolution" in China. We would qualify Ben-David's argument by noting that autonomy in science was a necessary but not a sufficient condition for the West's scientific revolution. Our concern in this paper is with institutional and role autonomy in science as a necessary condition for the emergence of modern science.

Two important preconditions for the institutionalization of scientific activity in Western Europe were established by the medieval period: the independence of the towns, and the differentiation of the "intellectual" role in the universities. In medieval Western Europe, as in ancient China, India, and Egypt, students interested in higher learning were attracted to famous masters. But the European towns in which the masters lived were corporations and independent of the King; foreign students were not, therefore, under the King's protection. The corporate organization of the European universities was encouraged by such things as the often violent confrontations between scholars and townspeople, and the failure of either the church or the state to regulate university affairs. This form of incorporation, while not unique to Western Europe, "attained a much greater importance there than elsewhere;" by the thirteenth century the university was an "autonomous, intellectual community relatively well-endowed and privileged" (Ben-David 1965, p. 19). Furthermore, Ben-David argues, the presence of different kinds of intellectual specialties within one autonomous organization fostered greater differentiation and gave the sciences their place in the universities. This process of differentiation, the establishment of an "intellectual" role, preceded the emergence of the academies and the establishment of the scientific role in Western Europe.

The institutionalization of science in Western Europe is marked by the shift of scientific activity from the medieval universities to the academies. The earliest of these academies, according to Dampier (1948, p. 149) was the Accademia Secretorum Naturae, founded in Naples in 1560. The first Accademia dei Lincei was founded in Rome (1603–1630), followed by the Accademia del Cimento (Florence, 1657–1667), The Royal Society (London, 1662–), and the Accademie des Sciences (Paris, 1666–). Among the earliest German academies was the Societas Ereunetica (Rostock, 1622–ca.1642). The Berlin Academy was founded by Leibniz in 1700. The academies, books and journals sponsored by the academies, and journals such as the *Journal des Savants* (Paris, 1665), and *Acta Eruditorum* (Leipzig, 1682), made possible an "uninterrupted scientific activity, more or less proportionate to the general growth of social resources (population and wealth) and relatively independent of the prevailing non-scientific culture . . . This contributed greatly to the development of the scientific identity and, beyond it, to the institutionalized role of the modern scientist" (Ben-David 1965, p. 49).[2]

There are, of course, earlier examples of scientific organization. Scientific activity was unified and academies founded under the Roman Empire (Ben-David 1964, p. 457); there were communities of "scientists" centered in Hellenistic Alexandria; and there was an Arabic "scientific community" that stretched from Spain to Persia by the thirteenth century A.D. (Needham 1949, pp. 6-7; Edelstein 1957, p. 121). But in terms of our definition of modern science, these must be considered protoscientific activities. They do not manifest the extensity of differentiation and autonomy, nor the continuity of interaction among scientists and scientific ideas characteristic of scientific activity in Western Europe beginning in A.D. 1500.

China, for all of its triumphs in science and technology, did not undergo a "spontaneous autochthonous" (Needham 1963, p. 139) scientific revolution.[3] This is illustrated, for example, by the failure of Chinese astronomers to develop a unified scientific system (Eberhard 1957, p, 66), the relatively "official" character of Chinese science within the state bureaucracy (Needham 1963, p. 124), and the intellectually inhibiting atmosphere of the schools of mathematics (Yabuuti 1963, p. 118). The general "lack of development" of Chinese science (Balazs 1964, p. 137) is noteworthy especially because of the advanced development of technology and industry in China.

The status of the scholar was high in China. But in China, as elsewhere, "the scholar or intellectual also was something else, usually a political or religious figure—a mandarin or a priest—or, in individual cases, a charismatic personality who was honored and supported as a unique phenomenon and did not earn his living as a member of an established intellectual occupation" (Ben-David 1964, p. 460). In Western Europe, by the seventeenth and eighteenth centuries, scientists had developed "a recognized role, or . . . 'identity';" they were "regarded as being quite different from philosophers and theologians; the overlapping of these activities in the same persons did not obscure the distinctive image or role of the scientist" (Ben-David, 1964, p. 464). The institutionalization of scientific activity facilitated the development of new symbols and the incorporation of rationalistic and universalistic norms generated by changes in the form of sustenance organization in Western Europe associated with the development of modern capitalism. This idea is discussed in more detail below.

Given conditions unfavorable to the institutionalization of science there was in China no opportunity for the Chinese to turn isolated examples of proto-typical theory construction and mathematization into a scientific revolution (Needham 1954–, v. I, pp. 146ff.; v.II, pp. 582ff.)

Ben-David (1965) stresses the significance of decentralized authority —ecclesiastical as well as political—in the emergence of modern science. The crucial point from a comparative perspective is that no centralized bureaucratic structure existed in Western Europe, nor did one emerge in conjunction with the development of conditions favorable to the institutionalization of science. Neither the church nor the state—nor any other "institutional sphere"—dominated the social terrain of Western Europe in the way that the "agrarian bureaucracy" dominated the history of China until modern times (Moore, Jr. 1966, Chapter 6, esp.

169ff.) It was this structural feature that prevented the emergence of new social classes—especially a merchant class (Moore, Jr. 1966, p. 174) —and of anything like the Protestant Reformation; just these social changes, according to Ben-David (1965), enhanced the status of science and led to the establishment of the scientific role.

To this point we have, for the most part, reviewed the basic framework from which our analysis originates. We turn now to (1) an analysis of authority patterns, social change, and scientific activity in China and Western Europe, viewed in terms of human ecology, and (2) a reinterpretation of the relationship between Protestantism, modern capitalism, and modern science. We begin by defining Western Europe and China as "ecological units." This simply acknowledges the interrelatedness in these two civilizational areas of populations, social organizations, environments, and technologies—manifested in a culturally bounded historical development generally recognized by students of European and Chinese history.[4] We can, of course, anticipate focusing on sub-units of these areas in the future for more detailed analyses, comparable to Ben-David's (1965) shifts in geographical focus as he outlines the stages in the development of the scientific role which finally emerges in Northern Europe.

SUSTENANCE ORGANIZATION AND SCIENTIFIC ACTIVITY

Sustenance organization in Western Europe was based on a farming economy that involved reliance on rainfall. In China sustenance organization was based on large-scale irrigation works and flood control. Rainfall farming or small-scale irrigation agriculture does not involve the patterns of organization that characterize large-scale irrigation agriculture; the control of large sources of water supply creates a technical task which involves an extensive division of, and control over, labor. Dams, reservoirs, aqueducts, and tunnels must be built and kept in repair, and, in addition to these large-scale "preparatory" operations, large-scale "protective" operations such as the building of dikes and embankments must be initiated to safeguard crops from periodic and excessive inundations. Further division of labor involves the operations necessary for the recruitment and staffing of the preparatory and protective functions. Wittfogel (1956) has analyzed the complexity of this division of labor.

Wittfogel notes that cooperation in hydraulic agriculture involves more than "digging, dredging, and damming," and problems of organization. More complicated questions arise: the number of persons needed, and where they can be found, the determination of quotas and selection criteria; selection, notification, and mobilization of persons.

Even simple water control operations require high levels of integrative social action; "In their more elaborate variations, they involve extensive and complex organizational planning" (Wittfogel 1957, p 26).

If irrigation farming depends, then, on the effective handling of a major supply of water, this large quantity of water can be controlled only through the use of an extensive division of labor. Moreover, to the extent that large scale irrigation agriculture requires an extensive division of labor, it further follows that the effective operation of such an agricultural system involves a high degree of coordination and social control. The tendency of such large scale systems to promote the centralization and concentration of authority has been noted by Weber (1920, pp. 298-99; 1927, p. 321) and Wittfogel (1957, pp. 26ff.).

In criticizing Wittfogel's hypothesis, Beals (1955, p. 54) asks whether the managerial functions implicit in irrigation may not be performed through local community leadership rather than a central authority. Indeed, Eberhard (1962, pp.32-45) has assembled evidence to show that the construction and maintenance of irrigation systems was much more dependent on the initiation and action of local authority than Wittfogel recognizes. The local authorities—landlords, or landlord families— were, however, products of the "Imperial system," with "academic degrees and the official contacts that such a degree made possible" (Moore 1966, p. 169; Moore's argument follows Lattimore 1960). The landlords exerted pressure on the State through their official contacts for the proper irrigation control necessary for growing quality crops. Provincial projects "were the work of provincial landlord cliques;" state projects were "the work of still more powerful cliques with a national vision" (Moore 1966, p. 169). All of this does not alter the fact that there was a central authority ultimately responsible for public works, maintaining peace, and collecting taxes (Moore 1966, p. 171).

It is also necessary to consider the extent to which conditions in a society permit intercommunity cooperation in public projects. Steward (1955B, p. 54) hypothesizes that "informal intercommunity cooperation" is possible in small-scale systems; increase in scale—in the size of dams, and the number of miles of canals—is accompanied by an increase in the demand for labor and in "managerial density." Eventually, volunteers are replaced by corvee labor, and temporary supervisors give way to state-appointed bureaucrats.

Lenski (1966, pp. 235-38) agrees that large scale irrigation systems have probably been a factor in the growth of "autocracy," but neither a "dominant" nor a "major" factor. The degree to which authority is centralized is influenced by such factors as size of territorial unit, quality of transportation and communication facilities, relative distribution of military skill and power in the population, and the nature of the laws of inheritance. While Lenski's thesis is plausible, we would argue, in contrast, that the influence of the factors he mentions is not independent of the tendency toward centralization of authority inherent in the practice of large scale irrigation. For example, to the extent that effective centralized authority involves first the accessibility of the directing agency to a subordinate population, a requisite for the emergence of centralization of control in large scale irrigation agriculture is the construction of extensive transportation and communication facilities. These facilities, in turn, increase the potential for further centralization of control and authority. The practice of large scale irrigation in China fostered the growth of centralized systems of navigation canals, roads, and postal systems which intensified still further the central government's monopoly on control over transportation and communication facilities (Wittfogel 1957, pp.31-40; Steward 1955A, p. 198). In contrast, the practice of rainfall agriculture in Western Europe was a factor in calling forth a decentralized system of authority in which local transportation and communication facilities were established long before governments undertook the construction of centralized systems of transport facilities (Pirenne 1937, pp. 86-88). This is manifest in the character of European feudalism. It was never uniform, never, in James Westfall Thompson's words, (1931, p. 698), "a meticulously differentiated society, a carefully graduated hierarchy rising to an apex in the king-super-suzerain, a neatly divided delegation of authority." Local communities in China did not enjoy the corporate autonomy of their European counterparts. European feudalism was unique in this respect not only with reference to China, but in relation to to all other feudal-type systems; Japanese feudalism, for example, entailed "loyalty to superiors and a divine ruler;" and in India, the concept of "free contract" did not arise (Moore 1966, p. 416). Only in Western Europe did the idea of collective immunity from the power of rulers emerge, "along with the right of resistance to unjust authority," and "the conception of contract as a mutual engagement freely undertaken by free persons" (Moore 1966, p. 415).

Are we to hypothesize that the necessity for large-scale water works in China was a determinant of the Chinese form of feudalism and of a centralized pattern of authority? And did rainfall agriculture in the West stimulate decentralization of authority and the Western form of feudalism? While these hypotheses are not misdirected they must be somewhat refined. Specifically, Wittfogel's "oriental despotism" hypothesis can be refined as follows: The tendency toward centralization of authority and "bureaucratic feudalism" in preindustrial societies is related to environmental conditions which necessitate the construction and coordination of large scale public works (Moore 1966, p. 416). The necessity of large scale public works in China was specifically the necessity for large scale irrigation.

The "democratic possibilities" Moore associates with Western feudalism—and the lack of such possibilities in China—are illustrated by the variations in capitalist activity in the West and in China. In turning to an examination of these variations we wish to draw attention to the parallel between the potential within China and Western Europe for institutional and role autonomy in capitalist activity and scientific activity.

SUSTENANCE ORGANIZATION AND CAPITALIST ACTIVITY

Ben-David (1965, p. 48) argues that the development of science in the European universities was stimulated by the emergence of a merchant or capitalist class whose interests were consistent with, and stimulated, an empirical, rational world view. An alternative hypothesis, however, is that the emergence of both "modern science" and "modern capitalism" was in response to the same set of ecological conditions. Those conditions which inhibited the institutionalization of scientific activity in China also, and concomitantly, inhibited the institutionalization of capitalist activity. We have discussed some of these conditions in the preceding section; a broader ecological perspective is introduced in the following discussion of capitalist activity in China and Western Europe.

The primary factors in the emergence and development of capitalist activity are competition and exchange in a "free," diversified market. The process of competition begins with the emergence of an aggregate demand in a population that exceeds the supply of resources (Hawley 1950, pp. 201-3). This is followed by an increasing homogeneity among the competitors. At the third stage of this process, "congestion" begins to selectively eliminate the weakest competitors. The fourth stage is the development of some form of differentiation according to territory, or intra-territorial specialization which entails dependence on but non-competition with those who command supplies. Thus, the differentiation and multiplication of functions in a society depends on competition for resources. In any such development, population size, environmental productivity and level of technology are conditioning factors.

Population size, for example, constitutes a limitation upon the extent to which functional differentiation can occur. Size limits the opportunities for specialization and also determines the extensity of specialization that can be supported by the population. Population size is affected by the productivity of the environment. A poor environment can support relatively few people, though the actual number is in part a function of the type of social organization they develop. Productivity, in turn, depends on the level of technology which in turn determines the efficiency of transportation and communication facilities; and these facilities affect the extent to which spatially separated activities can be inter-related. To the extent that population is isolated or independent of exchange relationships, its resources are limited to what may be ex-tracted from a local habitat. Isolation disappears as the development of technology, of economic surplus, and of transportation and communi-cation facilities promote and permit functional interdependence be-tween and among communities. Under conditions of exchange or interdependence a progressively larger environment is made accessible (Sjoberg 1964, p. 133).

The division of labor in China, while extensive, was not "functionally differentiated" along an exchange-of-surplus base. Whatever potential existed in China for such developments—and it appears to have been limited—local surpluses were commanded by the State for the rulers; there was no localized, independent development of and control over differentiated surpluses comparable to that which developed in Western Europe (Wittfogel 1957, pp. 67ff; Moore 1966, p. 175; and Marx, on the "Asiatic mode of production," 1965, pp. 70, 83, 91). The decentraliza-tion of authority in feudal Europe was a function of, and stimulated, regional exchange (Russell 1960, pp. 55-70) which, in addition to popu-lation increase and scarcity of land, contributed to the emergence of a class of private capitalists (Pirenne 1925, pp. 75-91). The free movement of the European merchants and their ability to share in the competition for power was dependent on the decentralization of authority, and on the associated conflicts amongst the landed nobility, the church, and the rulers (Pirenne 1937, pp. 54, 80; Moore 1966, p. 174).

It is widely recognized that China was isolated from intercultural exchange: "local and indigenous" factors conditioned the structure of Chinese life to a greater extent than was the case in the Near and Middle East, and Europe (McNeill 1963, p 238; Kracke 1953, pp. 4-5); and Lampard (1965, p. 538) has noted China's isolation from the "arc of urbanization that reached from the lower Nile in the west across the Mesopotamian heartland to the Indus Valley and Gangetic plain in the East." These factors were reinforced by the lack of an internal "urban market" (Moore 1966, pp. 178-80). These conditions can be considered principle obstacles to capitalist activity in China. Theoretically, it is important to determine the extent to which the Chinese environment could and did generate a differentiated surplus—a surplus of varied and inter-dependent goods and services—and the significance of such a surplus in the decentralization of authority, and the institutionalization of autonomous social activities. Here, of course, our focus is on capitalist activities.

It must be noted that competition results in a more elaborate division of labor that is functionally differentiated along an exchange-of-surplus base only to the extent that the production of a surplus has been institutionalized through some form of social organization. Students of the origin of urbanization usually start with the assumption that a precondition for the emergence and growth of urbanization or the cumulative differentiation of activity is a level of agricultural production sufficiently high to release a substantial part of the population from primary resource activities, and permit the involvement of those released in activities one or more levels removed from primary agriculture. But the accumulation of an agricultural surplus, while a necessary condition, is not a sufficient condition for functional differentiation in the elaboration of the division of labor. An appropriate form of social organization capable of accumulating the surplus is necessary (Keyfitz 1965).

Capitalist activity is oriented to profit through exchange. This requires (1) an environment which will yield not simply a surplus, but a differentiated surplus, and (2) social organization to accumulate and exchange the surplus (for profit); this implies differentiation of function among interdependent and more or less autonomous free "agents"— individuals or communities. Modern capitalism in particular was dependent on an ecological milieu conducive to the establishment of an autonomous role for merchants, or capitalists, and the institutionalization of their activities. The relation between these conditions and sustenance organization requires further explanation.

If the area of land available for cultivation and "environment" is held constant, the quantity of food available for any unit of population is a function of population size and the degree of farming knowledge and level of farming technology. Agricultural production in Europe was, until the eighteenth century, limited chiefly by the difficulty of restoring the fertility of the soil after cropping (van Bath 1963, pp. 9-14). There were at least three methods available for restoring soil fertility, but their use was limited (van Bath 1963, pp. 10-13). The difficulty of restoring soil fertility, in conjunction with population increases, led to population pressure on the food supply. This resulted in detaching increasing numbers of individuals from the land. These individuals represented potential occupants of roles ancillary to primary sustenance roles. Indeed, Pirenne suggests that the ancestors of the merchant class in Western Europe must be sought in the masses of individuals displaced from the soil (1925, pp. 80-81).

The fixation of the manorial regime in Western Europe during the ninth and tenth centuries (Thompson 1931, p. 722), furthermore, provided a form of social organization capable of generating the economic surplus requisite for support of activities one or more levels removed from resource extraction. Through various forms of manorial taxation and fiscal exaction the landed aristocracy secured the revenue with which to support an emergent merchant class (Thompson 1931, pp. 728-30).

From 1000 to 1700 the magnitude of population increase in China and Western Europe is estimated to have been about the same (Bennett 1954, p. 9). But while improved farming practices did not occur in Western Europe until the eighteenth century, improved techniques in China at a much earlier time resulted in increased yields to meet the needs of the growing population. The most significant development in land utilization and food production in China during the past millenium was, according to Ping-ti Ho (1959, pp. 169-71), brought about by the cultivation of early-ripening rice which began to be diffused extensively early in the eleventh century. The introduction of an early-ripening rice and development of early-ripening varieties appears to have been an important factor in stimulating the first long-term "revolution in land utilization and food production, based on the conquest of relatively well-watered hills" (Ping-ti Ho 1959, p. 171). During the greater part of the past thousand years, the food situation seems to have been much better in China than it was in Europe. Whereas the Europeans had to struggle with the problem of restoring soil fertility after cropping, the most striking feature of rice grown under irrigation is its indifference to soil fertility (Murphey 1957, pp. 191-92). This indifference accounts for the capacity of wet rice agriculture to stimulate the intensification of cultivation in response to a rising population. This supports the hypothesis that a differentiated surplus was not exploited by the Chinese; indeed, the exploitation of such a surplus (considering the nature and distribution of resources, and level of technology) appears to have been a uniquely Western prerogative.

The European situation we have described was an aspect of the process of urbanization in the West, a subject we turn to briefly next.

In ecological theory, general "cultural growth" is associated with "regional intercommunication and interstimulation" (Braidwood and Willey 1962, pp. 354-55). Communication and stimulation here imply not simply linkages but symbiotic linkages within a "naturally" differentiated area; the prerequisite for general cultural growth is the juxtaposition of diverse sub-environments or cultures symbiotically exploited through social interaction (Lampard 1965, p. 529). The process of urbanization is fed by population increases and the development of a surplus; technology injects "form and focus" into the process (Tisdale 1942, pp. 311-16). One of the necessary conditions for the Neolithic Revolution was technological innovation. The concentration of populations in towns was related to the productivity of the land (which defined the level of technology). Urbanization is thus a function of the exploitation and exchange of differentiated surpluses (e. g., Lampard 1965, p. 546); resulting interdependence between and among communities, and the increase in specialization are important not only in the history of general economic growth but in scientific development as well. Zilsel (1942) and White, Jr. (1963) provide a general overview of scientific development as a function of urban growth.

Between the second and fifteenth centuries A.D., China's general cultural growth can be "represented by a relatively slowly rising curve, noticeably running at a higher level, sometimes a much higher level, than European parallels" (Needham 1963, p. 139). The explanation for the exponential growth of European technology from the seventeenth century on, and the scientific revolution, must be based on an analysis of the prior and concurrent development of urban interdependence, and on the proliferation of transportation, communication, and exchange links in a relatively hospitable environment. By the seventeenth century the preconditions for economic and technological "take-off" had been established in Western Europe. The facilitating factors we have discussed did, of course, depend on processes of diffusion. The diffusion of technology, especially from China, appears to have been a critical process (Needham 1964, pp. 145-46). Theoretical support for this idea is offered by Meggers (1954, p. 822).

The structural consequence of the ecological constraints we have noted in China was that merchants and craftsmen were low in status, discriminated against, and relatively immobile within the rigid stratification system that characterized Chinese history from ancient to modern times (Eberhard 1962, pp. 5-50). In contrast to the merchant class that emerged in Western Europe from the ninth century onward, the Chinese merchants were interdependent with and subordinate to the scholar-officials, and subject to state supervision and regulation (Balazs 1964, pp. 23, 33, 70, 76). Among the most important distinctions was the fact, noted by Weber (1958, pp. 127-28), that the European merchants came to possess effective military power.

PROTESTANTISM, CAPITALISM, AND SCIENCE

The concurrent emergence of modern science, modern capitalism, and Protestantism made the search for mutual influences a natural research task. Weber, in *The Protestant Ethic and the Spirit of Capitalism,* suggested a relationship between ascetic Protestantism and modern capitalism. He noted too the strong influence of the natural sciences on the development of modern capitalism. Important stimulation for the development of the natural sciences was, in turn, provided by the interest of capitalists in their practical applications. The defining characteristic of modern capitalism, organized rationality, was dependent on the calculability of certain technical factors made possible by developments in mathematics and experimental method.

Given the basis provided by Weber for viewing ascetic Protestantism as instrumental in the emergence of modern capitalism, it followed from Merton's analysis that ascetic Protestantism was instrumental in furthering the development of science. Merton (1968, p. 628) argued that the Puritan ethic, an "ideal-typical expression of the value attitudes basic to ascetic Protestantism generally, so canalized the interests of seventeenth-century Englishmen as to constitute one important *element* in the enhanced cultivation of science." The glorification of God in His works, the welfare of society, rationality as a curb on the passions, the demand for systematic, methodic, and diligent work were, according to Merton, manifested in the pursuit of science (1968, p. 633).

In order to verify the general hypothesis that "the cultural attitudes induced by the Protestant ethic were favorable to science," Merton (1968, p. 637) deduced and tested the specific hypothesis that "if the Protestant ethic involved an attitudinal set favourable to science and technology in so many ways, then we should find amongst Protestants a greater propensity for these fields of endeavors than one would expect simply on the basis of their representation in the total population" (1968, p. 649). Merton's data indicated a "pronounced" association between Protestantism and scientific and technologic interests, "even when extra-religious influences are as far as possible eliminated."

The application of the Weber-Merton theses in the comparative study of science in China and Western Europe allowed the following conclusion: Protestantism, a doctrine of "rational mastery *over* the world" (Parsons 1949, p. 549), constituted "one important *element* in the enhanced cultivation of science" in Western Europe (Merton 1968, p. 33); Chinese civilization was dominated by Confucianism, a doctrine of "rational adaptation *to* the world" (Parsons 1949, p. 549) which inhibited organized scientific activity.

Parsons (1949, p. 541-42) has noted Weber's judgment that "in both China and India the combination of nonreligious factors was at the crucial time at least as favorable to capitalistic development as in the Western situation. Hence the strong probability that in this respect a principal differentiating factor with respect to capitalism lay in the religious element of the economic ethic." A parallel judgment concerning scientific development underlies attempts to attribute the inhibition of science in China, and its enhancement in Western Europe, to differences in the religious ethics of these two "civilizations." Weber argued that the Confucian ethical system fostered traditionalism in China, the "acceptance of an existing order, above all of the traditional religio-magical elements of it, whether state cult, ancestor worship or popular magic. Moreover, the ideal of the Confucian gentleman was a traditional status ideal, the basis of which was assimilation of a traditionally fixed body of literary culture, the classic;" the Protestant ethic, by contrast, was "a distinctly revolutionary force" (Parsons 1949, pp. 548-49). It is from this Weberian perspective that Merton views the instrumental influence of Protestantism on Western science. From this same perspective, Balazs (1964, p. 22) argues the widely accepted idea that "most probably the main inhibiting cause [for the lack of development in Chinese science] was the intellectual climate of Confucianist orthodoxy, not at all favorable for any form of trial or experiment, for innovations of any kind, or for the free play of the mind."

But the history of Protestant interference with the pursuit of science,[5] and the innovative contributions of the Chinese in science and technology documented by Needham, are enough to establish some doubts concerning the Weberian thesis. There is, however, a stronger argument for reconsidering that thesis. Eisenstadt (1968, pp. 25-27) has noted that Confucianism was an integral part of the "specific political framework of the Chinese empire." The fact that Confucianism developed and changed with the development of the state bureaucracy, and came to function as the legitimizing ethic of the empire, makes it reasonable to conceive Confucianism as a normative response to the form of sustenance organization in China (Balazs 1964, p. 7). One is tempted at this point to argue that Protestantism as an ethical system was a normative response to the form of sustenance organization in Western Europe. Actually it appears more reasonable to view rationalism-universalism as the normative response to the exchange based sustenance organization in Western Europe; rationalism-universalism is thus conceived as the conceptual equivalent of Confucianism. Protestantism, in this perspective, is interpreted as the particular response of the religious institution to rationalism-universalism which permeated to greater and lesser degrees all parts of Western European social structure beginning with the Renaissance.

As an integral part of the organization of the bureaucratic state, Confucianism might, indeed, be considered epiphenomenal in the inhibition of autonomous scientific activity in China. In fact, it is possible that even this interpretation over-emphasizes the influence of Confucianism. Needham (1969, p. 282), for example notes that "always alongside the Confucian veneration of the sages, and the Taoist threnodies about the lost age of primitive community, there flourished . . . other convictions that true knowledge had grown and could yet grow immeasureably more if men would look outward to things, and build upon what other men had found reliable in their outward looking." He notes further that "there is no Chinese century from which one could not cite quotations to illustrate the conception of science as cumulative disinterested co-operative enterprise" (Needham 1969, pp. 282ff.). The failure of such a conception to crystallize in China, and its successful institutionalization in Western Europe, can best be accounted for by examining conditions conducive to and inhibitive of autonomous social activities.

Thus, neither Merton's conclusion in 1938—that "the formal organization of values constituted by Puritanism led to the largely unwitting furtherance of modern science" (1970, p. 136)—nor his statement in 1970 that "before it became widely accepted as a value in its own right, science was required to justify itself to men in terms of values other than that of knowledge itself"—convey an appropriate appreciation of the relationship between scientific, capitalistic, and religious activities.

The distinction between the interpretation we are advocating and that associated with Weber and Merton is illustrated by Merton's (1970, p. xxviii) recent comment that "It would be satisfying to be able to answer the question, to estimate the proportion of cases which can be ascribed to the Puritanism science sequence and to the science Puritanism sequence." It is this "reciprocal influence" assumption our analysis has brought into question.

In addition to raising questions about the "reciprocal influence assumption," our analysis raises a related question concerning the "utilitarian hypothesis" and the emphasis on values and the roles of men in "evolving" science. Ben-David (1971, p.31) follows Merton (cf. 1970, p. xix) in this emphasis: "Before science could become institutionalized, there had to emerge a view that scientific knowledge for its own sake was good for society in the same sense as moral philosophy was. Something like this idea had apparently occurred to some natural philosophers. But in order to convince others that this was so, they had to show some moral, religious, or magical relevance of their insights. As a result, the scientific content of natural philosophy was either lost or concealed by the superstitions and rituals of esoteric cults." Ben-David's adherence to the utilitarian hypothesis is manifested in his emphasis on "needs," "demand," "social value," and "social interests." This strains his institutional approach (as it does Merton's) and leads him to focus on questions such as whether persons did or did not regard themselves as "scientists" (1971, pp. 45ff.). He argues that "the development of science depended on the determination of the minority who believed in science to fight for its general recognition openly and to express and develop its interest in science in public discussion and purposeful association (Ben-David 1971, p. 68). The more fundamental sociological questions, however, are (1) what stimulated such determinations, and (2) what conditions made institutionalization possible?[6]

We have argued from a comparative study of science in China and Western Europe that only in the latter setting did conditions exist which both stimulated scientific attitudes and concepts of a scientific role, *and* allowed for the institutionalization of science as an autonomous social activity. These conditions, from environmental factors to the structure of political systems, were in Western Europe conducive to the emergence of social activities in every institutional sphere with varying degrees of differentiation and autonomy; this occurred to a much lesser extent in China. Protestantism, modern capitalism, and modern science are thus conceived as parallel responses in different institutional spheres to an underlying set of conditions which promoted an exchange-based economy and political decentralization. The institutionalization of science, finally, is conceived to be a precondition for, and not in itself, the scientific revolution.

NOTES

1. The unique emergence of modern science has been the subject of much scholarly research and commentary. Weber (1958, pp. 13, 15-16) noted that "only in the West does science exist at a stage of development which we recognize today as valid." Price (1961, p. 3) writes that none of the other great civilizations followed a scientific path comparable to that followed by European civilization: "it becomes clearer from our fragmentary historical understanding of their case histories that none of them was even approaching it." See also Butterfield (1965, pp. 187-202); Ben-David (1964, pp. 455-76, and 1965, pp. 15-54); Needham, in Goldsmith and Mackay (1964, pp. 127-49) and in Crombie (1963, pp. 117-53). On periodization in the history of science there is universal consensus on dating the beginnings of the scientific revolution. The stages suggested by A. R. Hall (1962) are exemplary: "Rational science . . . is the creation of the seventeenth and eighteenth centuries. Since then dramatic achievements in understanding and power have followed successively. In this sense the period 1500-1800 was one of preparation, that since 1800 one of accomplishment and it is convenient to conclude this history of the scientific revolution with the early years of the nineteenth century for other reasons. Though profound changes in scientific thought have occurred since that time, and though the growth of complexity in both theory and experimental practice has been prodigious, the processes, the tactics and the forms by which modern science evolves have not changed."

2. On the shift of scientific activities to the academies and the impact of this shift on the scientific revolution see Zilsel (1957, pp. 273-74); Wolf (1959, pp. 8-9, 54-70); Dampier (1948, p. 149); Hall (1962, pp. 186-216); Johnson (1957, p. 328).

3. While Needham's phrase, "spontaneous autochthonous," expresses the fact that the scientific revolution occurred within the boundaries of Western Europe, it should not obscure the significance of external conditions and inputs. Needham clearly recognizes this when he suggests that the diffusion of Chinese technology to the West helped "to set the stage for the decisive break-through which came about in the favourable social and economic milieu of the Renaissance" (Needham, in Crombie 1963, p. 149).

4. On Europe as an "entity" see, for example, Needham (1963, p. 117); Price (1961, p. 2); Ben-David (1965, p. 17); Parsons (1951, pp. 339ff.); on China as an "entity" see, in addition to works by Needham, Eberhard, Wittfogel, and Balazs cited in this paper, Moore, Jr. (1966, chapter 4).

5. Candolle, in his *Historie des sciences et des savants depuis deux siecles* (Geneva, 1873), remarks that from 1535 to 1735 an authoritarian principle dominated Geneva, preventing any Genevan citizen from gaining real distinction in science. Concerning the Copernican revolution, Protestants as well as Catholics were confronted with the problem of reconciling the new astronomy with scriptures. Luther, and Melanchethon (author of *Physics,* published in 1552) reacted with hostility to the Copernican system. Tycho Brahe, who established the foundation for Kepler's work, rejected Copernicus out of respect for Scriptures. Kepler had to edit out a chapter on the reconciliation of heliocentrism with scriptures before he could get his *Mysterium Cosmographicum* into print (1596). He also had difficulties bringing out his account of the comet of 1607 due to the serious theological questions raised by members of the Lutheran University of Leipzig; see Russo (1963, pp. 300-301).

6. It should be clear that we think Ben-David has made an outstanding contribution to our understanding of the emergence and development of science. Of special note, as we have emphasized, is his identification of factors (e. g., decentralization) conducive to autonomy in science. We have, however, suggested an alternative to his stress on "social values and interests" in explaining the "takeoff into continuous accelerating growth of science" (1970, p. 169).

REFERENCES

Balazs, Etienne. 1964. *Chinese Civilization and Bureaucracy.* New Haven and London: Yale University Press.

Barber, Bernard. 1952. *Science and the Social Order.* New York: The Free Press.

Ben-David, Joseph. 1964. Scientific Growth: A Sociological View. *Minerva* 2 (Summer): 455-76.

_____. 1965. The Scientific Role: The Conditions of its Establishment in Europe. *Minerva* 4 (Autumn): 15-54.

_____. 1971. *The Scientist's Role in Society.* Englewood Cliffs, N. J.: Prentice-Hall.

Bennett, M. K. 1954. *The World's Food.* New York: Harper and Brothers.

Braidwood, Robert J. and G. R. Willey, eds. 1962. *Courses Toward Urban Life; Archeological Considerations of Some Alternatives.* Chicago: University of Chicago Press.

Butterfield, Herbert. 1965. *The Origins of Modern Science, 1300-1800.* Rev. ed. New York: The Free Press.

Crombie, A. C., ed. 1963. *Scientific Change.* New York: Basic Books.

Dampier, W. C. 1948. *A History of Science.* 4th ed., rev. and enlarged. Cambridge, England: Cambridge University Press.

Duncan, Otis Dudley. 1964. Social Organization and the Ecosystem. In *Handbook of Modern Sociology,* ed. R. E. L. Faris, pp. 36-82. Chicago: Rand McNally.

Duncan, Otis Dudley and Leo F. Schnore. 1959. Cultural, Behavioral, and Ecological Perspectives in the Study of Social Organization. *American Journal of Sociology* 65 (September): 132-46.

Duncan, Otis Dudley et al. 1960. *Metropolis and Region.* Baltimore: The John Hopkins Press.

Eberhard, Wolfram. 1957. The Political Function of Astronomy and Astronomers in Han China. In *Chinese Thought and Institutions,* ed. John K. Fairbank, pp. 33-70. Chicago and London: University of Chicago Press.

_____.1962. *Social Mobility in Traditional China.* Leiden: E. J. Brill, Ltd.

Edelstein, Ludwig. 1957. Recent Trends in the Interpretation of Ancient Science. In *Roots of Scientific Thought,* ed. P. Weiner and A. Noland, pp. 90-121. New York: Basic Books.

Eisenstadt, S. N. 1968. The Protestant Ethic Thesis in an Analytical and Comparative Framework. In *The Protestant Ethic and Modernization,* ed. S. N. Eisenstadt, pp. 3-45. New York: Basic Books.

Hall, A. R. 1962. *The Scientific Revolution 1500-1800.* 2d ed. Boston: Beacon Press.

Hawley, Amos H. 1950. *Human Ecology.* New York: The Ronald Press.

Johnson, F. R. 1957. Gresham College: Precursor of the Royal Society. In *Roots of Scientific Thought,* ed. P. Weiner and A. Noland, pp. 328-53. New York: Basic Books.

Keyfitz, Nathan. 1965. Political-Economic Aspects of Urbanization in South and Southeast Asia. In *The Study of Urbanization,* ed. Philip M. Hauser and Leo F. Schnore, pp. 265-309. New York: John Wiley.

Kracke, E. A., Jr. 1953. *Civil Service in Early Sung China, 960-1067.* Cambridge, Mass.: Harvard University Press.

Lampard, Eric E. 1965. Historical Aspects of Urbanization. In *The Study of Urbanization,* ed. Philip M. Hauser and Leo F. Schnore, pp. 519-44. New York: John Wiley.

Lattimore, Owen. 1960. The Industrial Impact on China, 1800-1950. In *First International Conference of Economic History,* Stockholm, August, pp. 103-12. Paris: Mouton and Company.

Lenski, Gerhard. 1966. *Power and Privilege.* New York: McGraw-Hill.

Marx, Karl. 1965. *Pre-Capitalist Economic Formations.* New York: International Publishers.

McNeill, William H. 1963. *The Rise of the West.* Chicago: University of Chicago Press.

Meggers, Betty J. 1954. Environmental Limitations on the Development of Culture. *American Anthropologist* 56 (October): 801-24.

Merton, Robert K. 1938. Science, Technology and Society in Seventeenth-Century England. *Osiris* 4: 360-632.

_____. 1968. *Social Theory and Social Structure.* Enlarged ed. New York: The Free Press.

_____. 1970. *Science, Technology and Society in Seventeenth-Century England.* New York: Harper Torchbooks.

Moore, Barrington, Jr. 1966. *Social Origins of Dictatorship and Democracy.* Boston: Beacon Press.

Murphey, R. 1957. The Ruins of Ancient Ceylon. *The Journal of Asian Studies* 16 (February): 181-200.

Needham, Joseph. 1954. *Science and Civilization in China* (successive volumes). Cambridge England: Cambridge University Press.

_____. 1963. The Poverties and Triumphs of the Chinese Scientific Tradition. In *Scientific Change,* ed. A. C. Crombie, pp. 117-53. New York: Basic Books.

_____. 1964. Science and Society in East and West. In *Society and Science,* ed. M. Goldsmith and A. Mackay, pp. 127-49. New York: Simon and Schuster.

_____. 1969. *The Grand Titration.* Toronto: University of Toronto Press.

Parsons, Talcott. 1949. *The Structure of Social Action.* New York: The Free Press.

_____. 1951. *The Social System.* New York: The Free Press.

Ping-ti Ho. 1959. *Studies on the Population of China 1368-1953.* Cambridge, Mass.: Harvard University Press.

Pines, S. 1963. What was Original in Arabic Science? In *Scientific Change.* ed. A. C. Crombie, pp. 181-205. New York: Basic Books.

Pirenne, Henri. 1925. *Medieval Cities.* New York: Doubleday-Anchor.

_____. 1937. *Economic and Social History of Medieval Europe.* New York: Harcourt, Brace and World.

Price, Derek J. de Solla. 1961. *Science Since Babylon.* New Haven and London: Yale University Press.

Russell, J. C. 1960. The Metropolitan City Region of the Middle Ages. *Journal of Regional Science* 2 (Fall): 55-70.

Russo, Francois. 1963. Catholicism, Protestantism, and the Development of Science in the Sixteenth and Seventeenth Centuries. In *The Evolution of Science,* ed. G. S. Metraux and Francois Crouzet, pp. 291-320. New York: New American Library.

Sjoberg, Gideon. 1964. The Rural-Urban Dimension in Pre-Industrial, Transitional, and Industrial Societies. In *Handbook of Modern Sociology,* ed R. E. L. Faris, pp. 127-59. Chicago: Rand McNally.

Steward, J. H. 1955A. *Theory of Culture Change.* Urbana: University of Illinois Press.

————. 1955B. Some Implications of the Symposium. In *Irrigation Civilizations: A Comparative Study,* ed. Julian H. Steward, pp. 58-78. Washington, D. C.: Pan American Union.

Thompson, James Westfall. 1931. *The Middle Ages.* 2 Vols. New York: A. Knopf.

Tisdale, Hope. 1942. The Process of Urbanization. *Social Forces* 20 (March); 311-16.

van Bath, B. H. Slicher. 1963. *The Agrarian History of Western Europe A. D. 500-1850.* New York: St. Martin's Press.

Weber, Max. 1920. Konfuzianismus and Taoismus. *Gesammelte Aufsatz zur Religionssoziologie,* Vol. 1. Tubingen: J. C. B. Mohr.

————. 1927. *General Economic History.* New York: Macmillan.

————. 1958. *The Protestant Ethic and the Spirit of Capitalism.* New York: Charles Scribner's Sons.

White, Lynn. 1963. What Accelerated Technological Progress in the Western Middle Ages? In *Scientific Change,* ed. A. C. Crombie, pp. 272-91. New York: Basic Books.

Wittfogel, Karl A. 1957. *Oriental Despotism.* New Haven: Yale University Press.

Wolf, A. 1959. *A History of Science, Technology and Philosophy in the 16th and 17th Centuries.* 2 Vols. New York: Harper Torchbooks.

Yabuuti, Kiyosi. 1963. Sciences in China from the Fourth to the End of the Twelfth Century. In *The Evolution of Science,* ed. G. S. Metraux and Francois Crouzet, pp. 108-27. New York: New American Library.

Zilsel, E. 1942. The Sociological Roots of Science. *The American Journal of Sociology* 47: 544-62.

WALTER HIRSCH

The Autonomy of Science*

THE PROBLEM

At the present time the assertion that "science is a social product" has found general acceptance, even among those who wish to disassociate themselves from a supposedly "Marxist" point of view.[1] A more problematical set of assertions and assumptions relates to the specific social conditions providing an optimal basis for the existence and flourishing of science. The prevalent view, at least among American sociologists and other scientists, is that "democratic" social structure provides the best possible conditions for the acceptance and implementation of the scientific ethos.[2] A minority view is held by George Lundberg, who asserts that "if we plot the course of scientific advances during the past two

Walter Hirsch, "The Autonomy of Science in Totalitarian Society," *Social Forces* 40 (1961): 15-22. Reprinted by permission of the publisher, University of North Carolina Press.
*This research was supported by a summer study grant from the Purdue Research Foundation. The bibliographical assistance of Dr. Hanna H. Meissner is gratefully acknowledged. The present paper is an extended version of one delivered at the meeting of the American Sociological Association, August 1960.

hundred years, the impressive fact is how little its main course has been deflected by all the petty movements of so-called 'social action,' including the major political revolutions."[3] More recently, Reinhold Niebuhr, who does not share Lundberg's frame of reference, writes that "on the whole, political despotism has no quarrel with pure science or with scientists who do not feel called upon to challenge the basic dogmas of the regime."[4] Historians of science have pointed out that important scientific discoveries have been made under political regimes and under conditions which are by no means "democratic." And lately the "sputnik effect" has brought about a questioning of the assumption that science must inevitably be backward in totalitarian societies.

The object of this paper is to investigate the proposition that certain types of scientific activity tend to be more "immune" than others to extra-scientific influences and controls. The proposition will be analysed in a setting where political controls are at a maximum, namely in the totalitarian societies of Nazi Germany and the Soviet Union.

The desirability of being able to generalize about "totalitarian" societies carries a price tag. How comparable are Nazi Germany and the USSR? There are obvious differences in their historical development, in their culture and social structure, in the development of science before the onset of totalitarianism, in the duration of the regimes, and in the idiosyncratic but influential role of a Stalin and a Hitler. We hope that, by keeping in mind the problems of comparability we shall be sensitive to both differences and similarities, and shall be able to do justice to both sociology and history. In addition, we shall be concerned with the implications of our findings for democratic societies.

ASSUMPTIONS AND HYPOTHESES

The hypotheses to be formulated rest on two major assumptions:

1. Totalitarian societies,[5] like other modern societies, depend on the fruits of science and technology for their economic well-being and military strength. In their turn, scientists and engineers depend on the holders of political power for their "conditions of existence," i.e. on funds and facilities for research in addition to the minimal requirements of survival, and on access to the available sources of knowledge and skill. These conditions also imply motivation on the part of scientists and engineers to perform tasks which may or may not be of their own choosing, i.e. the presence of rewards which can be manipulated by the political authorities.

2. Unlike other societies, totalitarian societies seek to implement their goals by maximum politicization of life, i.e. by stressing the rele-

vance of *all* behavior for the social goals, which have been largely determined by political authority, and by severely restricting individual choices considered antagonistic or irrelevant to these goals. The means for this include ideological indoctrination, centralized planning, the leadership principle, and the breaking up of "private" groups and institutions considered socially harmful.

Our two major working hypotheses are based on conventional categorical distinctions which appear throughout the relevant literature.[6]

1. The *natural* sciences are more autonomous, i.e. less subject to political controls than the *social* sciences. The rationale for this hypothesis is based on the following considerations:

a. By their nature, the social sciences are necessarily concerned with social values and are therefore of immediate interest to the ideologists of the regime, who must prevent the critical analysis on an "objective" level of the social goals that have been determined. The natural sciences on the other hand are not directly concerned with social goals; their findings must first be translated into value-relevant terms, as in the case of social Darwinism.

b. The natural sciences have accumulated knowledge which is more empirically testable; consequently it is more difficult to "distort" their findings in line with ideological needs. Ideologists, who wish to preserve the purity of their system, will hesitate to expose the potential weaknesses of the ideology in a conflict with the "hard facts" of natural science.

c. Natural science requires highly skilled experts while social science knowledge can be acquired more easily. Because of the esoteric knowledge and skills possessed by natural scientists, they are able to immure themselves behind an "invisible wall" through which nonexperts cannot easily penetrate.

What theoretical arguments can be adduced to negate the first hypothesis and to ascribe more autonomy to the social sciences? The plausibility of the argument may seem strained, but the need for "outrageous hypotheses" is as great in this area as in others:

a'. Since the results of natural science are more "fruitful" and immediately applicable in the form of technology than those of the social sciences, more effort will be expended to channelize the former into ideologically relevant goals. *Per contra,* the social sciences can be left alone, since their results are of minor consequence for the social goals.

b'. Since the object of totalitarian rule is to maximize control by manipulation of institutions and individuals, freedom to experiment should be accorded to social scientists, so that they may develop the most efficient methods for "human engineering."

2. Our other working hypothesis is that the *applied* aspects of science, including *technology,* are more autonomous than the *pure* or *theoretical.* The rationale favoring this is as follows:

a. Since the political leadership needs results, it cannot afford to meddle with the "experts." The luxury of ideological intrusion is then confined to the theoretical areas, where the outcome of scientific activity does not have immediate relevance for the society's well-being and the regime's power position.

a'. But there is a counter-argument: It is precisely in the applied sphere where political decisions have to be made, in terms of determining the most feasible goals, allocation of resources, etc. To the extent that these decisions are made on ideological grounds, applied science and technology would be less autonomous than pure or theoretical aspects of scientific activity.

TEST OF THE HYPOTHESES

We shall test our working hypotheses on the basis of widely scattered data from both Nazi Germany and the USSR. In view of the incomplete nature of the data a conclusive test is patently impossible at this time. Nevertheless, we may be able to come to some tentative conclusions and to ascertain the usefulness of the working hypotheses.

1. What is the evidence regarding the relative autonomy of the natural and the social sciences? On the ideological level this distinction is disparaged by Marxism and Nazism alike; terms like "idealist genetics" and "Jewish physics" indicate the totalitarian attempt to enforce the "social" significance of all scientific activity. However, Marxist theory recognizes differences in degree. Some aspects of science are located in the "superstructure," hence dependent on the underlying "relations of production," but other aspects, notably mathematics, are considered independent.[7] Nazi ideology does not contain any explicit distinction of this sort. In addition, we must, of course, look at the "behavioral" level, which may or may not be in accord with ideological presuppositions.

What was the initial impact of Nazi accession to power on scientific activity? Between 1932 and 1938, of the regular professors in the economic and social sciences, 54 out of 132, or 40 percent, vacated their chairs. It is not known what proportion was removed for either racial or political reasons and what proportion retired because of age, but 45 of the 54 were below age 60.[8]

Similarly, between 1931 and 1938, 3120 university and technical college teachers emigrated from Germany, comprising 8 percent of the total teachers. Of the emigrants 47 percent were in the economic and social sciences.[9] Clearly, the direct impact of political pressure was greater in the social science area. Many prominent German social scientists who did not leave the country became part of the "inner emigration." Among them can be counted Tönnies, Alfred Weber, v. Wiese, Vierkandt, Thurnwald, all of whom were not economically dependent on a university position. As one member of the profession who did emigrate put it, "They did not learn to lie, as those who remained in office necessarily had to."[10]

No comparable figures are available for the Soviet Union, and it is difficult to get a systematic picture of events during the early, chaotic days of the revolution. However, it appears that during the period of rapid industrialization introduced by the first five year plan, it was the economists and statisticians who were the first to suffer, because their projections and findings were not in accordance with the goals of the plan.[11] Subsequently the various purges became so extensive that it would be difficult to single out a given type of science as having suffered most from *direct* political controls. We must take recourse to different kinds of analysis, involving less direct pressures.

According to Wetter, who has written the most systematic analysis of the theory and application of dialectical materialism, as of 1954 the findings of "bourgeois" science were most acceptable to the authorities in the field of physics, less in biology, and least in anthropology and psychology.[12] From other sources we can supplement the list by adding mathematics and chemistry to the "acceptable" end of the continuum, and the other social sciences to the "unacceptable" pole. As is well known, sociology does not constitute an independent discipline in the USSR. A late, and authoritative definition of sociology by a professor at Moscow University reads, "Sociology in the Soviet Union constitutes a part of philosophy. Marxist sociology is historical materialism."[13]

In Germany, where sociology had become academically institutionalized before the Nazi revolution, its fruitful development was stopped. Some of its remaining practitioners tried to become mouthpieces for "Aryan" sociology or "race science," others retreated into formalism, still others turned to ethnology, folklore, and similar politically "safe" areas of research.[14] In German psychology, physiological psychology was least affected by the political change since it was considered "socially indifferent" (a paradoxical judgment, in view of the "racial" emphasis of Nazism), while the greatest ideological impact occurred in the field of personality, where strenuous efforts were made to provide a scientific validation of the superiority of "Aryan" character structure and "völkische" motivation.[15]

Another possible indicator of the degree of independence available in various sciences is its attraction for students. According to a visiting journalist, physics is the most "popular" department at Moscow University.[16] Obviously the popularity of a discipline is not proof of the lack of political interference. But there is other evidence of less political meddling in the "hard" sciences than in other academic pursuits at the present time. Top physicists like Kapitza and Tamm have gone on record against political interference and apparently have been instrumental in the recent decentralization of the Soviet Academy of Science —a reversal of a trend begun in 1950. The following figures for party affiliation, in 1947, among teachers in higher educational institutions indicate the relative political independence of the physical sciences: all fields, 34 percent; sociopolitical science and philosophy, 91 percent; physics and mathematics, 27 percent.[17] The most "independent" disciplines are the very ones where academic requirements are the most rigorous. This makes it unlikely that "popularity" of a discipline is simply a function of intellectual laziness. In this connection it is of interest to note the rating given by Nesmenaiov, the president of the Soviet Academy of Science, to the various fields in 1957: physics and mathematics—strong; theoretical mechanics and astronomy—weaker; biology and geology—weak and deficient; social sciences and language —very patchy.[18]

At present the natural sciences in the Soviet Union appear to be *relatively* free of political interference. (The apparent exception is genetics which we shall consider below.) Is there any evidence that under certain conditions there has been or may be less political interference in the social than in the natural sciences?

If such conditions exist at all, they tend to be very transitory. Thus in the early years of the revolution Lenin and other leaders believed that problems of social control would solve themselves "spontaneously." This belief left room for considerable experimentation in the fields of education and child psychology ("pedology"). However, during the early thirties the exigencies of planning required active intervention in all conceivable aspects of life, and experimentation became a victim of the need for individuals capable of keeping up with and surpassing the norms of production. The "production" of such individuals became a goal to be met by methods which were officially approved.[19] If, as often seemed to be the case, social science could not offer any direct aid in the battle of production—the application of psychology to industry, for example, was abandoned during this period—there remained the ideological function of reaffirming the "sacred" political goals and of "proving" that reality supported the wisdom of political decisions. It became the task of social science "not to challenge but to record, not to theorize but to rationalize, not to inquire but to reflect."[20] It is not surprising that Soviet social scientists are periodically criticized for not concerning themselves with acute problems of present Soviet society, instead taking refuge in "sterile academic" pursuits.

There are indications that the post-Stalin "thaw" has affected the social sciences to some degree, but it seems unlikely that it will produce substantial changes in the social function of these disciplines.[21] On the basis of the available data, we see no essential differences between Nazi Germany and the Soviet Union in respect to the hypothesis under discussion.

2. We shall now consider our second hypothesis regarding the relative autonomy of pure and applied science. Our investigation leads us to conclude that the conventional distinction employed has little theoretical utility. As is well known, some of the most "useful" scientific discoveries have been unanticipated by-products of theoretical research.[22] On the ideological level the Marxist dictum of the "unity of theory and practice" stresses the circular connection between the two types of activity,[23] and Nazi political leaders, although often contemptuous of "pure" research, did recognize its importance for building the economic and military power of the Third Reich.[24] Nevertheless, it may be argued, in spite of the conceptual shortcomings of the distinction, it is often used by those who allocate research funds, whether on a private or a socialist level. What evidence do we have in terms of the conventional dichotomy?

In the period immediately following the Soviet revolution it was the engineers, statisticians, pedologists, and other practitioners who first felt the application of political controls, while academics were left in relative peace.[25] What is the current situation? There are a number of areas whose immediate practical application appears relatively remote to the political authorities, e.g. cosmology, and in which scientists are required merely to exercise "negative conformity," i.e. not to explicitly question postulates such as materialism.[26] However, political needs of various party echelons as well as of scientists looking for advancement or rehabilitation may encourage search for heresy in these areas as well. Thus philosophers and theoretical physicists were charged with the task of deciding whether quantum physics, relativity theory, and cybernetics were compatible with the eternal truths of *diamat*. The results of the inquiry fluctuated between "yes," "no," and "perhaps," but the end result was in most cases the ideological acceptability of the "bourgeois" theories.[27] In the field of psychology, research in sensory physiology and perception appears to be remote from immediate practical application and is relatively free of political interference.[28] What accounts for autonomy in these areas?

One reason may be the recognition by the political authorities of the potential long-range usefulness of the theoretical areas. This seems to be the case with mathematics and nuclear physics for military purposes and development of new power resources, and with cybernetics for application in electronic calculators. Another possible reason is the studied avoidance on the part of theorists of discussing the ideological implications of their theories, leaving this thankless task to philosophers, who in turn are unable to grasp the esoteric ideas which they are supposed to judge. The latter then tend to avoid this challenge, the result being that the theories remain "value free."[29] We cannot concur with Vucinich's generalization that "the more abstract a science the less it is open to ideological influence."[30] Apart from the doubtful usefulness of the concepts of "abstract" and "concrete," this generalization disregards "functional" criteria such as the attribution of practical utility to a field on the part of the political authorities.

Turning to Nazi Germany we find that an early observer noted the anti-intellectual stance of Nazi ideologists, derived from their typically lower-middle class origin, which caused them to "distrust speculation and admire practical science." Symptomatically, philosophy was no longer required as a subject for most Ph.D. aspirants, and sciences "not readily utilizable tended to become neglected."[31] But this is an overstatement of the case: The existing German intellectual tradition with its penchant for "speculation" was by no means entirely eliminated even during time of war, and among the political leadership there were men who recognized the utilitarian aspects of theoretical research. As late as 1943 the head of the planning office in the Reich Research Council, who doubled as a member of the "cultural department" of the Gestapo(!) prevailed upon Hitler to issue a decree releasing 5000 scientists from the armed forces.[32] When the American physicist, Goudsmith, entered Germany with the Allied troops he was "surprised at the amount of pure physics which was done during the war."[33] Apparently he had expected all German physicists to be engaged in rocket research and the like. Leslie Simon, an American engineer on a similar mission, opined that "the research establishments of the German air force were the most magnificent, carefully planned, and fully equipped that the world had ever seen."[34] The Technical Academy of the air force "gave magnificent support *in basic research* to the many organizations that were engaged in technical research and development work," which incidentally closely parallelled that of the Allies, in spite of mutual isolation.[35] To be sure, German air force research was not characteristic of that sponsored by the army or navy. It depended on the dynamic role of its head, a civil servant retained from the Weimar republic, who organized research on the basis of independent institutes whose leaders were scientists rather than mere administrators. There prevailed "a minimum of arbitrary government and the substitution . . . of stimulus, inspiration, and encouragement by a competent head."[36] The air ministry "rejected both pressures from representatives of private business and political intimidation."[37] Evidently there were countervailing forces opposing the Nazi ideologists. A good example may be found in the complaints of members of Rosenberg's ministry to their boss that the war effort was being hurt by a clique which was disparaging modern physics as a "Jewish invention."[38]

While it might be difficult to find a counterpart to this in the USSR, it can be argued that the Nazis accorded at least as much autonomy to their air force research as the democratic USA did to research in the field of atomic physics (Manhattan project) during World War II. In both Russia and Germany the *instrumental uses of theoretical science* were recognized by the political leadership, and a relatively high degree of autonomy was and is considered functional for the pursuits of totalitarian goals. At this point we must face the enigma of the role of genetics in the USSR. Paradoxically, genetics has been the most documented and discussed aspect of Soviet science until the last few years, and at the same time is considered the least typical by many students of Soviet science.[39]

Since Lysenko's coming to power pure research in genetics has been taboo, because of his alleged demonstration that it is a complete waste of scientific resources in the face of his success in the battle against nature with "Lamarckian" methods. How was it possible for Lysenko to maintain his political power as long as he did if his scientific theories are not valid? How can the Soviet rulers afford to leave the direction of agricultural production to a charlatan in view of the need to maximize agricultural production?

The most plausible explanation is that Lysenko and Co. were able to pull the wool over the eyes of the political leadership. A number of practices which did not work, such as "vernalization" of grain were dropped without fanfare, while liberal use was made of methods and seeds developed on the basis of Mendelian theory, without, of course, giving credit to "bourgeois" genetics and its "mystical, mythical, and actually non-material gene of foreign and rather strange origin."[40] The technique is essentially the same which Lysenko accuses western scientists of using when he says that "Morganism [i.e. Mendelian genetics] as a theory is being developed [in the USA] for its own sake while practical farmers go their own way."[41]

A PROPOSED ANALYTICAL SCHEME

The phenomenon of Lysenkoism, which spread far beyond the confines of plant and animal breeding, is not easily reconcilable with a rationalistic model of a social system where "efficiency" is at a premium. Another conception of totalitarianism stresses the sheer drive to "power," coupled with a quasi-religious adherence to long-range goals.[42] Still other models stress power as the only "real" factor, reducing ideology to a mere rationalization, *à la* Pareto. It seems to us that all these models are unduly simplistic and reductionistic, and that a meaningful analysis must take into account the functions of these and other variables in a specific historical and social setting.[43] Thus, Dobzhansky, in a review of Zirkle's *Evolution, Marxian Biology, and the Social Scene,* rightly criticizes the author for asserting that "environmentalism is so important to the Communists that they would preserve it at all costs."[44] Zirkle's own data tend to negate this thesis. Apparently he is not aware that in the twenties it was Lamarckism, not Mendelism, that was taboo in the Soviet Union, and a thorough study of the status of genetics prior to the rise of Lysenko documents that "the widespread tendency to regard the texts of Marx and Engels as the chief determinants of the conflict [on genetics] is very much mistaken."[45] Lysenko maintained himself in power not simply through his adherence to "Marxist dogma," but because he was able to convince the political leadership that his methods were most effective to implement the goals of increased production. He, or rather his ideological mentor Prezent, was able to translate his power position into acceptable ideological jargon.

The case of Lysenkoism points up the weakness of the conventional distinction between "pure" and "applied" science for our purposes. The "theoretical" aspects of Lysenkoism have an "applied," i.e. political function for both Lysenko and the party leadership. Lysenko was able to maintain the autonomy of his system by dint of political power. In the long run it is unlikely that he and his school will be able to do so, but the same limitation applies to other scientific areas as well.[46] Both theoretical and applied science can and have become instrumentalized to meet nonscientific needs which are legitimized in ideological terms. This is true of "democratic" as well as totalitarian societies.[47] The problem we should be concerned with is, what effects does the *instrumental use* of science, be it "pure" or "applied," have on its autonomy and on its functions? Ultimately we must analyze the generally accepted (in the western world) assumption that autonomy is *always* functional. As a step towards this we propose the following analytical scheme and the associated typology.

Type 1 represents the least desirable situation from the standpoint of a given scientific discipline. Its activities are controlled, but it enjoys no official support since its aims are considered inimical to the goals of society. The obvious result is the eventual extinction of this discipline. Examples are Freudian psychoanalysis in Nazi Germany and Mendelian genetics in Soviet Russia.

TABLE 1.   Modes of Relationship Between Science and Political Institutions

TYPE	INSTRUMENTAL		NONINSTRUMENTAL	
	Control	Support	Control	Support
1	Yes	No		
2	Yes	No		
3			No	Yes
4			No	Yes

Glossary:
Instrumental: Considered important or harmful for the furthering of political, economic, military, or other social goals of the society, rather than being a value per se.

Control and Support: Measures taken by agencies of the state to minimize or maximize scientific activity.

Type 2 depicts the situation most typical of the natural sciences in totalitarian societies. They are supported *and* controlled by the state. This represents the optimal choice for scientists under the given political limits, and it is here where are fought the battles for "minimal autonomy within the system."[48]

Types 3 and 4 represent the noninstrumentalized scientific activities. Type 3, without either state support or control, is dependent upon the private sector for its existence and largely irrelevant for totalitarian societies. In other societies the existence of private interests will determine whether such activities will go on at all and to what extent. As an example, we may cite archaeological research in the Near East undertaken by American universities and compare it with geological research in the same areas sponsored by oil companies. The latter falls into the instrumental category, of course. Type 4, finally, represents in a sense the ideal from the standpoint of democratic ideology, insofar as it values "liberal" education as an end in itself.

Crude as it is, this typology should enable us to go beyond the conventional categories previously used. We have already noticed that totalitarian regimes recognize the need for autonomy of science under certain conditions; on the other hand we are aware of the growing political control of science in democratic societies. Both kinds of societies are operating under increasingly similar needs and pressures of economic and military expansion. The kinds of questions to be asked should be along the following lines: Under what conditions is scientific autonomy functional and for whom? Why and how does a given science become instrumentalized? How and when do scientists resist encroachment on their autonomy, with what degree of success, and with what effects on their work? We have cited some of the data which are needed to answer these questions in the case of Nazi Germany and the Soviet Union, and there are more available.[49] To the extent that these questions can be answered by the methods of social science they are of import not only for a clarification of what happened in history, but even more so for what will happen.

NOTES

1. Robert K. Merton, *Social Theory and Social Structure* (Glencoe, Ill.: Free Press, 1957), p. 533.

2. Merton, *op. cit.;* Bernard Barber, *Science and the Social Order* (Glencoe, Ill.: Free Press, 1952), Ch. 3.

3. *Can Science Save Us?* (New York: Longmans, Green, 1947), p. 48.

4. *New Leader,* 40 (November 25, 1957), pp. 7-8.

5. We do not consider it essential at this point to discuss the problem of the definition of totalitarianism. For our purposes it is sufficient that Nazi Germany and Soviet Russia are generally regarded as totalitarian, and that our assumptions are applicable to both. Cf. Carl J. Friedrich, ed., *Totalitarianism* (Harvard, 1954). Our view is that modern totalitarianism is a phenomenon *sui generis,* characteristic of industrialized societies. Consequently we are reluctant to discuss possible analogous situations, such as the control of science by the Catholic church (e.g. the case of Galileo), or the still current controversy over evolution.

6. e.g. Barber, *op. cit.,* pp. 94-100.

7. David Joravsky, "Soviet Views on the History of Science," *Isis,* 46, pp. 3-13, (1955).

8. Helmuth Plessner, *Untersuchungen zur Lage der deutschen Hochschullehrer* (Göttingen: Vondenbroeck and Ruprecht, 1956), p. 283f.

9. Christian von Ferber, *Die Entwicklung des Lehrkörpers der deutschen Universitäten und Hochschulen* (Göttingen: Vondenbroeck and Ruprecht, 1956), pp. 143-146.

10. Rene König, "Die Situation der emigrierten deutschen Soziologen in Europa," *Kölner Zeitschrift für Soziologie und Sozialpsychologie,* 11; p. 120 (1959).

11. Lazar Volin, "Science and Intellectual Freedom in Russia," in *Soviet Science* (Washington, D.C.: American Association for the Advancement of Science, 1952), p. 90f.

12. Gustav A. Wetter, *Dialectical Materialism* (New York: Praeger, 1958), p. 487.

13. "Teaching of the Social Sciences in the USSR," *International Social Science Journal,* 11, No. 2, p. 178 (1959).

14. Heinz Maus, "Bericht über die Soziologie in Deutschland, 1933–1945, *Kölner Zeitschrift für Soziologie und Sozialpsychologie,* 11, pp. 72-99 (1959).

15. F. Wyatt and H. L. Teuber, "German Psychology under the Nazi System," *Psychological Review,* 51, pp. 229-247 (1944).

16. Patricia Blake, "Russia: The Scientific Elite," *Reporter,* 17, (November 14, 1957), pp. 17-19.

17. Alexander G. Korol, *Soviet Education for Science and Technology* (New York: Wiley, 1957), p. 292.

18. Leopold Labedz, "How Free Is Soviet Science?" *Commentary,* 25, p. 476 (June 1958).

19. Raymond A. Bauer, *The New Man in Soviet Psychology* (Harvard, 1952), pp. 180-181.

20. Alexander Vucinich, *The Soviet Academy of Sciences* (Hoover Institute: Stanford University, 1956), p. 44.

21. Norman Birnbaum, "Science, Ideology and Dialogue," *Commentary*, 22, pp. 567-575 (December 1956); Ronald L. Meek, "Conversations with Soviet Economists," *Soviet Studies*, 6, pp. 238-246 (1954–55): For more optimistic views, see Jean Piaget, "Some Impressions of a Visit to Soviet Psychologists," *International Social Science Bulletin*, 8, No. 2, pp. 401-404 (1956), and Arvid Brodersen, "Soviet Social Science and Our Own," *Social Research*, 24, pp. 253-286 (1957).

22. Bernard Barber, *op. cit.*, p. 95ff.

23. Barrington Moore, Jr., *Terror and Progress USSR* (Harvard, 1954), Ch. 4.

24. Cf. the references below to Simon and Goudsmith.

25. Korol, *op. cit.*, p. 269f.; J. S. Joffe, "Russian Contributions to Soil Science," in *Soviet Science, op. cit.*, p. 61.

26. N. W. Mikulak, "Soviet Philosophic-Cosmological Thought," *Philosophy of Science*, 25, pp. 35-50 (1958).

27. Wetter, *op. cit.*, Ch. 5; P. S. Epstein, "The Diamat and Modern Science," *Bulletin of the Atomic Scientists*, 8, (August 1952) pp. 190-194.

28. Ivan D. London, "Toward a Realistic Appraisal of Soviet Science," *Bulletin of the Atomic Scientists*, 13, (May 1957) p. 169ff.

29. *Academic Freedom Under the Soviet Regime* (Munich: Institute for the Study of the History and Culture of the USSR, 1954) p. 13.

30. Vucinich, *op. cit.*, p. 64.

31. Edward Y. Hartshorne, Jr., *The German Universities and National Socialism* (Harvard, 1939), p. 110f.

32. Samuel A. Goudsmith, *Alsos* (New York: Henry Schuman, 1947), p. 188f.

33. *Ibid.*, p. 79.

34. Leslie A. Simon, *German Research in World War II* (New York: Wiley, 1947), p. 12.

35. *Ibid.*, p. 37. Italics supplied.

36. *Ibid.*, p. 58.

37. *Ibid.*, p. 95.

38. Leon Poliakov and Josef Wulf, *Das dritte Reich und scine Denker* (Berlin: Arami, Verlag, 1959), pp. 98-103.

39. For documentary sources, see Conway Zirkle, *Death of a Science in Russia* (Philadelphia: U. of Pennsylvania Press, 1949). The same author's recently published *Evolution, Marxian Biology, and the Social Scene* (Philadelphia: University of Pennsylvania Press, 1959), contains an analysis of the role of Marxist ideology which is sharply criticized in a review by Th. Dobzhansky in *Science*, 129, p. 1179f (May 9, 1959). Wetter, *op. cit.*, p. 456ff, provides valuable material on the impact of Lysenkoism on other disciplines.

40. The Soviet scientists Perov, quoted in Pamela N. Wrinch, "Science and Politics in the USSR: The Genetics Debate," *World Politics*, 3, pp. 486-519 (1951).

41. *Ibid.*, p. 496.

42. Z. Brzezinski, "Totalitarianism and Rationality," *American Political Science Review*, 50, pp. 750-763 (1956).

43. For good examples of this type of analysis, see Raymond R. Bauer, Alex Inkeles, and Clyde Kluckhohn, *How the Soviet System Works* (Harvard, 1956), and Barrington Moore, Jr., *op. cit.* Nothing comparable has been done for Nazi Germany, possibly because of ideological bias. Thus, Bauer characterizes "Nazism and Fascism . . . as almost unequivocally anti-rational and anti-intellectual" (in Friedrich, *op. cit.*, p. 156), while in the Soviet Union "much of what looks like far-fetched dogmatism is usually far-reaching pragmatism" (*ibid.*, pp. 150-151).

44. See note 38.

45. David Joravsky, "Soviet Marxism and Biology Before Lysenko," *Journal of the History of Ideas*, 20, pp. 85-104 (1959).

46. One obviously relevant variable is the nature of the political power structure. For other examples, which are neither totalitarian nor democratic, see Philip G. Frank, "Non-Scientific Symbols in Science," in Lyman Bryson, ed., *Symbols and Values* (New York: Harper, 1954), pp. 341-348.

47. Kurt P. Tauber, "Science and Politics: A Commentary," *World Politics,* 4, pp. 432-436 (1952).

48. Barrington Moore, Jr., *op. cit.,* p. 153.

49. For example, Carl J. Friedrich and Z. B. Brzezinski, *Totalitarian Dictatorship and Autocracy* (Harvard, 1956), p. 271; Moore, *op. cit.* p. 129, 143; H. J. Berman, "The 'Right to Knowledge' in the Soviet Union," *Columbia Law Review,* 54, pp. 749-764 (1954); Vucinich, *op. cit.,* p. 73, 87; Goudsmith, *op. cit.,* p. 191; Wrinch, *op. cit.,* p. 495; Friedrich, *op. cit.,* p. 145; Korol, *op. cit.,* p. 278; *Academic Freedom Under the Soviet Regime, loc. cit.,* p. 91; Robert Jungk, *Brighter than a Thousand Suns* (New York: Harcourt, Brace, 1958) p. 91.

KENNETH PRANDY

Technologists in Industrial Society

This study has two main intentions. One is as a contribution to the theory of social stratification, the second as a contribution to the sociology of science . . . Our present concern is with the latter. Scientists and engineers have been chosen for study because it is believed that they are a particularly important occupational group, or number of related groups, about whom information tends to be sadly lacking. In the main part of the study technologists (for the sake of simplicity this term will often be used to cover all scientists and engineers) are dealt with in one particular area only, that of work, and in the context of the concepts of class and status. As an introduction to this, however, it seems desirable to undertake a more general discussion of the place of technologists in modern industrial society—how the various occupational groups have developed and the ways in which they have become involved in the process of politics and the system of power.

Reprinted by permission of Faber and Faber Ltd. from *Professional Employees: A Study of Scientists and Engineers* by Kenneth Prandy, (London: Faber and Faber, 1965), pp. 15-29.

Not so very long ago it might have been necessary to offer some justification for the statement that technologists are a particularly important group in modern society. More recently, and especially since the last war, the need for justification has considerably diminished; indeed the statement is now a commonplace, and if anything the tendency is for the importance of this group to be overstressed. The growing science-based industries, and perhaps even more the new scientific methods of warfare, have brought about this change.

It is natural that an interest in the relationship between science and society should reflect itself also in a greater interest in the scientists and technologists themselves. Much more important, however, has been the influence of popular concern with this group. In large measure this concern dates from the very recent past, in fact from the Second World War or, to be more precise, from the dropping of the first atomic bomb. This event demonstrated in spectacular fashion the decisive influence that science could exert on society, and it came as a shock as much to politicians and intellectuals as to the ordinary man in the street.

> Until the revelation of Hiroshima, Congressmen, like most laymen, had little reason to be much concerned with either science or scientists. The majority of them undoubtedly shared the popular conception of science as the well from which material benefits flowed in an endless stream symbolized by the familiar picture of the man in a white smock holding up a test tube to the light. As for scientists, most politicians seemed to view them either as useful tools for increasing the productive resources of industry, or as impractical visionaries and eccentric crackpots ... The atomic bomb changed this situation completely, forcibly thrusting science and scientists into the forefront of politicians' focus of attention ... Detonation of the bomb drove into people's consciousness the realization, hitherto understood by only a few laymen, that science was a major social force.[1]

So writes an American author, who goes on to describe the different images, or stereotypes, that politicians have of scientists. Many, he believed, "seemed to regard scientists in much the same way that primitive men regard their magician priests." Following a similar analogy another writer concluded from a study of public opinion about science and scientists that "if the inquisitive observer watches the worshippers and the more casual passers-by, he notices respect and appreciation, but little real curiosity and interest, and he can overhear a certain amount of distrust and apprehension expressed in subdued conversations."[2]

This lack of curiosity and interest can be detected at two levels. There is not only the ignorance of the "common man," for whom science is something far beyond his understanding, but what is worse, the studied ignorance of those learned in the classical disciplines and the humanities. C. P. Snow has made a similar point in his strictures on the division between the two cultures.[3] Many people have taken his words to heart, but there has been in any case a greater desire to know about science, together with more effort on the part of scientists to explain and popularize their work. The intellectual opposition to science is undoubtedly diminishing, but the danger is that it is changing into an uncritical worship. The scientist has become a glamorous personality, and the scientific career has also in the eyes of many become equally attractive. There has long been a highly influential current of thought which holds that anything practical is by nature inferior and degrading, and although its influence may be declining it is still strong. Science now is accepted, but only science of the pure sort. The very term used, with its high value loading, is indicative of the attitude. Pure, basic science, so the idea runs, is eminently useless and must therefore be decent—a fit occupation for a gentleman. If this is a caricature of influential opinion, the scientists themselves are partly to blame for it. They have emphasized the importance of basic research—science for its own sake—partly in order to maintain control over their own subjects, since clearly only they are competent to judge the value of such research. Once they attempt to be practical, however, they surrender judgement and therefore also direction into the hands of laymen.

At the same time, since scientists as a group are not gentlemen of leisure but have to earn their living, they always have to temper their zeal for basic research with reminders that this may eventually have results of tremendous practical importance, and they point to atomic physics as an awesome example. However, they can be excused these occasional lapses since, after all, their intentions are honourable. The full force of ignorance, misunderstanding and social disapproval is reserved for the practical men, the technologists, especially the engineers. Even though science has now become an attractive career, engineering, on the whole, has not,[4] and despite the present great demand for higher education there are quite often unfilled places in engineering departments not only in the technical colleges but even in the universities. Quite apart from other consequences these attitudes are likely to create many dissatisfied scientists, since opportunities for pursuing basic research are limited. Indeed, the distinction between pure and applied research is becoming increasingly difficult to uphold in practice.

One indication of the inferior position of engineering relative to science is the difference in social background of the practitioners. These differences in fact go back to the early periods of the two disciplines and the different ways in which they arose. Whereas science was strongly associated with the activities of gentlemen amateurs,[5] engineering has always been a practical matter, only recently becoming an academic discipline (or rather cluster of disciplines). Traditionally, the training of engineers has been largely "on the job." Nevertheless, although the differences existed they did not prevent a great deal of interest in practical problems by scientists. With the rise of professional scientists and a greater theoretical emphasis in science, associated to some extent with its pursuit within the universities, this practical interest diminished, and applied science, technology and engineering, assumed a definitely inferior position, looked down upon as much by the new pure scientific culture as by the old literary one. Having been largely rejected by the universities technologists have had to be content almost until the present day with second-class educational institutions. Even now many engineers have learned their skill by some form of apprenticeship with part-time further education, and only about one-half are graduates. This contrasts markedly with the situation in science, where it is only comparatively recently that the Royal Institute of Chemistry, for example, has provided a means of professional qualification outside the university. There can be little doubt that in Britain these differences in training are both a reflection of social background and a determinant of the relative social prestige of scientists and engineers.

The whole problem of social attitudes towards science and engineering is clearly of more than academic interest. Ours is an industrial society, and if it is to maintain or improve the material standards of its members it is essential that industry becomes ever more prepared to use the skills of scientists and engineers in the right ways. It also means that these people should be produced in the right quantity and of the proper quality.

This is no new problem in Britain. It is only at the present time, if at all, that there has been a sufficient awareness of the need and an attempt made to ensure that enough trained technologists are provided. Despite the early lead in industrialization (or perhaps because of it) and the undoubted brilliance of gifted individuals, there has been a constant lag in the introduction of new industrial techniques. In the nineteenth century France and Germany, envious of Britain's lead, were much more willing to develop the industrial potential of new scientific discoveries. Although Cardwell attributes the first step in the invention of applied science to France, there is no doubt that the second, and by far the greater, was made by Germany. That country had no very great advantage, for example, in the discovery of aniline dyes, perhaps the opposite, but they showed themselves far more willing to found a new industry on the basis of this discovery. A major reason for this was that, as Cardwell says, there was "behind the industrial scene the great educational system of the country,"[6] which ensured the supply not merely of gifted individuals but also of the large number of trained technologists needed to use and to develop new techniques.

The chemical industry, and in the early stages particularly the dye industry, is the first example of the widespread application of science. Naturally, the industry was not completely neglected in Britain, but it undoubtedly lagged behind that in Germany. In large part this was a result of the quite inadequate provision for technical education. As even *The Times* once said, "the chemical industry owes nothing to the historic educational institutions of this country"[7]—nor, one might add, much to any non-historic ones. There is a good deal of truth in the assertion that "here we tried to start chemical industries practically without chemists."[8] Another, associated reason was the attitudes held by those who made the decisions in industry. It was not, for example, until 1892 that the first research laboratory was set up in the chemical industry, and even then was seen as a revolutionary step, consisting as it did of "half a dozen chemists, a general handyman and a confidential clerk."[9] It may be noted that both the Chief Chemist of the concern which set up the laboratory and the first applicant for a post in it were Swiss.

Nevertheless, although the chemical industry in Britain lagged behind that in Germany, its growth was quite substantial. Most of the improvements which brought about the advance were made by experienced manufacturers, often themselves trained in chemistry, rather than by research chemists. Chemicals as a modern science-based industry only developed, in Britain at least, when the twentieth century was already a couple of decades old. In 1902, for example, the British Association estimated from a survey that there were some 1,500 chemists in British industry, of whom only 225 were graduates.[10] Nor can the blame for these low numbers be entirely laid on industry. If industrialists saw little point in recruiting chemists, the chemists themselves, it seems, were equally averse to working in industry—"trade" was still very much *infra dig.*

After the First World War the pace of expansion quickened, and between 1920 and 1938 the number of research chemists and the amount spent on research quadrupled. Even so there was some unemployment amongst chemists. Industry was willing to take the first-class research men, but had little use for the remainder. It took the Second World War and post-war conditions to create a "shortage" of chemists.

The electrical industry, in all its aspects, was more obviously science-based from the beginning. The earliest industrial application of electricity, that of telegraphy, created only a fairly small demand for technologists. Subsequent expansion in this field has been steady, but even with the development of broadcasting the demand has not been very great. It is only since the Second World War that the growth of the electronics industry has stimulated a vastly increased need not only for telecommunications and electronic engineers (developed from telegraph and radio engineers) but also for physicists, who were now employed for the first time on a large scale in industry.

The other branch of the electrical industry, with its two sides of the generation of power and the machinery and appliances that make use of it, has been much more important in the demand for technological manpower. Electricity generation, as Ashworth says, "was probably the most notable late nineteenth-century example of the influence of technology in the creation of new industry,"[11] although here again development was neither smooth nor rapid—"in this combination of missed chances and great, novel achievements the history of power equipment seems very typical of the history of the techniques of British industry generally in this period" (i.e. around the turn of the century).[12] As early as 1907 14 per cent of the output of the engineering industry was electrical. Electrical engineers were demanded especially in the generating side, but were also needed by manufacturing industry for development and design work.

In the history of the growing demand for scientists and engineers the Second World War can be clearly seen as a watershed. In the latter part of the nineteenth century discussion about technical manpower centered on the two aspects that not enough men were being trained and that industry was too slow to use those who were available. During the first third of the present century training improved considerably, so much so that despite the great increase in the employment of technologists there was, if anything, a certain amount of underemployment. But since 1945, despite a great increase in the numbers trained, the stress is on the shortage of qualified personnel. There can be little doubt that demand has grown substantially. The science-based industries have become much more important as a result of wartime and post-war developments—electronics, aircraft, petro-chemicals and atomic energy particularly—and other more traditional industries have also seen the need for scientists and engineers. At the same time the role of the government as an employer has also grown in importance. The stimulus for this advance has come partly from an increasing rate of technological innovation,[13] partly from full employment, but in part also from certain political considerations, to which we shall return.

There is now undoubtedly a far greater concern about scientific manpower and the use made of it. There exists a permanent Advisory Council on Scientific Policy with a Committee on Scientific Manpower, which has now issued several reports, and which has been incorporated into the sphere of competence of the Minister for Science. Despite the official pronouncements of this body, however, arguments still rage over the question of whether there is a shortage of scientists and engineers. The reports seem to show that there are some areas where supply is short, but that on the whole it is not great and that, moreover, by 1970 there should be if anything an excess.[14] This opinion is by no means widely shared, however, and *The New Scientist,* for example, tends to see it as merely another example of British complacency over scientific matters. Economists, also, have entered the discussion, some pointing out that the price paid for technologists has risen in response to excess demand, though with a lag,[15] others maintaining that the "shortage" is more apparent than real, and that what is meant is that there is actually an "unmet need," a much more difficult concept.[16]

This problem of unmet need brings us back to the question of the use made by industry of science. There may be little room to criticize the science-based industries themselves, but many of the others seem very loath either to engage in research and development or indeed to make use of what has already been done. Much of the impetus in setting up the Research Associations has come from the state, which is constantly trying in many ways to persuade industry of the value of science through the D.S.I.R. and other bodies. The Trend Report[17] is but the latest example of this.

The reason for this in the past has been the fear of technological backwardness which would put the country at a commercial disadvantage compared with other countries, particularly Germany and the United States. The motives, however, have not been purely economic. In the early days of the chemical industry, Germany was seen as the main competitor, and there can be little doubt that a large component in the fears aroused was political, the threat of dominance. The United States, perhaps for other reasons as well, was never feared in quite the same way, even though that country was even more superior technologically. At the present time, although some emphasis is laid on our position *vis-à-vis* the German and American competitors, the main point made is that Britain and "the West" must not fall behind in the "race" with Russia. Economics are inevitably involved in this way in international politics, and since science now plays so important a role in industry it means that science, and thereby scientists, are also brought into the same international political arena.

However, science does not merely play a greater role in industry, its importance in the techniques of war is as great or even greater. Thus there have been two reasons why science has become deeply involved in politics; the old commercial one, with its political overtones, and the new one of the era of scientific warfare, with the overshadowing importance of nuclear weapons. It is of some significance that the chairman of the Committee on Scientific Manpower is Sir Solly Zuckerman, who is also Scientific Adviser to the Ministry of Defence. As he has said himself: "This war was the turning point. Whereas previously scientists were seen, according to the interests of the observer, either as dedicated scholars, or as the source of invention, or as the technical guardians of the social services on which an urban civilization depends, today they also appear in a number of new guises—as the backbone of national defence; as pioneers of outer space; and even as the councillors of presidents and prime ministers."[18]

The "even" in the last phrase is significant. In a world in which science has now presented us with the means of our own total destruction, to say nothing of its less spectacular gifts, an eminent scientist expresses an element of surprise that politicians should take advice from scientists. The statement raises a very important problem. If science is so vital for modern society, for either its "life" or its "death," and if scientists control the development of science and technologists its application, what part do scientists, as individuals, play in making decisions about our society? Are they becoming, as Veblen prophesied, the elite of our society, or is their role merely one of being "even" councillors of presidents and prime ministers? Do they, in short, wield power; do they even exercise very much influence?

Our knowledge about this important problem is lamentably slight, but there are some suggestive indications. It has been pointed out, for example, that in 1961, of 20 members of the cabinet 1 was a scientist (in the widest sense), of 22 permanent heads of government departments 2 were scientists, and of 20 heads of national newspapers none were scientists.[19] These figures hardly suggest very much involvement in the centres of power in Great Britain. Even more illuminating, perhaps, is the comment of an "eminent" Fellow of the Royal Society, quoted by Anthony Sampson. He said of this supreme scientific body that "they had the choice after the war of remaining a mutual admiration society, or really taking part in the control of science. They chose the former. They threw away the handles of power."[20]

Nevertheless, scientists are not quite without influence. In his study of the relationship between government and science in the U.S.A., Price argues that many policy decisions depend upon research already completed and that "in the long run this system, or lack of system, gives a great deal of influence in public affairs to men whose positions enable them to maintain a comprehensive view of new scientific developments," that is, "scientists who are leaders in their professional societies and research councils."[21]

Some of the differences between this point of view and that presented in the preceding paragraph can be explained by the fact that one refers to Britain and the other to America, but not all. It would seem that Price's argument does not stand up very well to some of the facts that he himself presents—for example, that whereas in 1938 only one-fifth of federal money spent on science was for military research, in 1953 nine-tenths was used for this purpose. Men other than scientists surely made the decisions which brought about this state of affairs.

Of course scientists are used to giving advice. C. P. Snow in his Harvard lectures has described the parts played by Tizard and Lindemann in determining strategy during the Second World War.[22] More recently there have been others, such as Sir William Penney and Sir Solly Zuckerman, who has already been mentioned.

It may thus be true that a small number of scientific advisers are influential, but as we have suggested, two questions remain. First, how influential, and second, how representative are they? Snow's lectures are a plea for scientific advice to government to be far more representative, and the story of Tizard and Lindemann is a warning of how a scientist can become immensely influential, even in non-scientific matters,[23] whilst being quite out of touch with general scientific feeling. No single person can be a good scientific adviser, if only because he cannot separate his expert judgements from his own opinions on policy. In his excellent discussion of this problem[24] Werner Schilling states that "the scientist, in short, is not likely to orbit the centers of political power emitting upon request 'beeps' of purely technical information. He will inevitably be pulled into the political arena."

This author nevertheless sees the role of the scientist as little different from that of any other expert in (American) government. He describes the various predispositions to which scientists as advisers are particularly prone. They have, it seems, a predilection towards naïve utopianism or naïve belligerency ("impatient optimism" in Snow's words), towards dealing with whole problems rather than with their allotted fragments, towards quantum jumps rather than piecemeal improvements (e.g. developing new weapons instead of improving old ones), towards "technology for its own sweet sake," towards a "sense of paradise lost," and a belief that "science serves mankind." These are seen as a result of the character of their expertise, although any individual scientist may exhibit none of them.

Schilling concludes, with the example of the development of radar in Britain, that "British scientists and science were in the final measure but ready tools. They were good tools, but the use to which they were put was the result of the kind of ideas the military men had about war. The contributions that science and technology will bring to international politics will largely turn, not so much on the particular arrangements of scientists in the policy-making process, but on the purposes of statesmen and the theories they have about the political world in which they live."

For the most part, then, scientists seem to exert influence only as tools, as technical-problem-solving machines, as superior, or perhaps rather, inferior computers. The more government relies on one individual the less true this will be. The more scientists as a whole are consulted the more likely it is to be the case. The scientists themselves seem to be content to be represented either by a very few individuals, as witness the attitude of the Royal Society, or to be used as tools of someone else's policy. This traditional attitude of the divorce of values (the basis of politics) from technical expertise suffered a setback in the early days of atomic weapons, when, for example, many scientists within the Manhattan project felt a troubled concern. "It raised no moral question about the rightness of their own actions in the realization of the atomic bomb, but it insisted that their will be consulted about its applications."[25] This basic concern soon meant that "problems had to be pursued into areas of social life which might have appeared earlier to be unconnected with the interests of responsible scientists." However, this scientists' movement, and the Pugwash conferences which are another manifestation of a similar feeling, have never had the backing of the mass of technologists, nor has their influence on government policy been very great.

Many more scientists are used, and more money spent on research, for defence work than for any other activity. If their influence in this area is weak, it is no less so in others. There is doubt about the influence of scientists in the "office of the Minister of Science," and further doubt about the influence of the Minister himself.[26] Outside the government and within industry the position is no better and may be worse. The technologist has a good chance of reaching the ranks of management, and in certain cases even the board of directors, but there is a widespread feeling that their numbers are far too few and their influence far too small.

How much influence technologists have is a question of fact; whether it is too much or too little is largely one of value, though it depends also on the factual question of how best to arrange matters so as to obtain a particular desired end in government or industry. From whatever point of view one approaches the problem, however, the necessity of studying technologists and of increasing our store of knowledge about them can hardly be disputed.

The decisions of engineers, says Merton, "do not merely affect the methods of production. They are inescapable social decisions affecting the routines and satisfactions of men at work on the machine and, in their larger reaches, shaping the very organization of the economy and society." "The central role of engineers as the General Staff of our productive system only underscores the great importance of their social and political orientations: the social strata with which they identify themselves; the texture of group loyalties woven by their economic position and their occupational careers; the groups to whom they look for direction; the types of social affects of their work which they take into account—in short, only by exploring the entire range of their allegiances, perspectives, and concerns can engineers achieve that self-clarification of their social role which makes for fully responsible participation in society."[27] He suggests that it is the task of the sociologist to carry out this exploration.

NOTES

1. Harry S. Hall, "Scientists and Politicians," reprinted in B. Barber and W. Hirsch, *The Sociology of Science,* Free Press, 1962, p. 269.

2. Stephen B. Withey, "Public Opinion about Science and Scientists," reprinted in Barber and Hirsch, op. cit., p. 159.

3. C. P. Snow, *The Two Cultures and the Scientific Revolution,* the Rede Lecture 1959, Cambridge U.P., 1959.

4. For example, those taking up engineering have lower "A" level results than those taking up science. See *Technology and the Sixth Form Boy,* Oxford University Department of Education, 1961.

5. See D. S. L. Cardwell, *The Organisation of Science in England,* Heinemann, 1957, p. 12.

6. Ibid., p. 136.

7. In 1902. Quoted in D. W. F. Hardie, *A History of the Chemical Industry in Widnes,* I.C.I. Ltd., 1950, p. 176.

8. Quoted in Cardwell, op. cit., p. 147.

9. Hardie, op. cit., p. 176.

10. See R. M. Pike, *The Growth of Scientific Institutions and Employment of Natural Science Graduates in Britain 1900–1960,* unpublished M.Sc. thesis, London, 1961, Chapter 1.

11. W. Ashworth, *An Economic History of England,* Methuen, 1960, p. 79.

12. Ibid., p. 86.

13. According to one writer the growth is exponential with a doubling every 10–15 years. See D. J. Price, "The Exponential Curve of Science," reprinted in Barber and Hirsch, op. cit., p. 517.

14. Advisory Council on Scientific Policy, Committee on Scientific Manpower, Statistics Committee, *The Long-Term Demand for Scientific Manpower,* Cmnd. 1490, H.M.S.O., 1961.

15. K. J. Arrow and W. M. Capron, "Dynamic Shortages and Price Rises—the engineer-scientist case," *Quarterly Journal of Economics,* vol. 73, May 1959, pp. 292-308.

16. See, for example, the report of the speech by Professor Jewkes at the 1957 meeting of the B.A.A.S. in the *Economist,* vol. 192, pp. 845-6.

17. Committee of Enquiry into the Organization of Civil Science, *Report* Cmnd. 2171, H.M.S.O., 1963.

18. "Liberty in an Age of Science," *Nature,* 18 July 1959, quoted in A. Sampson, *Anatomy of Britain,* Hodder and Stoughton, p. 509.

19. Ibid., p. 510.

20. Ibid., p. 514.

21. Don K. Price, *Government and Science,* New York U.P., 1954, pp. 28-29.

22. C. P. Snow, *Science and Government,* Oxford U.P., 1961.

23. In economic questions, for example. See R. F. Harrod, *The Prof,* Macmillan, 1959, p. 179.

24. W. R. Schilling, "Scientists, Foreign Policy, and Politics," *American Political Science Review,* vol. 56, June 1962, pp. 287-300.

25. E. Shils, "Freedom and Influence: Observations on the Scientists' Movement in the United States," reprinted in L. J. Gould and E. W. Steele, *People, Power and Politics,* Random House, 1961, pp. 635-40.

26. Sampson, op. cit., p. 527.

27. R. K. Merton, "The Machine, the Worker and the Engineer" in *Social Theory and Social Structure,* Free Press, 1957, p. 533.

PART II

Science and
Social Organization

The basic theme in part two is the conflict between the norms of scientific research and the need for bureaucratic administrative controls in industrial organizations, and in complex organizations generally. Box and Cotgrove discuss a familiar aspect of this theme: the incompatibility between professional and bureaucratic authority. The authors show how mechanisms of occupational choice and selection may alleviate conflicts and strains associated with such incompatibility. The thesis about conflict between professional and occupational "needs" is modified to take into account the degree of commitment to values associated with professional occupations. Differential attachment to the values of autonomy, disciplinary communism, and a scienctific career is the basis for identifying public, private, and instrumental scientists. Their identification of types of scientists is a significant departure from conventional conceptions of science as a monolithic system.

In their recently published *Science, Industry and Society* (1970), Box and Cotgrove extend the analysis presented in the paper reprinted here. Their research, exploratory and sometimes methodologically crude, is nonetheless an important addition to the literature on scientists in industry. It is worth noting that in their book (but not in the article) they suggest participative management opportunities for self-actualization as a strategy for more productive accommodation of scientists to potentially stressful industrial roles.

Kornhauser's essay is an examination of the "inherent" structural strains between science and industry which are most clearly manifested in the industrial research organization. The strains reflect conflicts between the need for autonomy in science and for coordination in industry. Kornhauser explores ways in which these strains are accommodated in setting goals, controlling research, providing incentives, and establishing responsibilities for the utilization of research. Sociological factors in the tension between science, professionalism, and industrial organization are identified and means for accommodation are discussed and critically evaluated. The crucial problem, according to Kornhauser, is how to affirm the power of applied science without undermining the pursuit of "scientific understanding of the natural order."

Beer and Lewis discuss "the second scientific revolution," the institutionalization and professionalization of science in the twentieth century. Relationships between staff and management in research departments, subtle pressures in support of research stressing potential application, imbalances between (1) military and civilian research, (2) research in big and small business, and (3) basic, academic research and applied civilian and military research are discussed in terms of a basic theme: the convergence of academic, government, and industrial research, and a concomitant blurring of distinctions between basic and applied research, and between academic and nonacademic institutions. In conceiving the development of a science for managing science, Beer and Lewis anticipate that increasing understanding of the process of discovery will be correlated with an increasingly uniform pattern of management. This trend, they suggest, "will be accelerated by governmental guidance and regulation as the nation seeks both to enhance and balance its scientific effort." Recalling the "science and society" arguments of J. D. Bernal and other "Marxist-socialist" members of the "Invisible College" in the 1930s, and considering the current "science for the people" movements, it is necessary to inquire whether a science of science management is possible without a radical restructuring of society. Beer and Lewis are critical of various "imbalances," but do not inquire whether these imbalances are manifestations of the trends they describe and extrapolate. Just a few years prior to the emergence of radical caucuses in the scientific professions, and a variety of social movements in science, Beer and Lewis were writing that "class solidarity and common political action have never developed among scientists and engineers." Recent "action" in science reflects, in part, a consciousness of and response to the convergence of dysfunctions in professionalization-bureaucratization, and may be considered actual, or proto-typical and potential counter-processes in science (cf. Brown 1971; Allen 1970).

The first three articles in this section focus attention on the incompatibility between professional and bureaucratic authority (Box and Cotgrove), structural strains between science and industry (Kornhauser), and the possibilities for a science of science management (Beer and Lewis). They pose a general question concerning the amount of administration necessary in complex organizations. Norman Kaplan raises just this question as a result of some surprising findings in his exploratory studies of research administration in the U.S. and the U.S.S.R. Following a study of research administration in the U.S., Kaplan interviewed directors, deputy directors, department heads, and other scientists in thirteen medical institutes located in Moscow, Leningrad, and other Soviet cities. The crucial questions raised by Kaplan's comparative analysis were: (1) "Why does the American administrator occupy such a superordinate position in the research organization relative to his Soviet counterpart?" and (2) "Why does the Soviet administrator occupy such a subordinate position in the research organization compared with the American administrator?" Kaplan discovered that the Soviet administrator, while more willing and more likely to engage in administration than his American counterpart, is able to spend more time than his counterpart on his own research. He suggests that there is "less administration" in the Soviet research institute. Several reasons for this possibility are discussed, and the assumption that the level of administrative activity in research and other organizations is near or at the minimum necessary for co-ordination, control and communication in an adequately functioning organization is challenged.

REFERENCES

Allen, J., ed. 1970. *Scientists, Students, and Society*. Cambridge, Mass.: Massachusetts Institute of Technology Press.

Box, Steven and Stephen Cotgrove. 1970. *Science, Industry, and Society*. New York: Barnes and Noble, Inc.

Brown, Martin, ed. 1971. *The Social Responsibility of the Scientist*. New York: The Free Press.

STEVEN BOX
STEPHEN COTGROVE

Scientific Identity, Occupational Selection, and Role Strain[1]

The growing application of science and technology to production involves the increasing employment of scientists in industrial organizations. Industry will soon be the main employer of scientists. Nearly 70
per cent of all chemists engaged in research and development work are
to be found in industrial laboratories.

The research reported here arises out of the now familiar theme that
the employment of professionals in organizations gives rise to problems
which derive from the conflict between professional and bureaucratic
authority.[2] The professional demands autonomy in the exercise of his
expert knowledge. But at the same time, he is expected to comply with
the demands of the organization which employs him, and in particular
is subject to the authority of administrators.

The scientist, like other academics, is likely to experience special
difficulties. His intellectual product is knowledge (unlike the practising

Steven Box and Stephen Cotgrove, "Scientific Identity, Occupational Selection and Role
Strain," *British Journal of Sociology* 17 (March 1966): 20-28. Reprinted by permission of
the publisher.

engineer, accountant or architect). His professional reputation depends on the evaluation of his contribution to knowledge by his peers, and this in turn compels publication.[3] Industry, however, frequently restricts publication, at least until findings have been adequately covered by patents, with all the hazards to recognition which this involves.[4] Furthermore, scientists in industry experience restrictions on professional autonomy in the choice of research goals, the termination of projects and the application and utilization of findings.[5] Moreover, there is a lack of congruence between the rewards and incentives desired by the scientist in the form of autonomy and public recognition, and those offered by industry, where the highest rewards usually go to those who switch from a career in science to administration.[6]

The thesis as so far stated assumes that all those with qualifications in science have identified themselves with the values of science and are equally committed to pursue scientific goals such as publication. Although some of the more sensational literature makes such assumptions,[7] there is a growing recognition that academics, professionals and scientists may in fact be differentiated according to the extent to which they are attached to professional, scientific and organizational goals. Various hyphenated neologisms have been employed, such as "cosmopolitans-locals," "professionals-organizationals," "science oriented-company oriented" to characterize different types of attachment qualified scientists can have towards professional values.[8] Recently these either-or dichotomies have come under criticism. Glaser[9] has suggested that a social actor could combine both an organizational and a scientific orientation; whilst Goldberg[10] presents some evidence to suggest that the dimensions are independent, and hence if each is dichotomized, a four-fold classification of scientists is essential.

This paper will firstly attempt to construct a typology of scientists based on fresh research data. But once it is recognized that there are varying degrees of attachment to scientific values, and that in this sense, there are different types of scientist, the problems of *role strain*[11] and *organizational tension*[12] take on a new perspective. In particular, we will attempt to demonstrate how the mechanisms of occupational choice and selection may reduce the conflicts between scientists and organizations, and the resultant strains.

SCIENTIFIC IDENTITY

The specific clusters of values and beliefs which constitute the ethos of science have been identified by Merton[13] and elaborated by many writers since.[14] Three clusters of values are of central importance to our present research: autonomy, disciplinary communism and personal commitment. The professional element in science claims autonomy based upon personal expertise. In addition, if knowledge is to be cumulative and verifiable, communication of findings must not be hindered by non-scientific barriers. In this sense, science is universal, cutting across barriers of class and nation. Finally, dedication to the pursuit of knowledge takes precedence over other goals, such as material gain, and requires substantial personal disinterestedness.[15]

We have attempted to measure differential commitment to each of these values by means of a questionnaire which confronts the respondent with a series of choice situations. The questions involve a choice of job offering greater research autonomy, but requiring geographical mobility;[16] a job offering slightly higher material rewards, but involving restrictions on publication; and a choice between a career in science, or a career switch to administration, with more salary and status. Each of these choices is made more complicated by adding other factors which weigh against attachment to the particular value of science.[17] Respondents were asked to indicate degree of agreement with a series of such statements on a five-point scale.

The response to these questions by undergraduates[18] suggests a three-fold classification of scientists, rather than a dichotomy. The evidence further suggests that values are linked rather than being independent. Thus, if a respondent indicated that he attached importance to publishing results (disciplinary communism), it invariably meant that he was also attached to the values of autonomy and personal commitment. However, support for the last two did not necessarily mean a high value placed on publication. Nor did those who value autonomy necessarily have a strong commitment to a career in science. From these findings, we have constructed the following typology of scientists, according to their attachment to the three main clusters of values:

1. *Public (extrinsic)*—has identified with the profession of science and will attach maximum importance to publication and communication as a means of achieving recognition as a scientist. In this way he can sustain an identity of himself as a "scientist." To maximize publication, he will seek a fairly autonomous position, and will pursue research with a high degree of dedication.

2. *Private(intrinsic)*—for whom the greatest excitement is in the job itself. The solution to the problem is everything. His main concern will be to work under conditions which enable him to get the best results. But the public world of science is not his reference group.

3. *Instrumental*—has acquired the knowledge and skills of science but does not seek public recognition as a scientist. He will use his expertise

for occupational advancement, but will be prepared to abandon a career in science.

This classification is represented schematically in Figure 1.

FIGURE 1.   Types of Scientist

| | ATTACHMENT TO VALUE OF | | |
	Autonomy	Commitment	Disciplinary communism
Public	+	+	+
Private	+	+	−
Instrumental	±	−	−

This classification's "public" and "instrumental" resembles previous dichotomies, such as "professional-organizational," but the "private" type is not a mixed type, or a combination of both types. It is distinct theoretically, and in its career correlates. It is this "private" type which is of particular importance in the discussion of role strain in industrial research laboratories.

OCCUPATIONAL CHOICE AND SELECTION

There are, then, significant differences in the extent to which scientists attach importance to freedom to publish, freedom to choose research projects and are committed to a career in science. Our research among undergraduates also shows that they believe industry and the universities to differ considerably in conditions of employment. The overwhelming impression is that industrial laboratories offer work which is better paid, has superior social and welfare benefits and provides superior technical supports, such as equipment and technicians. University laboratories were thought to be better for autonomy and for professional conditions, such as freedom to publish and freedom to attend scientific meetings. Moreover, these images of employment conditions were the same, whatever the "identity" of the scientist and whatever his intended job. But what is important is the different types of scientist attached different degrees of importance to various job conditions. It is to be expected then, that public scientists (who attach importance to publication) would be more likely to seek employment in a university or government laboratory, which are seen as providing more favourable conditions for a career as public scientist than industry. This expected association can be seen in Table 1.

However, even if a graduate has stated his preference for a work location he still has to be selected. In the spring and summer of 1964, we carried out a series of interviews in ten chemical or pharmaceutical companies. One specific interest we had was to obtain a description of the method of recruitment into research staff, with particular reference to the personal qualities sought. The typical recruitment pattern which emerged from these interviews was that a firm sent talent scouts to universities towards the end of the academic year. These scouts locate students who indicate an interest in an area of research which is also of interest to the particular company. These students are then invited to the company for a day or two. There they are interviewed and vetted by both research managers and research colleagues, particularly section leaders under whom they may be working. The interesting aspect of these interviews was that they were searching for potential executive material rather than pure research ability. It was made clear that "academic types" and people who would not fit in were to be avoided. The conclusion we drew was that industrial research management overwhelmingly favours the "instrumental" and the "private" types of scientist.

TABLE 1.   Type of Scientist and Organizational Choice

TYPE OF SCIENTIST	ORGANIZATIONAL PREFERENCE			
	Industry	University/ Govt. Lab.	% diff.	N = 100%
	%	%		
Public	28	72	+44	61
Private	46	54	+ 8	39
Instrumental	51	49	− 2	43

But further selection may also take place in industry. Public scientists may be allocated to (or seek transfer to) work at the basic end of the research spectrum, which in some firms may mean substantial autonomy. Our data suggest that some such selection does occur, although of those public scientists in industry in our sample one-third were, in fact, employed in development or service work (Table 2).

ROLE STRAIN

If such choice and selection mechanisms were able to operate effectively, we could expect them to minimize role strain[20] resulting from a lack of congruence between the needs and interests of the individual and the demands of the organization. But the fact that industry employs such a high proportion of all chemists employed on research and development work means that some graduates who would prefer to work in university or government will find that they are working in industry because this is the only job open to them. Amongst this minority, role strain is likely to be felt more acutely, although it will also be present in "private" scientists if they find that industry does not afford them sufficient autonomy, and if they discover that in order to obtain a reasonable status and salary a switch to a non-scientific career is essential. It is above all the *public* scientist and, to a lesser extent, the *private* scientist among whom such strain can be expected. As Table 3 shows, it is, in fact, the public scientists who most frequently report dissatisfaction[21] with publication and patents policy, and with supervision and autonomy. By contrast, the instrumental scientist expresses a positive satisfaction with many items which are major sources of complaint for the public scientist.

TABLE 2.   Position on the Research Spectrum and Type of Scientist[19]

TYPE OF SCIENTIST	POSITION IN RESEARCH SPECTRUM			
	Basic/Applied	Development/Service	% diff.	N = 100%
	%	%		
Public	71	29	+42	35
Private	34	66	−32	124
Instrumental	32	68	−36	84

This is further illustrated by comments in reply to our open question on role strain. One of our respondents, for example, denied any sense of strain. "I do not believe that industry employs scientists in order to advance the 'frontiers of science', but in order to make money. If scientists do not like this approach, they may do university research, but if they work in industry they should accept their contractual obligations." Another considered "most scientists in this organization work for their salaries rather than science."

Public and private scientists, however, may well find themselves in industrial employment in which they experience strains and frustrations. One such replied: "In this hard commercial world, basic research is rarely allowed to continue unless its ultimate commercial value is clearly apparent. In our own case, extremely interesting work on cellulose structure was stopped, to the annoyance of the chemist involved, because it was leading us in an academic direction. In another case, an

TABLE 3. Type of Scientist and Dissatisfaction with Various Job Items

JOB ITEM DISSATISFACTION	TYPE OF SCIENTIST[22]		
	Instrumental	Private	Public
The amount of influence research workers have in choosing their research projects	.04	.09	.30[*]
Research workers' influence over their hours of work	−.16	.15	.23
The way in which projects are terminated	−.10	.02	.29
Attention given by management to research personnel suggestions and complaints	.09	.14	.26
Publications policy	−.12	.07	.38
Patents policy	−.10	.33	.46
The degree of supervision over your main work	.05	.30	.37

[*]A positive correlation indicates that the more a respondent is like that type of scientist, the more dissatisfied he is with the job item provided in his company.

individual who had developed a method of coating, of no commercial value, refused to devote his attention to anything else, and had eventually to be discharged." Another senior chemist reported that he was "very annoyed" at the delays to publication which resulted from the company's policy of insisting that all work must be completely covered by patents before publication. Such delay, he felt, might cause duplication and might blunt one's claim to recognition for scientific contributions. It is not surprising that this scientist scored very high on our identity test.

CONCLUSION

The thesis that there is conflict between the needs of professionals and the organizations which employ them requires some modification. An important variable is the degree of commitment to those clusters of values which have been assumed to characterize members of professional occupations. Scientists, for example, differ in the importance which they attach to the pursuit of scientific and professional goals such as publication and autonomy. Three main types may be conceptualized (public, private and instrumental) on the basis of differential attachment to the values of autonomy, disciplinary communism (e.g. publication) and commitment to a career in science.

Both individuals and organizations are selective. Science students are aware of the differing extent to which scientific needs may be met in various types of employment, and seek a job where the needs to which they attach importance are likely to be satisfied. Moreover, industry selects those whom it thinks will fit. But such selection mechanisms operate imperfectly. Both public and private scientists are to be found in development work, and it is public scientists in industry who experience greatest strain, as evidenced by their dissatisfaction with publications and patents policy, and with supervision and autonomy.

Further research is now in progress to investigate the effects of role strain on the productivity of scientists.[23]

NOTES

1. This paper is a rewritten version of a paper first delivered by Dr. S. Cotgrove at the British Sociological Association Conference, April 1965. The researches reported here are part of an inquiry financed by a grant from D.S.I.R. (now S.R.C.).

2. For a work incorporating much of the literature, see W. Kornhauser and W. O. Hagstrom, *Scientists in Industry,* Berkeley: California University Press, 1962. See also H. A. Shepard, "Nine Dilemmas in Industrial Research," *Admin. Sc. Qrt.,* vol. 1, 1956, pp. 295-309; Opinion Research Corporation, *The Conflict between the Scientific Mind and the Management Mind,* Princeton: Opinion Research Corp., 1959; S. Marcson, *The Scientist in American Industry,* Industrial Relations Section, Princeton University, 1960.

For a brief summary of these and other relevant researchers, see S. Cotgrove and S. Box, "Scientists and Employers," *New Scientist,* 7 May 1964.

3. For the classic formulation on publication and career contingencies of scientists, see R. K. Merton, "Priorities in Scientific Discovery; A. Chapter in the Sociology of Science," *A.S.R.,* vol. 22, 1957, pp. 635-59. See also T. Caplow and R. McGee, *The Academic Market Place,* New York: Basic Books, 1958, ch. 4, for a study of publications and careers in the academic world. For a further development of some of Merton's ideas, see B. G. Glaser, *Organizational Scientists: Their Professional Careers,* Indianapolis: Bobbs Merril, 1964.

4. For recent evidence on industrial publication, see National Science Foundation, *Publication of Basic Research Findings in Industry,* Washington, 1962. The evidence shows that only 14 per cent of companies in their sample ($n = 174$) allowed publication of all basic research, whilst 61 per cent allowed some or none of their basic research to be published. Since development work is nearer production and would probably be of more use to competitors, the figures for nonpublication must be substantially higher for scientists engaged in this level of work. For other evidence, see W. Kornhauser, op. cit., pp. 73-80.

5. "Scientists versus Administrators: An approach towards achieving greater understanding," *Pub. Ad.,* vol. 37, 1959, pp. 213-57.

Tarkowski and Turnbull contrast the enthusiasm of the research scientist and his often intense irritation at frustrations and delays with the caution of the administrator who may even deliberately avoid contamination by enthusiasm.

6. W. Kornhauser and W. O. Hagstrom, op. cit.

7. See, for instance, W. H. Whyte, *Organization Man,* Cape, 1957, esp. chs. 16, 17, 18; C. W. Mills, *White Collar,* Oxford University Press, 1951, esp. chs. 6, 7.

8. For the literature on these hyphenated neologisms, see D. Marvic, *Career Perspectives in a Bureaucratic Setting,* Michigan Governmental Studies, No. 27, Ann Arbor, Bureau of Govt., Inst. of Pub. Ad., University of Michigan, 1954; H. L. Wilensky, *Intellectuals in*

Labour Unions, Free Press, 1956; A. W. Gouldner, "Cosmopolitans and Locals: Towards an Analysis of Latent Social Roles," *A.S.Q.,* vol. 11, 1957-8, pp. 281-306, 444-80; T. Caplow, op. cit., p. 85.

9. B. G. Glaser, op. cit., ch. 2. This is not a genuine reformulation since the organization which he examined was one devoted to basis research; hence no clear conflict existed between the organization's value system and science. For another example of professional and organizational orientation being combined, see P. Blau and R. W. Scott, *Formal Organisations,* Routledge and Kegan Paul, 1963, ch. 3.

10. L. C. Goldberg, *et al.,* "Local-Cosmopolitan: Undimensional or Multi-dimensional," *A.J.S.,* vol. 70, 1965, pp. 704-17; H. A. Shepard, "The Values of a University Research Group," *A.S.R.,* vol. 19, 1954, pp. 456-62.; H. A. Shepard, "Basic Research and the Value System of Pure Science," *Philosophy of Science,* vol. 23, 1956, pp. 48-57; F. Reif, "The Competitive World of the Pure Scientist," *Science,* vol. 134, 1961, pp. 1957-62.

11. W. J. Goode, "A Theory of Role Strain," *A.S.R.,* vol. 25, 1960, pp. 483-96.

12. For an exposition of this concept see R. K. Merton and E. Barber, "Sociological Ambivalence," in E. A. Tiryakian (ed.), *Sociological Theory, Values, and Socio-cultural Change,* Free Press, 1963, pp. 91-121.

13. R. K. Merton. "Science and the Democratic Social Structure," in *Social Theory and Social Structure,* 2nd ed., Free Press, 1957, pp. 550-62.

14. B. Barber, *Science and the Social Order,* Allen and Unwin, 1953; S. S. West, "The Ideology of Academic Scientists," *IRE Transactions on Engineering Management,* EM-7, 1960, pp. 54-62; R. G. Krohn, "The Institutional Location of the Scientist and his Scientific Values," *IRE,* EM-8, 1961, pp. 133-8. See also W. Kornhauser, op. cit., ch. 1.

15. That these values are not just abstract logical derivations, but are characteristics of highly esteemed scientists, has already been demonstrated in the literature.
See, for instance, A. Roe, *The Making of a Scientist,* New York; Dodd, Mead, 1953; B. T. Eudison, *Scientists: Their Psychological World,* New York: Basic Books, 1962.

16. For a discussion and report of research on "public" scientists and geographical mobility, see M. Abrahamson, "Cosmopolitanism, Dependence-Identification and Geographical Mobility," *A.S.Q.,* vol. 10, 1965, pp. 98-106. His evidence suggests that "cosmopolitanism" and mobility are positively related.

17. For example, "Research scientist Smith, who is now in his early thirties, has been working for a company for five years. He is offered an administrative position with more status and higher salary, but with little scope for further research. (a) Smith, who is single, accepts the offer. (b) Smith, who is married with no children, accepts the offer."

18. A questionnaire was sent to third year undergraduates in chemistry at most London University colleges (internal and external). The total N involved in the analysis was 166, 83 per cent of the 200 students originally contacted.

19. This is a preliminary table from our survey of chemists in industry. The 243 chemists represent a 66 per cent response from 368 originally contacted in six chemical companies.

20. There are, of course, other mechanisms operating to reduce role strain. See, for example, Avery, "Enculturation in Industrial Laboratories," *IRE Transactions on Engineering Management,* EM-7, 1960, pp. 20-4. See also S. Marcson, op. cit.; M. Abrahamson, "The Integration of Industrial Scientists," *A.S.Q.,* vol. 9, 1964, pp. 208-18.
For studies particularly stressing administrative methods of resolving some of the conflict, see, for instance, R. M. Hower and C. D. Orth, *Managers and Scientists: Some Human Problems in Industrial Research Organisations,* Division of Research, Graduate School of Business Administration, Harvard University, 1963; E. Raudsepp, *Managing Creative Scientists and Engineers,* New York: MacMillan, 1963; R. Livingston and S. Milberg (eds.), *Human Relations in Industrial Research Management,* New York: Columbia University Press, 1957; T. Burns and G. M. Stalker, *The Management of Innovation,* London: Tavistock Publications, 1961.
See W. Kornhauser, op. cit., pp. 195-208; T. R. Laporte, "Conditions of Strain and Accommodation in Industrial Research Organisations," *A.S.Q.,* vol. 10, 1965, pp. 21-38.

21. A variety of criteria have been suggested as indices of "role strain." W. M. Evan, "Role Strain and the Norm of Reciprocity in Research Organisations," *A.J.S.,* vol. 68, 1962, pp. 346-54 (employed labour turnover, absenteeism, late arrival and early departure). We have taken felt dissatisfaction, times importance attached to an item as an index of "role strain."

22. For this table different indices of "public," "private" and "instrumental" scientist were employed. Respondents were asked the following question: "People obtain satisfaction at work from different sources. Please look at the following three statements, and say how far you agree with each. Code answer 1 = complete agreement; 2 = agree; 3 = uncertain; 4 = disagree; 5 = complete disagreement.

1. One of my main satisfactions comes from the interest and sometimes the excitement of solving scientific problems.

2. One of my main satisfactions is from seeing the results of my research efforts incorporated into a company product which sells well.

3. One of my main satisfactions comes from publishing a paper which is well received in the scientific world.

Answers to each are used respectively as an index of "private," "instrumental" and "public orientation."

23. D. Pelz and F. M. Andrews, "Organizational Atmosphere, Motivation and Research Contribution," *Amer. Behv. Sc.,* Dec. 1962, pp. 43-7; N. Kaplan, "Some Organisational Factors Affecting Creativity," *IRE Transactions on Engineering Management,* EM-7, 1960, pp. 24-9; D. C. Pelz, "Some Social Factors related to Performance in a Research Organisation," *A.S.Q.,* vol. 1, 1956, pp. 310-25; M. I. Stein, *On the role of the Industrial Research Chemist and its relationship to the problem of creativity* (unpublished mimeograph).

For a view suggesting that the evidence on lower productivity among industrial research scientists is suspect, see J. R. Hinricks, *Creativity among Industrial Scientific Research,* American Management Ass., Management Bulletin No. 12, New York, 1961.

WILLIAM KORNHAUSER

Strains and Accommodations in Industrial Research Organisations in the United States

Fundamental problems arise as a result of the transformation of research from an activity conducted primarily by separate individuals on a relatively small scale to an activity increasingly carried on by professional groups in large organisations. This change has produced a wide gap between older images and newer forms of scientific organisation, and strains between scientists and their sponsors. The massive employment of scientists in large research organisations has opened up new opportunities for the extension of scientific knowledge and values. But it also has created new problems for scientists.

Scientific research must be integrated with other functions if it is to contribute to industrial goals. However, scientists have their own requirements for furthering their creative work above all protection of their autonomy as scientists. Science needs autonomy to realise its purposes, but industry needs coordination to achieve its goals. There are,

William Kornhauser, "Strains and Accommodations in Industrial Research Organisations in the United States," *Minerva* 1 (Autumn 1962): 30-42. Reprinted by permission of the author and the publisher.

therefore, *inherent strains* between science and industry that find their most concrete expression in the industrial research organisation. These strains tend to give rise to new relationships to accommodate the need of science for autonomy and the need of industry for organisational coordination. These strains and accommodations in the industrial research organisation shall be examined in this paper with respect to (1) goals of research, (2) controls over research, (3) incentives for research, and (4) responsibilities for the utilisation of research.

GOAL-SETTING IN INDUSTRIAL RESEARCH

Strains between science and industry emerge first of all in the formulation of research goals. the scientific enterprise seeks understanding rather than utility, technical excellence rather than operating ease, creativity rather than routine. Specialised organisation, on the other hand, is designed for utilitarian ends; it places a premium on orderly and predictable action. Thus, *any* combination of science and organisation, including even one within a university research laboratory, induces some tension among these divergent purposes. But of course the strain between science and organisation is much greater where the organisation is designed for purposes other than the furtherance of science *per se.* It is the industrial or governmental laboratory that generates considerable conflict over goals for research.

The academic training of scientists is rooted in a tradition of deeply probing research and the advancement of knowledge. The experience of industrial managers, on the other hand, initially disposes them to use professional specialists as mere technicians for routine operations. The clash between the scientific and industrial perspectives, as well as the uncertainties inherent in research, have led to considerable variations in objectives of industrial research. In some firms research is used primarily as a service to manufacturing; while in others it is used primarily for the advancement of knowledge in areas of special interest to the organisation. There are numerous possibilities for the use of research within the limits set by the nature of the firm's products and technology. As a general proposition, it appears that the functions of industrial research units tend to increase over time beginning with service work and subsequently adding research directed toward the improvement of the product and then some exploratory research. The very act of establishing a research unit within a firm, and of hiring increasing numbers of scientists with advanced degrees, inevitably engenders a commitment to more fundamental research.

A central problem in developing research policies in industry is the balancing of pressures on research groups to bend their energies toward the achievement of immediate advantages for the company, on the one hand, and the protection of the integrity of the research, on the other. Where research is turned into a mere service operation for manufacturing, its distinctive competence to contribute to long-range goals will be rendered ineffective. Furthermore, the security of the research unit will be seriously impaired, so that it will lack the basis for developing a more important set of goals in the future. Much depends on the opportunity for the research staff to free itself from immediate control by other parts of the industrial organisation. When research is clearly separated from other operations, and when the several functions of research are distinguished from one another, by separate organisation, the integrity of research goals gains protection. In this manner, the strain between scientific and industrial objectives is reduced.

Research units have been established apart from production units, with a director of research often on the same level of authority as the head of production. The research unit itself, in this style of arrangement, has undergone considerable differentiation. Basic research increasingly is seen as performing an important exploratory function that requires separation from other research functions. Applied research as a source of long-term innovations also requires protection from pressures to satisfy the organisation's immediate needs. Broad programmes in applied research are recognised as being something different from product development and technical service.

The assignment of the different functions of research to separate units helps to ensure that each will be effectively performed. This does not resolve all conflict between the interests of the research staff and the immediate concerns of management. The enterprise also seeks the integration of research functions, in order that research and non-research divisions may benefit from one another's special competence and to coordinate their efforts for the achievement of the organisation's goals. As a consequence, conflict between professional and organisational goals becomes translated into tensions among units performing different research functions, as well as between research and other parts of the enterprise. But the multiplication of strains among increasingly specialised units is likely to be more beneficial to the enterprise than would be the polarisation of conflict between research and operations.

It lies in the very nature of industrial research that high standards of intellectual excellence will continue to be threatened by organisational pressures for quick and easy solutions. Since governmental and industrial sponsorship of research for practical ends looms so large today, the central problem is not whether the scientific community will fail to adapt itself to practical needs, but whether this adaptation will impair the creativity of science. The greatest burden is on the university as the major custodian of free inquiry; but government and industry also have a major stake in the preservation of free inquiry for the future quality of men and ideas upon which they will depend.

SOCIAL CONTROL IN INDUSTRIAL RESEARCH

A second area of strain between science and industry centres in the *modes of control* over research. To be effective, the organisation requires systems of control appropriate to its goals and to the dispositions and skills of its participants. Each class of participant is likely to have certain goals of its own, commitment to which will clash to some extent with the organisation's objectives. A fundamental problem of social control is the accommodation of these conflicting goals in order to achieve a balance between the initiative and compliance of the organisations' members.

An important problem for an industrial enterprise is the determination of this balance in light of the expanding role of scientific laboratories and professional research personnel. This poses the dilemma of control by professional groups *versus* administrative control. This dilemma is concretely encountered in the selection of scientists, the organisation of research groups, the choice and assignment of research projects, the regulation of research communication, and the administration of research.

Professional groups in industry generally do not possess the power of independent professional groups over recruitment of new members. One consequence of management's control over hiring is that scientists and engineers often are selected for their organisational abilities rather than for their professional competence. This may mean only that they evidence a lively interest in bending their research talents to the specific interests of the company. Or it also may mean that some of them will be chosen for their promise as future managers. Most large firms in particular seek managerial personnel from their research divisions. However, since good scientists often do not make good organisation men, and *vice versa*, recruitment practices generally represent different kinds of compromises. Firms which emphasise intellectual qualifications, such as research originality, tend to recruit more from universities, to look for men with advanced degrees, and to use their senior scientists for recruitment, whereas those firms stressing organisational criteria, such as administrative skills, tend to recruit more from other companies, to look for men with industrial experience, and to rely primarily on management judgment in hiring.

The research staff in industry has to give up organisation strictly according to professional scientific specialities, since industrial firms need to coordinate diverse activities according to their own specialisation of products and processes. But industry also has had to accommodate professional scientific lines of organisation, in order to call forth the sense of responsibility and the initiative of its scientific staff. Many large industrial laboratories organise departments on the basis of scientific disciplines, but also use task groups composed of diverse specialists.

The tradition of science as a profession teaches that the research worker must be free to work on problems of interest to him and to follow new leads as they emerge from his work, if he is to make the maximum contribution to the knowledge of his subject. Industry has curtailed this freedom in work in order to increase the chance of securing scientific results that can be commercially exploited. At the same time, industry must depend on the scientist's initiative; hence various compromises are sought between the organisation's control over tasks and the individual scientist's freedom. Thus, industrial laboratories frequently invite scientists to suggest research projects subject to management approval, allot a small amount of the time for research of the scientists' own choosing, and provide more freedom of choice for senior scientists and those engaged in basic research.

The science profession's commitment to freedom of communication in research has been sharply curtailed by the organisation's desire to control technical communication to prevent its potential exploitation by competitors. Individual scientists also compete with one another for professional achievement and recognition, and therefore seek more liberal publication policies. Scientific secrecy may be of limited value to the organisation in any case. Characteristic accommodations are to delay but not prohibit public communication of research results, and to allow early publication of research that has no immediate commercial application.

The scientific community places a very high value on its freedom from outside control, and on the autonomy of the individual scientist. The ideal form of professional control is advice and consultation among colleagues, leaving the final responsibility for professional judgment to the individual. This procedure is subject to considerable pressure from the organisation. For the characteristic mode of organisational control is a structure of authority in which orders are given by superordinates to subordinates. In so far as scientific activity is conducted in large groups for practical objectives, professional norms of control have had to be modified. At the same time, the character of scientific work and of professional norms has required modification of the usual organisational hierarchy. The management of professional work requires the adaptation of professional norms to organisational requirements. This has been facilitated through the creation of the role of scientist-administrator.

The scientist-administrator must seek to accommodate the view of scientists that only professionally trained persons are capable of judging scientific work, and simultaneously to accommodate the organisational need for a hierarchy of authority. A characteristic way the scientist-administrator meets these pressures is to rely on advice rather than orders in matters directly relating to professional judgment, notably in the formulation of specific research problems and procedures; and to operate unilaterally on administrative matters, such as the scheduling and coordination of work.

Since the role of scientist-administrator is an accommodation of divergent norms of control, it harbours considerable inner tension. If the scientist-administrator is selected for his scientific ability alone, top management will be concerned that he may fail to represent adequately the organisation's interests to the men under him; and the scientists may be frustrated by his lack of administrative skill. If, on the other hand, the role is filled by an individual who is competent in administration but inconsequential in science, the scientists may lack confidence in his scientific judgments and resent his superior status in the organisation. The scientist-administrator must please top management or run the risk of failing to gain satisfactory support for research. But he also must meet the expectations of the scientists or forfeit their cooperation and respect.

The role of scientist-administrator has been discussed thus far as if it were a single position. But there is actually a hierarchy of positions of research administration. The direction of adaptation of the scientist-administrator tends to vary in accordance with his rank; the higher the position, the more controlling are the norms of management and the organisation; and the lower the position, the more responsive is the scientist-administrator to the norms and demands of his scientific profession. Since the lower positions entail the supervision of scientists on matters closest to their professional interests—such as their research assignments and procedures, and the assessment of the quality of their research—the greater responsiveness of men in this role is of special importance in meeting professional scientific standards. Since high-level scientist-administrators are viewed by the research staff as having primarily "administrative" responsibilities—including the formulation of broad policy, the selection of new personnel, the drawing up of a budget, and so on—their greater use of executive authority is generally more acceptable to the staff.

The differentiation of research administration by kinds of research functions is another mechanism of accommodation in the control of professional activity. There are fewer administrative problems in basic research, and basic research is the area of strongest professional concern for the autonomy of the scientist. As a result, basic research units are subject to less supervision than are applied research units; and applied research units are granted greater autonomy than units engaged in product development or technical service.

In sum, the strain between autonomy of scientific research and the exigencies of bureaucratic control is accommodated by the creation of new roles for research administration. Administrative matters are controlled on the basis of hierarchical principles of authority, while matters regarded by professionals as the primary responsibility of the individual are more subject to multilateral determination through collegial consultation and consensus. Thus organisational controls are relied upon to a greater extent in the sphere of general policy, in research areas close to operations, and by top research directors, whereas professional scientific criteria are used more extensively in research assignments and procedures, in more basic research areas, and by first-line research supervisors.

In spite of these accommodations, great organisational pressures on professional autonomy remain. There is a continual strain in industry toward the transformation of professional scientific concerns into administrative issues. As a result, professional scientific autonomy requires outside support, principally from professional scientific associations. In certain professions, notably medicine and law, the professional association has succeeded in exercising considerable support for individual autonomy. But this has not been true in industrial science. To be sure, technical societies have greatly increased their interest in the professional qualifications and status of their members. They also continue to perform their central functions of facilitating scientific communication, assessing and recognising scientific contributions, and reinforcing professional identification. But the adaptations of professional societies to the very large-scale employment of their members in industry have thus far been very limited. As a consequence they have developed little competence to support the professional autonomy of industrial scientists and engineers. One major reason for this is the very great influence of those members who are also industrial executives: the leadership of societies in which the bulk of the members work in industry tends to be dominated by engineers and scientists who are industrial executives. For this and other reasons, these societies tend to avoid dealing with matters where there is a conflict of interest between professional scientific employees and their employers.

Professional unionism has been one response to the massive employment of engineers and scientists in industry, and to the relative ineffectiveness of the professional scientific society in adapting to this new condition. But unionism is primarily a method of countering one set of organisational controls (management) with another set of organisational controls. For this reason, unions of engineers and scientists have failed to establish themselves as representatives of professionalism, and especially of professional autonomy. This failure is indicated by the acute conflict of purpose and identity within professional unionism, concerning such matters as the relative importance of professional interests *versus* employee interests, the relationship of unionism to professional societies and to non-professional unions, and the legitimacy of orthodox union tactics of collective bargaining and strikes. The failure of professional scientific trades unions is also evidenced in the recent precipitous decline in their membership: at the present time only a very small percentage of all industrial engineers and scientists are union members.

In view of the inadequacies of professional societies and unions, new efforts may be expected on the part of representatives of professional science and engineering in their effort to restrict organisational pressures on professional autonomy.

INCENTIVE SYSTEMS IN INDUSTRIAL RESEARCH

A third source of strain between science and organisation lies in their different *systems of incentives.* The extensive employment of scientists and engineers in industry and government opens up new types of careers on a hitherto unprecedented scale. This new opportunity for the growth of professional employment also generates new pressures on professional loyalties. The process of training developed in the scientific profession requires a far-reaching and deep devotion, and a commitment to continue in scientific work for a life-time. But organisations that employ professionally-trained persons seek to use some of them in positions that are marginal to or outside of the strictly professional scientific endeavours. Furthermore, industrial firms try to arouse and maintain loyalty to the enterprise, and to make these relations decisive whenever they conflict with others. As a result, the professional scientist-employee is pulled between his primary loyalty to the profession of science within which he is trained and the firm for which he works. The individual, fresh from professional training, enters organisational employment committed to a professional carreer, but subsequently he may develop aspirations for a career in the organisation which entails the renunciation of his career as a professional scientist.

The scientist or engineer in industry—and government—undergoes powerful temptations to direct his ambition toward administrative posts —for that is where the greater power, prestige, and income in the organisation lie. Organisations need scientists and engineers in administration to give research effective direction and coordination. They also need some specialists at the highest level of management if their special concerns are to receive proper consideration in the general policy of the enterprise. Industrial organisations in particular have the additional need for scientists and engineers in operations. In the absence of persons trained in science and technology directly attached to operations, research and technical information will be poorly utilised. In short, organisations need scientists and engineers for general administration and the utilisation of scientific and technical work, as well as for production of technical results themselves.

For this purpose, organisations employ incentives to attract some scientists and engineers to general administration and policy. Such incentives are inherent in the structure of the firm. The distinctive problem for the firm is its poverty in incentives for strictly professional scientific work. This deficiency endangers attachment to professional scientific standards and loyalties, and it is therefore of special concern to professions that have many members in industry. The scientific and technical professions respond by pressing for the introduction and increase of professional incentives, such as opportunities for publishing research findings, participating in meetings of professional societies, continuing professional training through subsidy by employers for further study, pursuing duties that fully utilise professional training and competence, and advancing in salary and status while continuing strictly professional scientific work. Along with such individual incentives, the viability of a strong professional tradition depends on the quality of the research organisation and management's receptivity to new ideas.

The willingness of a firm to adopt professional incentives varies with the importance of professional scientific contributions for the achievement of its goals. Thus, organisations which depend on technological innovations for their competitive position and future growth need to provide professional incentives in order to attract highly competent scientists and to give them a strong motivation to do their best work for the firm. If the effectiveness of incentives varies according to the kind of contribution the organisation seeks to elicit from its professional employees, then the need for different kinds of contributions requires a differentiated system of incentives. This also means the establishment of the possibility of a multiplicity of careers for professional people. The separation of professional scientific from administrative careers helps to meet the needs of the organisation for both professional and managerial services. It also helps to ensure that the professional commitment will not be overwhelmed by organisational pressures. In so far as the traditional system of incentives in industrial firms becomes preponderant, motivation for a career in science or engineering will be weakened. Industrial leaders are becoming aware of this, and recently have been experimenting with "professional ladders," whereby scientists and engineers can secure advances in salary and status without taking on administrative duties. Instead of greater administrative authority, they are rewarded with greater freedom to engage in their scientific and technological specialties. Whether or not such an innovation can help solve the problem, it does show an awareness of its existence.

Professional science and engineering also are showing increasing recognition of the legitimacy of a multiplicity of careers for their adherents. This is indicated by the recognition of research administration as a valued activity, and the leadership exercised by scientist-administrators in professional societies. It might lead in the future to the development of special training programmes for the administration of science, paralleling the formation of post-graduate schools of business administration, hospital administration, and so on. At the same time, professional science and engineering continue to reserve principal recognition for strictly professional achievements, which are, of course, achievements in research. The purely scientific professions have demonstrated a greater capacity to develop a professional commitment than have the various engineering specialties. The integrity of science and engineering depends on their capacity to sustain the pre-eminence of the professional commitment in the future.

RESPONSIBILITIES FOR THE UTILISATION OF RESEARCH

A fourth source of strain between science and industrial organisation lies in the allocation of responsibility for the utilisation of research results. Scientists in Western society generally have sought sharply to circumscribe their responsibilities, in the belief that the scientist is responsible only for increasing knowledge of the laws of nature and of their possible applications but not for the actual use made of that knowledge. This claim has served the important function of protecting the independence of science. The tenacity with which it has been held suggests that the scientific community has been more concerned to protect its independence than to enlarge its influence. However, the immense practical consequences of modern science have begun to awaken in the scientific community an incipient belief that scientists and technologists have some responsibilities for the use of their discoveries and inventions. The "splendid isolation" of science has been rent asunder, and the scientific community is beginning to be aware of how much it is at the centre of practical affairs. This development presents great opportunities for science to extend its values, including free inquiry, the disciplined use of the mind, and the critical spirit. But it also contains great dangers to science. A report of the American Association for the Advancement of Science suggests that "there is some evidence that the integrity of science is beginning to erode under the abrasive pressure of its close partnership with economic, social, and political affairs." For example, "unseemly claims of priority may be encouraged (and) premature reports of new scientific discoveries . . . may be permitted to acquire a semblance of credibility."[1]

The scientific community has not been able to counter these pressures by denying responsibility for the utilisation of scientific work, since scientists are employed in large numbers expressly for purely industrial or industrial-military ends. The specification and implementation of the responsibility of the research group for the manner in which its work is utilised by the industrial firm which employs it encounters severe difficulties. Consideration of this problem brings us full circle, since the delineation of these responsibilities is inextricably intertwined with the definition of the goals of research. Even though the *utilisation* of research results is closely connected with their *creation,* the facilitation of the former might inhibit the latter. Thus, if the optimal condition of creativity in research is a high degree of professional scientific autonomy, the effective use of research may require close integration of the research function with the ultimate user of research products. Research sponsored by industry tends to be highly interdependent with market strategy, and therefore often is conducted within the industrial establishment. This greatly limits the autonomy of research units, and in consequence may discourage research creativity to the detriment of effective innovation over the longer run. Scientists in universities and independent research institutes possess greater freedom to tackle new problems; but the distance between them and potential users of their work increases the risk that research findings will not be relevant or effectively utilised.

There are serious obstacles to the extension of the responsibilities of research units for the use of the results of their research. These obstacles are thrown up by scientists as well as by highest management and the operating departments. Scientists tend to be much more interested in discovering new knowledge than in imparting it. They tend to avoid the role of educator in industry for much the same reasons that research-minded scientists in universities often regard teaching as a minor duty.

Members of operating departments, on the other hand, want quick and easy solutions to their technical problems; they tend to be impatient with the niceties of scientific procedure and disturbed by large-scale innovations. As a result the work of the research staff is highly vulnerable to ineffective use or even abuse. The misuse of research can readily undermine the conditions conducive to the future growth of science. When business firms make false claims to the scientific validation of industrial products, or premature claims to new industrial discoveries, no denial by scientists that this is not their responsibility can provide sufficient protection for the reputation of scientists for an unimpugnable integrity.

Some of the factors that tend to impair the influence of scientists on the utilisation of their research are the limited authority of industrial scientists and engineers, the resistance of non-research units to innovations, and the isolation of research groups from the potential users of their work. Characteristic mechanisms that diminish conflicts over the utilisation of research can be subsumed under three headings.

First, industrial research groups have sought to institutionalise the pedagogical role of the scientist. Expert scientific authority, lacking the same control of financial punishments and rewards that is possessed by executive authority, is dependent on the willingness of the professional scientist to assume the role of teacher and on the willingness of the executive to be taught. But industrial employment, owing to differences in function, status, and power, typically attenuates the authority of research scientists and engineers in comparison with that of independent practitioners of these professions or members of independent scientific and technological institutions. The views of research groups may be ignored or not accepted as authoritative. Research scientists and engineers in their turn, if they manifest a strong professional orientation, may be more interested in gaining freedom in research and status in the profession than in informing the executives of industrial firms of potential utilities and disutilities in applying their findings. Though also serving other purposes, information and liaison services conducted by research units within a firm establish channels and procedures which simultaneously make it more difficult for workers to evade their pedagogical responsibility. They also help to strengthen the authority of research workers.

Second, the strategic deployment of scientists and engineers throughout a firm tends to reduce resistances to research resulting from conflicting perspectives on innovation. One way in which compromise arises is through training and experience in both research and non-research functions. Hence, the horizontal and vertical movement of scientists and engineers between research and other segments of an enterprise provides management and operating units with people whose views help to bridge the gap between research and the rest of the organisation. Mobile individuals thereby help to integrate the scientific and the executive parts of the firm. Mediating, compromise-promoting perspectives are in fact inherent in certain roles. The training, recruitment, and task of engineers lead to their strong commitment to industrial values, notably technical progress. Thus, they are often capable of mediating between research and non-technically trained personnel, although this role is more effectively performed by some types of engineers than others. Companies in which, for example, engineers are used in large numbers to staff operating units and to fill management positions are in a better position to utilise research than are those in which trained engineers are segregated and used only or primarily in staff departments.

Third, restrictions on the autonomy of research units help to keep open channels of contact and communication, and to increase the influence of the users of research on the activities and policies of the research unit. For example, a research organisation may accept or have imposed on it the responsibility for providing technical services, even if their functions are organised in separate units. In addition to providing services that are desired and appreciated by the users of research, these activities provide channels through which the research needs of production and sales departments are communicated to the laboratory.

Social arrangements that transform scientific activity into an institutional source of innovation add to man's power to control the environment. This is not the fundamental purpose of science: knowledge and power are very different values. But if power has not been the central objective of scientific activity, and cannot become so without undermining the integrity of science and of the larger cultural system of which it is an important part, the scientific community has nevertheless to acknowledge and deal with the largely unanticipated fact of its enormous capacity to change the environment. This power is fully demonstrated in highly industrialised nations, and it exercises a magnetic influence on the leadership of numerous underdeveloped countries even now seeking modernisation by adapting for their own needs the scientific techniques of the West. The challenge to the scientific community is to find ways of affirming the power of science in application to practical affairs while keeping faith with the fundamental quest for scientific understanding of the natural order.

NOTES

1. "Science and Human Welfare," *Science,* CXXXII (1960), p. 69.

JOHN G. BEER
W. DAVID LEWIS

Aspects of the Professionalization of Science

The institutionalization and professionalization of science that has taken place in the past century has justly been called the second scientific revolution. Seen historically, this revolution has resulted from the fusion of rapidly maturing scientific disciplines with western organizational and administrative techniques, enabling large numbers of scientists with varying interests and abilities to be marshaled for massive projects of research and development. In the process, the separation which once existed between science and technology has been narrowed and bridged, and science has come to exert a major influence upon economic growth. At the same time, the very methods and objectives of science have been deeply affected, for they are no longer primarily private or personal but have a large social component.

The first scientific revolution introduced to the western world a new way of looking at phenomena, with an emphasis upon accuracy of

John J. Beer and W. David Lewis, "Aspects of the Professionalization of Science," *Daedalus* 92:764-84. Reprinted by permission of the publisher.

observation, quantification of data, verification of results, and useful prediction. Although many of these values were shared by those who took part in the technological revolution which laid the foundation for the modern industrial age, the scientist and the engineer or mechanic proceeded along parallel rather than convergent paths. Only when the full utility of science became evident to capitalistic promoters or state officials were the two streams of development combined.

The consequences of this mutual interaction upon society in general have been both numerous and momentous, and they need not be described here. With regard to science and its practitioners, however, two results should be recalled. First, because of the way in which advances in scientific knowledge can be transmuted into tangible and even awesome results, the professional scientist today enjoys financial support and public recognition in high degree. Since World War II, he has gained considerable influence with regard to the formulation and execution of public policy and a strong voice in the councils of many large industrial corporations. Second, as science and technology are drawn into ever closer cooperation, the chief types of institution in which professional scientists are employed—colleges or universities and industrial or governmental laboratories—share an increasing number of common characteristics. The researcher who is employed in the one may be engaged in work which is quite similar to that performed by a scientist in the other.

This situation differs greatly from that which prevailed only a century ago, for professionalized science manifested itself first in educational institutions and only somewhat later in private and state laboratories devoted to increasing agricultural or industrial productivity and efficiency. Before 1900, most professional scientists were teachers in universities, technical institutes and trade or secondary schools. Relatively few were employed in industry, with the major exceptions of the dyestuffs and electrical fields. In some countries, such as Great Britain and France, there was an aristocratic prejudice against abandoning the quest of pure truth for the pursuit of financial gain, stemming in part from a tradition of scientific research conducted by amateur gentlemen of private means. For the scientist unlucky enough to have to work for a living, the main alternative to teaching was government service. Even here jobs were difficult to secure, there being few if any government research laboratories and only a handful of technical or supervisory jobs available in arsenals, mines, observatories, public health services, geodetic surveys or civil engineering projects.

If many scientists looked with disfavor upon industrial employment, manufacturers often had serious misgivings about establishing close relationships with scientists. Although some academically trained researchers had well-developed business instincts—for example, Charles

Martin Hall and William Henry Perkin—the entrepreneur typically thought of the scientist as a person too committed to abstract scholarship to be directly useful in a profit-seeking enterprise. In addition, the tradition of trade secrecy which existed in many factories clashed with the desire of scholars to disseminate the results of their research. To those who managed industries based upon empirical methods and employed artisans imbued with a craft viewpoint, the scientist might also appear as a menace to established routine and job classification, and hence as a potentially antagonizing influence with regard to the working force. Finally, and perhaps most important, academic science had only begun to be translated into practical terms so clear cut and potentially lucrative as to entice business interest. More often the nonacademic inventor, with largely commercial motives, managed to produce significant innovations by making empirical use of scientific principles already fifty or more years old. In the absence of theoretically trained industrial scientists ever ready to pounce upon new academic discoveries, the lag between initial advance and ultimate application thus remained large.

Although most technical and industrial innovation was still in the hands of empirical inventors during the late nineteenth century, several of the academic sciences, such as thermodynamics, electromagnetism, optics, chemistry, bacteriology and biology, had now reached a point of development at which the existence of large and well-classified accumulations of data permitted the formulation of highly useful and comprehensive theoretical systems which could be exploited in a commercial and technological sense. As the practical and pecuniary potential of these new syntheses and advances came more clearly into view, particularly in the dyestuffs and electrical industries, the purse strings of private donors and governmental bodies were opened to finance improved scientific education and to establish well-equipped research laboratories.

These developments ultimately produced more and more trained researchers and a steadily growing number of positions for which they could be hired. There was an initial tendency for most scientifically trained university graduates to be brought back into expanding programs of technical education, but the proportion gradually decreased as those who entered industry demonstrated their value—indeed their indispensability—to formerly skeptical entrepreneurs. Although the timing and rate of scientific professionalization differed from one industrial nation to another, similar trends were taking place throughout the world. The nearly complete displacement of the amateur scientist by his teaching colleague was ultimately followed by a sharp growth in the number of industrial researchers, who displaced the empirical inventors and who in time came to outnumber the academic scientists.[1]

It is of interest to note that as science became professionalized late in the nineteenth century, the division between basic and applied research became institutionalized. Basic science remained almost exclusively the province of the university. In industrial laboratories, entrepreneurs seized upon scientific processes or techniques which could lead swiftly to profitable results. Product testing, for example, was one of the first ways in which science demonstrated its utility to the manufacturer, thanks to the proliferation of ingenious analytical instruments and to the need of the industrialist for precise measurements and specifications in the volume production of goods made with interchangeable parts or standardized ingredients. The application of science to improving efficiency, eliminating unnecessary steps, conserving raw materials and finding various uses for by-products was also of great benefit to entrepreneurs who were trying to cut costs in the face of competition and declining price levels. By serving in such capacities, scientists could bring quick and tangible returns at a relatively small financial outlay to a great number of industries.

Positions involving mere testing and analysis had a tendency to become stultifyingly routine, and they were consequently unattractive to scientists with first rate capabilities. Other circumstances also created nagging annoyances which many professionally trained men found difficult to tolerate. Although eager to derive profit from science, most industrialists wanted to keep research budgets as low as possible. In addition, as we have previously mentioned, they wanted to guard against antagonizing craft-oriented employees, to preserve factory stability and to keep knowledge of certain processes from leaking out to potential competitors. It was not uncommon, therefore, for an industrial scientist to be consigned to a poorly equipped back room or an out-of-the-way corner of a plant and strictly forbidden to trespass in other departments. In some cases, manufacturers found it possible to train raw factory hands to make simple tests and routine analyses, and so to limit their scientific staffs to a few professionals who could be counted on absolutely to preserve trade secrets.[2] To such entrepreneurs, it seemed cheaper and less dangerous to buy patents from the outside than to run the risk and expense of large and continuous development programs of their own.

Because of this cautious approach, much applied research had to be done outside the factory if it was to be carried on at all. One type of institution which sprang up partly in response to this situation was the independent commercial research laboratory, typified by the firm of Arthur D. Little, founded in 1886. In addition, certain universities and academic scholars, while retaining a primary emphasis upon basic inquiries, became engaged in applied research. Such faculty members as

Michael Pupin of Columbia University, who devised a loading coil for the effective transmission of impulses in long-range telephony, found it profitable to invent as well as to theorize.[3] Professors more and more frequently entered into formal consulting arrangements with manufacturing concerns, served as expert witnesses in patent litigation, and steered their students into research projects having a bearing upon industrial problems. In time, universities themselves established institutional ties with industry, thanks to the efforts of such men as Robert Kennedy Duncan, a chemistry professor at the University of Kansas. Believing that the research chemist was placed in a stultifying position in industrial employment, where he was likely to do one routine testing operation over and over, Duncan came to the conclusion that most manufacturers had neither the background nor the education to direct the efforts of scientists. The solution was for the scientist to remain at the university and to have the industrialist come there for help. Duncan persuaded academic officials at Kansas to start a program of industrial fellowships under which graduate students could work at the university on projects suggested and financed by businessmen. He later moved to the University of Pittsburgh, where his ideas were implemented at the Mellon Institute. By 1917, the industrial fellowship idea, which had also been put into effect in Germany in the late nineteenth century, had spread to universities and technical institutes in such countries as Canada, Great Britain, Australia, Finland and Japan.[4]

Although their ideas about the undesirability of placing scientists in factories thus led to the establishment of an important type of institution, the industry-oriented university research institute, such men as Duncan failed to realize that the condition of the scientist in a nonacademic habitat did not necessarily have to be as stultifying as they depicted it. The small manufacturing concern, highly competitive and cost-conscious, continued to rely on routine testing, and farmed more demanding projects out to other institutions. On the other hand, a large industrial organization, heavily capitalized and holding a dominant market position under cartel or oligopoly conditions, could provide a reasonably satisfying atmosphere for the creative researcher under wise and percipient executive leadership. This type of situation, lacking in most manufacturing plants, eventually came to exist in such fields as the dyestuffs and electrical industries. In Germany, for example, such firms as Meister, Lucius & Brüning and the Badische Anilin- und Soda-Fabrik established genuine research organizations clearly distinguishable from the one- or two-man operation or from the mere testing division.[5] Similar developments occurred in the German and English electrical industries as new leaders succeeded men like Werner Siemens, who had retarded his company's development by rejecting the alternating current system as unworkable, by insisting that the laboratory work mainly on his own pet problems, and by manifesting an excessive skepticism toward academic theoreticians.

In the United States, where the emergence of large chemical empires was delayed until World War I, the electrical industry had the distinction of bringing the first really modern industrial laboratory into existence. After Thomas A. Edison had relinquished control of the electrical enterprise associated with his name, and Charles Coffin became the dominant figure in the newly constituted General Electric Corporation, conditions were brought about under which such academically trained theorists as Charles P. Steinmetz could make outstanding careers within the confines of an industrial organization. Furthermore, the exploitation of alternating current phenomena and the need to develop highly complex equipment for electrical transmission and use made it absolutely necessary for the industry to hire the best men available for research work. In 1900, the General Electric Laboratory was established at Schenectady, New York, under the direction of Willis R. Whitney, and an outstanding program of experimentation and creative effort was begun. Other large American firms followed in short order as Du Pont established explosives laboratories in Wilmington, Delaware and Gibbstown, New Jersey; Eastman-Kodak set up a research program in photography under C. E. K. Mees at Rochester, New York; and A.T.&T. began a centralized laboratory in New York City through the efforts of John J. Carty.

At such laboratories, both in Europe and in the United States, research work was placed under the direction of men who were purely scientific in their interests and who managed to introduce much of the university atmosphere into the industrial situation. Under such scientists as Carl Duisberg in Germany and Whitney in the United States, academically trained personnel were encouraged to undertake basic research projects, to participate in conferences modeled upon university seminars, to maintain regular contact with academic consultants, to write papers for professional journals and to attend scholarly meetings. In laboratories of this type advanced research was clearly distinguished from development: extensive libraries were provided; up-to-date equipment was made available; auxiliary specialists were hired; and well-organized programs of recruitment and training were undertaken. With respect to the last, industrial laboratories began to exert effective influence upon the curricula of universities and technical institutes to obtain the types of staff members which they desired.

Thus, through the efforts of men like Duncan on the one hand and various industrial laboratory officials on the other, and despite obvious differences in motivation, manufacturing organizations and universities were brought into a mutually helpful relationship. Basic science remained the chief concern of academic institutions, but many industrial researchers, like Irving Langmuir of General Electric, carried on fundamental explorations of their own. Applied research and development predominated in the activities of industrial laboratories, but some university scholars and students also worked on projects of this type. In either case a direct link was provided between science and technology as basic discoveries were systematically explored at the higher theoretical levels of applied research, adapted to industrial use in pilot plants and ultimately exploited commercially. In addition to serving as the source of most of the fundamental achievements and a considerable number of practical applications, the university also provided virtually all of the scientific personnel of the laboratories, thus heavily influencing the quality of work which would be done in such institutions. In turn, industrial organizations endowed professorial chairs; financed university facilities; established scholarships, prizes and loan funds; and placed representatives on academic boards of trustees. As a result of this interaction, educational and business institutions have become parts of a continuous spectrum of scientific effort.

The closeness of this relationship has given rise to a number of problems affecting professional scientists in both the academic and the industrial fields. Fears are frequently expressed that the capacity of the university to advance human knowledge is being impaired by an undue emphasis upon explorations in areas of immediate usefulness. While some educational institutions have not hesitated to seek financial reward by engaging in applied research on toothpaste additives, rodent killers and vitamin compounds,[6] others have made earnest attempts to confine the efforts of faculty scientists to investigations of a fundamental nature. This, however, is not easy at a time in which more and more research at the academic level is being financed through grants from private organizations or governmental agencies. There is no doubt that many basic projects are promoted in this manner, because of an awareness on the part of the donors that serendipity does occur and that fundamental advances can lead to profitable results in unsuspected fields. But the faculty member often still finds that support for research comes more readily if, when writing his proposal, he is careful to stress the potential practical applications of his ideas and does not propose to explore too far beyond the existing frontiers of science. Thus a subtle pressure is detected which could, if unchecked, pose a danger to progress in fundamental science itself.

Of more immediate concern to many professional scientists, however, is the problem of fitting vocational hopes and aspirations into the spectrum of effort just described. One of the most powerful traditions established in the first scientific revolution was that of the liberty of the scientist to pursue truth according to the dictates of his curiosity. Because professionalized science first arose in an academic environment, this tradition was reinforced by the university scholar's desire for academic freedom in teaching and research. In addition, as a member of a profession the scientist shares with physicians, lawyers and other practitioners of specialized knowledge the desire to have the standards of his work set by his peers rather than by persons outside his field of competence. In short, he wants to be a colleague, not an employee. In this respect, the vocational choices available to him frequently fall short of his desires.

Although the sociologist Simon Marcson in particular has emphasized the degree to which the university scientist of today is caught up in an employee situation,[7] the professional who desires a maximum of freedom in his work and an opportunity to expand basic knowledge will normally realize these objectives more readily in an academic position than in an industrial laboratory. Evidence gathered by such analysts as Francis Bello indicates that the most gifted and creative young scientists usually gravitate to the universities or toward employment with a very few business corporations noted for allowing great latitude to talented research personnel. As enrollments burgeon at institutions of higher learning, more and more academic positions are becoming available; but it is obvious that colleges and universities could not begin to absorb all the products of graduate training, let alone the great numbers of those gaining baccalaureate degrees. Most scientists, therefore, must find employment in other fields. The popular stereotype of the scientist as a professor thus fails to correspond with actual facts; he is more likely to be a researcher with an industrial firm or a government agency. If one wishes to understand the pressures and problems facing the typical scientist today, one must examine the conditions that exist in such types of environment.

Despite the many ways in which the industrial laboratory has come to resemble the university situation in the twentieth century, the fledgling researcher may find in this area of employment conditions for which he may have been unprepared by his academic training. Like any business organization, the manufacturing firm exists in order to make a profitable return on the investments of its owners, and its research efforts are conducted with this aim in mind. As we have seen, this does not mean that fundamental research will be entirely neglected in favor of applied science, nor that the industrial laboratory will be subjected to the same type of efficiency analysis and cost accounting used with relation to other departments of the same firm. It does mean that the thought of ultimate application can never be wholly absent from any investigator's mind, and that some basic advances will have to lead to profitable results if a basic research program is to justify itself to management. It also means that most of the scientists employed in industry and most of the money spent for their work will be channeled into the development of applications for known principles.

The industrial research director is well aware that he is dealing with scarce and highly trained manpower imbued with professional aspirations and unwilling to settle for the status of the ordinary employee. As a man of scientific training himself, he also knows that genuine creativity can exist only in a situation in which staff members are impelled by inner drives, rather than by exterior compulsion, to penetrate the unknown. Top management too is normally cognizant of such special circumstances, so that in most large firms the laboratory divisions are allowed a considerable degree of autonomy. However, there are limits beyond which permissiveness cannot go, as the engineer C. E. Skinner learned in the period 1916–1921, when as first director of the Westinghouse research organization, he tried to establish a laboratory in which there was little or no pressure for immediate or short-range results.[8] Far more percipient in this regard was Willis Whitney of General Electric, who unfailingly gave priority to requests for help on problems submitted by the corporation's various operating departments.[9] The successful research director therefore plays a mediatory role between the executive echelons of the company and the professional staff, reconciling if possible the expectations and viewpoints of each.

This far from easy task is usually accomplished by a slow process of suggestion and persuasion through which the scientist is motivated in a relatively indirect manner to carry out company wishes. In at least one contemporary American laboratory analyzed by Simon Marcson, the similarities between university and industrial practices are stressed when a prospective staff member is interviewed. Most new Ph.D.'s in science are imbued at least in some measure with the academic ideal of pursuing knowledge for its own sake, and they express a desire to be permitted to do some independent research. Industrial recruiters are usually able to assure the prospective staff member that facilities and some time are available for this purpose, and they point with pride to the laboratory's discoveries, the articles published in professional journals and the national reputation of the staff. But after the young man is hired, a period of adjustment occurs in which he gradually becomes aware of what his superiors expect of him. In Marcson's words, "Acculturation takes place as the laboratory and the recruited scientist learn from each other and change each other." However, "The laboratory . . . effects more changes in the scientist than vice versa."[10] The very atmosphere and work load of most industrial laboratories are so strongly weighted toward obtaining prompt, tangible and profitable results that the recruit's ideal of independent, basic research is put to a severe test.

At this point the majority of the new men abandon or modify their earlier values and ambitions. They do so for a number of reasons. They may discover that the pressure of assigned projects is so great or the effort to please superiors so consuming that insufficient time is left in the 8:30 to 5:00 day to pursue individual research interests. Staying beyond 5:00 P.M., although perhaps officially encouraged by the company, is generally not practiced by the staff, and hence only the most strongly motivated will stay to work overtime on pet projects. Or, as happens frequently, the new researcher may discover that assigned research is equally taxing and fascinating, and fully as satisfying, as his thesis research had been. Incidentally, that thesis topic is likely to have been suggested by his professor. Hence at this point in his career the young scientist, his idealistic commitment to scientific self-direction notwithstanding, has possibly never initiated a research project of his own, and thus finds industrial research not so different from what he knew at the university as he might originally have thought. Further, he may discover that a considerable amount of self-reliance and ingenious initiative is required in solving assigned problems, and that his work for industry often requires greater tenacity and elegance of execution than were needed in academic research. For at the university virtually any result represents a publishable article, which is the tangible product and measure of academic productivity; but in many industrial laboratories the only acceptable result is one which works, and works simply and well enough to be commercially feasible.

In fact, many of the new staff soon become so absorbed by the challenge of rendering an idea commercially feasible that they forget all about their one-time ambition of making some great contribution to basic science. All too often we hear only of scientists leaving industry for freer research environments elsewhere, when there are probably at least as many or more who leave large laboratories to develop on their own the practical potential of ideas which they obtained while working for industry or government.

Those among the new staff who persevere in their determination to contribute to basic science can usually find some time (one fifth or more in some industrial laboratories) for this purpose; but initially they may need to work overtime on such projects if these are to move along satisfactorily. Should the results prove of interest to the company, the researcher may obtain permission to devote more, or possibly all of his time to his project. Beyond that always exists the chance of being given the support of an entire research team. If, on the other hand, such a scientist fails to convince his superiors of the commercial potential of his personal project, he may feel sufficiently restricted in his effort to quit the company for the academic world or for another research organization more favorably inclined toward his own field of interest. This, however, does not seem to happen too often. Most scientists in industry learn to live with the frustrations and opportunities of their jobs. They usually understand the business framework within which decisions concerning industrial research must be made, and they are aware of the fact that other research environments, including universities, are no less free of tensions and disadvantages.

Aside from the usually superior monetary rewards proffered by industry, many scientists find satisfactions in industrial research which may be lacking in university life. Some, who have little interest in or aptitude for teaching, may welcome freedom from lecture preparation and working with students. Others find the auxiliary services available in industrial laboratories superior to those afforded by graduate assistants and other personnel on the academic scene. Some large companies may be better able to provide expensive equipment than any but the most well-endowed educational institutions. According to at least one laboratory director, it may actually be possible for a manufacturing firm to provide a scientist having marked personality idiosyncracies with a more congenial setting than would be attainable in an academic institution, where such a person would have frequent contacts with students and be obliged to take part in committee meetings or other functions.

Here again must be emphasized the wide variety of objectives and opportunities which exist at some industrial laboratories, particularly the three hundred largest ones which together perform 80 per cent of all American industrial research. Even though the industrial laboratory can make more effective use than the universities of scientists whose talents are rather ordinary, industry is no haven for scientific mediocrity, for it must also attract highly imaginative and superior persons who are among the best that the graduate schools can offer, and retain their services if at all possible. Thus when one examines the techniques used in managing research personnel one encounters a wide variety of administrative practices, *ad hoc* relationships, written and unwritten rules of behavior and informal, but none the less real, channels of influence. The highly structured chain of command which typifies the industrial enterprise is present, but it is considerably modified in practice because of the special problems which are encountered in operating a hybrid organization staffed by specialized individuals who are both self-motivated and disciplined, professionals and employees.

It is difficult to generalize in the face of such a situation as this, but two scientists connected with the Bell Telephone Laboratories have presented a cogent analysis of three basic relationships which develop between staff members of a research department and the management under which they work.[11] At the lowest level is the *artisan-master* relationship. Here the employer knows exactly what he wants and has a concrete idea of how this can be obtained. The scientist engaged in routine testing and analysis fits into this pattern, which exists at any large laboratory and may still be the dominant type of situation prevailing in a small research department run by a highly cost-conscious company which farms out most projects of a higher order, if indeed it has any. At the next level is the *professional-client* relationship. Here the autonomy of management is abridged because those who hire the services of specialists do not really know how a given project is to be consummated, how much time it will take, how much equipment will be needed and how the results will affect the market strategy of the company. Presumably most industrial scientists with advanced credentials—master's or doctor's degrees—will fall into this category. Finally, there is the *protégé-patron* relationship similar to that which once existed between members of the nobility and artists like Michaelangelo, musicians like Haydn, and scientists like Galileo. Perhaps the earliest example of such a person in American industrial research was Charles P. Steinmetz; at a later date, Irving Langmuir filled such a role at General Electric. Working under minimal supervision, protégés may discover basic phenomena of great potential profitability and considerably enhance the growth possibilities of the corporations which employ them. They may also bring the firm great prestige, some of which can be transmuted into advertising value. Langmuir, for example, won a Nobel Prize in 1932 for his studies in the behavior of surface films. Steinmetz brought General Electric such renown that a company official once estimated his advertising value alone at $1,000,000.[12] Thus even the highest form of brilliance does not escape the commercializing tendency in the industrial situation.

The rewards for proficiency and achievement in industrial research are substantial, but often for the professionally minded scientist they are too exclusively monetary and unaccompanied by honorific titles which denote changes in status. There is a tendency in the business world to bestow new job titles only when persons are given new responsibilities or elevated to positions of increasing power, and not merely for growth in capability within the same type of work. Even when industrial scientists are put in charge of group projects this is frequently done on an informal basis and with no titular change. This is justified by laboratory officials on two grounds: first, that research groupings are only temporary and that flexibility must be preserved if personnel are to be shifted about as the needs of the company dictate; second, that the spirit of teamwork in a laboratory may be impaired if too many distinctions are made. Plausible though these reasons may be, they fail to satisfy the status aspirations of persons whose thinking is heavily imbued with academic standards of recognition.

Cognizant of these circumstances, some industrial corporations have begun to experiment with ranking systems which demonstrate particularly well the manner in which university science departments and company laboratories are becoming more and more alike in certain respects. One leading American chemical firm, for example, has established two scales of titles—two ladders of achievement—one for scientists who remain in actual research and the other for those who go into administrative positions. The recruit is normally classed as a "Research Chemist," corresponding roughly with the grade of "Instructor" in the academic world. If he stays in research work, he may rise to "Senior Research Chemist," "Research Associate," and ultimately "Research Fellow," the pinnacle of recognition on this ladder. If on the other hand he becomes a research administrator, he advances through such grades as "Supervisor," "Senior Supervisor," and "Manager." If his executive abilities are particularly marked, he may eventually become a "Laboratory Director" or proceed into the highest managerial ranks of the company. The obvious distinction between the two ladders, of course, is that the one is almost purely honorific while the other involves formally recognized progress in power.

In order to hold its most creative researchers, the government too has found it necessary since World War II to create civil service classifications that reward the research scientist at a rate parallel and commensurate with the scientist-administrator.[13] Government, however, has found it difficult to make such changes in personnel policy as quickly as has industry, so that it has tended to lag behind in its capacity to attract and retain scientists. Despite vigorous action by the research agencies and the Civil Service to extend the salary scale, particularly at the top, to expedite hiring procedures and to maximize the researcher's freedom of action and communication, many scientists still leave or refuse government employment because they consider the salaries and the opportunities for advancement inferior to those of industry.[14] Less often than formerly, however, are they quitting because of unsympathetic management practices and red tape. Job security, superior research facilities, and fringe benefits which in former times tended to offset the disadvantages of government employment are on the whole no longer superior to what industry now offers, though they remain better than what academic institutions can manage to hold out. The fact that much governmental research is contracted out to private industry on a cost-plus-fixed-fee basis has often permitted industry to outbid government laboratories for scientific personnel, and hence has added to the government's recruitment difficulties.[15]

Yet, despite all these handicaps, the government has managed to diminish or offset them so as to increase appreciably the fraction of the nation's scientific manpower which it employs. It has been particularly successful in attracting and holding biological and social scientists, for whom there is a less desperate demand than for mathematicians and physical scientists. In mathematics, the physical sciences and engineering the government has competed effectively with industry for recent university graduates, but less successfully at higher levels. Indeed in many cases government laboratories have acted as training centers for industry, which lures away promising scientists just as they reach full productivity. When an agency loses such an individual, it often raids other government laboratories to fill the vacancy.

Clearly the point is at hand when the government must reevaluate its manpower policy and scientific priority system if it is to maintain any kind of order and balance in its research activities particularly as it plunges deeper and deeper into the gigantic space programs now underway. Furthermore, since governmental science by its very size now vitally affects the direction of all scientific activity in the country, such a manpower policy must be conceived within the framework of a plan encompassing the whole nation's scientific endeavors. Thus far, no such plan is in the making. Vast federal research programs continue to battle each other for men and funds while all of them together compete with industry and the universities.

Such competition would be more salutary if the supply of scientists and engineers were not already scarce in relation to projects needing their services. By and large, projects involving military space research are now outbidding those oriented toward the commercial sector of the economy or toward the extension of basic knowledge. The National Aeronautics and Space Administration's own research centers and industrial subcontractors alone could easily absorb all of the 30,000 new engineers and scientists who will become available in 1963. Of the 400,000 American scientists and engineers currently engaged in research and development 280,000 are employed in government-sponsored projects such as defense and space, while the remaining 120,000 work in industry on civilian objectives. Likewise, over 71 per cent of the $15 billion total expended in 1962 for research and development in the United States was spent by the government for military and related projects.[16] The major redeployment of our scientific manpower made possible by these enormous government appropriations has not allowed a proportionate strengthening of other sectors of the scientific front and has given rise to much concern.

During the late 1940's and the 1950's, it was often argued that, far from retarding innovation in the civilian sector of the economy, the shift toward massive military research would not only produce the necessary weapons but also simultaneously generate a flood of peace-time applications far surpassing in quantity what existing industrial laboratories could produce with purely commercial backing. We have now learned that this is not the case. Thus far, investigations with military potential in the 1940's and 1950's in fields such as radar, atomic energy and space exploration have not made as great a contribution to the gross national product, despite the vast resources devoted to their development, as a number of far more modest research programs specifically aimed at the civilian market in such areas as agriculture and polymer chemistry. The fact is that the technical adaptation of scientific principles to such civilian needs as transportation, housing, food, clothing and education is a very complicated process requiring much time and labor, and quite different in nature from the equally complicated development of modern weapons. The arms race plus the current shortage of scientific manpower often preclude the simultaneous pursuit by defense-oriented laboratories of promising civilian applications. Thus, despite the fact that our overall national research expenditures tripled between 1947 and 1960, the rate of economic growth actually dropped during that period from a fair 3.7 per cent per annum to a sluggish 3 per cent. Officials of the Department of Commerce now point to the imbalance in our research effort toward military ends as an important cause for the lackluster economic progress made by the United States in recent years, warning us that such vigorous world competitors as the Netherlands, Germany, Japan and Sweden have all been spending about 1.25 per cent of their gross national product on civilian research to our 0.80 per cent.[17]

Another imbalance in our scientific effort, aggravated by the manpower shortage and by the spiraling cost of research, is that which exists between small and big business. As we noted, of 300,000 manufacturing companies in the United States, approximately 300 perform 80 per cent of industrially sponsored research. These same 300 companies account for 60 per cent of the sales of all manufactured products and for 61 per cent of the total manufacturing employment. While these companies spend an amount equal to 2.75 per cent of their sales on research and development, the remaining small firms can afford an average of only 0.9 per cent, or about one third as much.[18] If small business is to remain vigorous, and if it is desirable to try allocating to all sectors of our economy a more or less comparable share of the nation's scientific resources, new measures must be taken. These will possibly involve increasing cooperative research by several firms or by entire industries, assisted where necessary by government, as has been done with considerable success in Europe since World War I.

To these imbalances between military and civilian research and between scientific research in big and small business must be added the imbalance existing between basic and academic research on one side and applied civilian and military research on the other. Furthermore, disparities arise between the various sciences as a consequence of some gaining public favor while others fall into neglect. Thus the magnitude of the problems facing modern science in using its limited resources wisely becomes apparent.

Until recently, the institutional structure of science was built piecemeal and haphazardly as each government agency or business corporation sought to devise, more or less empirically, that organizational pattern which it felt would best meet its immediate needs. Now, as these originally separate units have grown more interdependent scientifically, while at the same time competing more vigorously for scarce manpower, there has arisen a need on the part of each to make the most of its existing research staff, and on the part of all to coordinate their efforts for maximum overall benefit to each other and to society. Yet progress along both these lines is considerably hampered by a dearth of abundant, accurate data relating to the science of managing science, a discipline still in its infancy but which hopefully, when mature, will help to guide the formulation of scientific policy as economics now guides decision-making in the business world.

It is also highly likely that in the future, research institutions will engage the services of efficiency analysts and science management experts trained in scientific fields, to peer over the shoulders of researchers, to evaluate records and team performance and to act as consultants to scientific task forces during all phases of their research work from inception to final evaluation. They may at first be received with smirks and later with resentment by working scientists, who will declare that discovery is an unscrutable, highly personal and necessarily inefficient act; but at that point, the scientists themselves might do well to recall that the artisan and the factory foreman a half century ago accorded them a very similar reception. As the process of discovery becomes more clearly understood, it seems probable that its management will assume a more uniform pattern, and that this trend will be accelerated by government guidance and regulation as the nation seeks both to enhance and balance its scientific effort.

Indeed, a considerable standardization already exists among big laboratories. Differences between academic, government and industrial research are dwindling as all three types of institution have become active along the whole spectrum of research from basic, through applied, to routine testing, though admittedly each still applies the weight of its effort to different parts of the spectrum. All three types of institution are also in the business of scientific education and extension, both in training their own staffs and in spreading appreciation and knowledge of science among the wider public. The director of the Atomic Energy Commission's Oak Ridge laboratory, Alvin Weinberg, recently advocated converting that institution into a bona fide graduate school in atomic science to help universities train more scientists while at the same time to bring the benefits of a genuine graduate school atmosphere to the Oak Ridge laboratory.[19] He explained that this was not as drastic a step as one might suppose at first, because the laboratory was already deeply involved in scientific education, conducting seminars at and away from Oak Ridge, granting summer research fellowships to graduate students and university professors, cooperating closely with the Massachusetts Institute of Technology in training nuclear engineers and taking part in other related activities. Other government laboratories, such as those of the Department of Agriculture, the National Institute of Health and the Bureau of Standards, have similar programs, as does industry, although to a lesser degree. Scientists from these laboratories often teach specialized graduate courses in nearby universities and frequently take part in academic seminars. Likewise, professors are often in very close touch as consultants with industrial or governmental research.

In fact, as is well known, there has been a marked tendency in recent years to build new research institutions in close proximity to existing ones, especially near large universities. Towns such as Princeton, Berkeley, Ann Arbor, and Cambridge (Massachusetts) have become enormous centers for scientific research. The main advantage in thus concentrating research facilities is to have easy access to other experts and to allow the constant trend toward specialization to proceed apace unhampered by loss of contact with parent sciences and with other specialties. If the present trend continues, the majority of our scientists will find themselves working in such communities, in environments so organized as to enable each specialist to serve as many of the local research institutions as his capacity permits. An industrial scientist, for instance, could direct research for his own firm, act as a consultant for a government laboratory and teach at a nearby university. Should government desire to organize some crash program, he and others could be quickly marshaled for this task.

From the very beginning of the second scientific revolution, when academic and, somewhat later, industrial science were first institutionalized, we have noted a trend toward blurring the lines between basic and applied research and between academic and nonacademic institutions. The professional scientist of stature, like his counterparts in the professionally older fields of medicine and law, practices, teaches, consults, writes and keeps learning more about his discipline. While on the one hand he has necessarily had to surrender some freedom of action upon becoming an employee of a large institution, he has still retained a considerable amount of self-direction by making the most of society's extraordinary dependence upon his scarce and highly specialized services.

He has, however, not gone so far as Thorstein Veblen urged him to go in the years immediately following World War I. At that time, Veblen sought to organize a seizure of industry and government by engineers and industrial scientists, whom one of his disciples called technocrats.[20] The Technocracy was to come into being by the simple expedient of a paralyzing strike during which these specialists would withhold their indispensable know-how. Nothing came of the movement, for Veblen failed to arouse that class consciousness among the technocrats which he felt sure already existed in latent form. Class solidarity and common political action have never developed among scientists and engineers, since those with administrative ambitions and abilities have found their striving toward the pinnacles of power unprejudiced by their technical background. By climbing up the bureaucratic ladder, rather than by revolution, they have steadily increased their number and influence in the high places of industry and in such government agencies as the National Aeronautics and Space Administration, the Atomic Energy Commission, the National Institute of Health, the National Science Foundation, the Bureau of Standards and the many research divisions of the Departments of Agriculture and Defense.

It is in some ways surprising that the technocrats have not thus far achieved an even greater control of our society. Businessmen, lawyers and professional politicians continue to predominate in those key positions where the weightiest decisions affecting our national well-being must be made. That this is so may be attributed to the fact that all these professions require the skill of persuasion, and beyond that a keen insight into the ways of men and organizations. Among scientists and engineers such skill and insight often remain underdeveloped. By personal inclination, training and early job experience they are more likely to focus on scientific and material problems. Yet as the institutional character of science continues to grow, and conversely as the scientific aspects of the work of more and more institutions increase in significance, researchers will wish to sharpen their managerial and cooperative capabilities. As they develop this aspect of their profession, we can look for a still greater influx of scientists into positions of high influence than has occurred to date.

NOTES

1. In England, for example, 70 to 75 per cent of a total of 550 graduate chemists were employed in teaching at the beginning of the twentieth century. This percentage steadily decreased, especially after 1918, as graduate scientists who entered industry proved their value to entrepreneurs. In Germany, the shift in favor of industrial scientists occurred earlier; by 1900, 4000 chemists alone were working for business and manufacturing firms there. There is a need for detailed statistical studies to be done for the United States. See D. S. L. Cardwell, *The Organisation of Science in England: A Retrospect* (London: 1957), pp. 135, 155, 160.

2. See especially Arthur D. Little, "The Chemist's Place in Industry," *Journal of Industrial and Engineering Chemistry,* II (February, 1910), 63-66, and Alexis F. du Pont, "The Du Pont Company and Francis G. du Pont, With Special Reference to the Years 1887 to 1900 Inclusive," in Allan J. Henry (ed. and comp.), *Francis Gurney du Pont: A Memoir* (Philadelphia: 1951), I, 14-16.

3. Raymond Stevens, "Little, Arthur Dehon," *Dictionary of American Biography,* XXI, 500-501; N. R. Danelian, *A. T. & T.: The Story of Industrial Conquest* (New York: 1939), pp. 98-100.

4. See William A. Hamor, "Duncan, Robert Kennedy," *Dictionary of American Biography,* V, 511-512; Robert Kennedy Duncan, "Temporary Industrial Fellowships," *North American Review,* CLXXXV (1907), 54-61, and "On Industrial Fellowships," *Journal of Industrial and Engineering Chemistry,* I (1909), 601; Raymond F. Bacon, "The Object and Work of the Mellon Institute," *Journal of Industrial and Engineering Chemistry,* VII (1915), 343-344; Report of R. F. Bacon in *Science,* XLV (1917), 399-403.

5. John J. Beer, "Coal Tar Dye Manufacture and the Origins of the Modern Industrial Research Laboratory," *Isis,* XLIX (1958), 124-125.

6. See "Patents and Profs," *Wall Street Journal,* February 7, 1961, pp. 1,8.

7. Simon Marcson, "Decision-Making in a University Physics Department," *American Behavioral Scientist* (December, 1962), p. 37.

8. W. Rupert Maclaurin and R. Joyce Harman, *Invention and Innovation in the Radio Industry* (New York: 1949), pp. 172-176.

9. On this and other aspects of the development of the General Electric Laboratory, see Kendall Birr, *Pioneering in Industrial Research; The Story of the General Electric Research Laboratory* (Washington, D.C.: 1957), and Laurence A. Hawkins, *Adventure Into the Unknown: The First Years of the General Electric Research Laboratory* (New York: 1950), *passim.* For a biography of Whitney, see John T. Broderick, *Willis Rodney Whitney: Pioneer of Industrial Research* (Albany: 1945).

10. Simon Marcson, *The Scientist in American Industry: Some Organizational Determinants in Manpower Utilization* (Princeton: 1960), pp. 70-71.

11. Bruce F. Gordon and Ian C. Ross, "Professionals and the Corporation," *Research Management* (November, 1962), pp. 493-505.

12. F. Russell Bichowsky, *Industrial Research* (Brooklyn, N.Y.: 1942), p. 120.

13. [Bernard Barber], *Science and the Social Order* (New York: 1952), p. 234.

14. Earl W. Lindveit, *Scientists in Government* (Washington, D.C.: 1960), pp. 35-54.

15. At the time of Lindveit's publication, about half of industry's research and development expenditures were derived from federal government contracts. *Ibid.,* p. 8.

16. J. Herbert Hollomon, "Science, Technology, and Economic Growth," *Physics Today,* XVI (1963), 42, 44. Unfortunately, Mr. Hollomon does not indicate what proportion, if any, of these scientists and engineers are academically employed.

17. *Ibid.,* p. 38.

18. *Ibid.,* p. 40.

19. Alvin Weinberg, "The Federal Laboratories and Science Education," *Science,* CXXXVI (1962), 27-30.

20. Thorstein Veblen, *The Engineers and the Price System* (New York: 1944), p. 71. This chapter was first printed as "The Captains of Finance and the Engineers" in *Dial,* 66 (June 14, 1919).

NORMAN KAPLAN

Research Administration and the Administrator: U.S.S.R. and United States

In a recent paper I described the newly emerging role of the research administrator in the U.S. and tried to analyze a number of conflicting definitions and problems that stem primarily from the organizational structure in which the role is embedded.[1] During the summer of 1959, an attempt was made to compare these findings on the American research administrator with the situation of the Soviet research administrator. Some preliminary results of this comparative study are reported in this paper.[2]

After a brief description of the study in the U.S.S.R., I will outline a typical large-scale Soviet medical research organization. The administrator is located in this structure and his role is then described and compared with that of his counterpart in American research organizations. Although there are many similarities between these two roles in the two societies, some basic differences emerge, which are of potential significance for both the concrete study of research organization and for

Norman Kaplan, "Research Administration and the Administrator: U.S.S.R. and U.S.," *Administrative Science Quarterly* 6 (June 1961): 51-72. Reprinted by permission of the author and the publisher.

organizational theory generally. In a later section of the paper, some of the factors that may account for this basic difference are explored. Finally, some implications of this analysis are discussed.

DESCRIPTION OF THE STUDY

One of the most important objectives of the study in the U.S.S.R. was to obtain data on the organizational structures and practices in research institutes that would permit comparisons with results previously found in the U.S. The study was therefore restricted to research institutes in the medical field, and especially those concentrating on cancer research, so as to examine roughly similar types of organizations engaged in roughly similar activities in both the U.S.S.R. and the U.S.

In all, I interviewed the director or deputy director, as well as a number of department heads and other scientists, in thirteen medical institutes located in Moscow, Leningrad, and elsewhere in the U.S.S.R.[3] Interviews were frequently conducted in a mixture of English, French, and German, as well as Russian. Sometimes we relied on interpreters almost entirely, and in general, either lay or scientific interpreters were almost always available. A qualitative interview guide was used, and on the whole the cooperation in answering questions very specifically was exemplary. Most of the interviews lasted a minimum of two hours, and many were much longer. In a few instances it was possible to conduct several interviews with the same person on successive days.

Most of the organizations visited were under the jurisdiction of the Academy of Medical Sciences, and the majority of these were concerned primarily with cancer research.[4] The smallest institute had over two hundred people while the largest had over a thousand research workers including auxiliary staff. In size, scope, and nature of specific research activities, these institutes were not unlike many to be found in many parts of the U.S.

STRUCTURE OF A RESEARCH INSTITUTE

As one might expect where most of the institutes studied are under the jurisdiction of a single organization, namely the Academy of Medical Sciences, the basic structure tends to be the same in most of the institutes.[5] Differences were, of course, encountered but these appear to be related primarily to differences in size and especially to differences in emphasis with respect to clinical operations. In this section the basic outline of the structure encountered in most research institutes is described in general terms. No claim is made that this structure is typical of all medical research institutes in the U.S.S.R., let alone all scientific research institutes. My interviews lead me to believe, however, that the deviations and differences which may exist in other research institutes are not basic ones. This will necessarily be an exploratory account, since the primary purpose here is to locate the role of research administration and the administrator.

The director is the chief executive of the research institute and has over-all responsibility for the conduct of the research program and the maintenance of the research institute and its staff. He is appointed by the Presidium of the Academy of Medical Sciences for a three-year term which is renewable indefinitely. Directly below him in the organizational hierarchy is the deputy director or vice-director and typically the title contains the phrase "for research." He assists the director, acts for the institute in his absence, and has primary responsibility for the conduct and co-ordination of the scientific program of the institute. Below the vice-director are the departments into which the institute is divided with the department heads or chiefs reporting directly to the vice-director. The number of departments as well as their composition depends upon the size of the institute and the scope of its program. Below the department heads, one is likely to find a number of laboratories with the laboratory chiefs reporting directly to the department heads.

The basic outline of this type of structure is very familiar and certainly resembles that of most larger medical research institutes in the U.S. and many European countries. Parenthetically, it might be noted that I saw only one organization chart at all the institutes visited, although most of the directors with whom I talked were quite willing to help me draw one up.

Two other elements are always present in the organizational structure and should be described in some detail. The first is the Scientific Council (*Soviet*) which is nominally responsible for the over-all research plan of the institute, evaluating progress of the institute and of individuals, and in general dealing with any organizational or scientific problems that may arise. The director of the institute is the chairman of the Council which is made up of all or most of the department heads. The Party is represented formally on this Council by the secretary of the local Trade Union of Scientific Workers who is normally one of the regular scientists on the staff. Senior scientists who may not be department heads may also be on the Council. In addition, at least two eminent scientists, usually in related fields, but always from other institutes, are also members of this Council. The total number of members varies, of course, according to the size of the institute, and most of the ones about which information is available vary from about twelve to about thirty-five members. The frequency of meetings varies from institute to institute, but in general there are regularly scheduled meetings once or twice a month although they may occur as often as once a week.

The Council appears to combine in a single group the functions normally incorporated in two separate groups in most scientific research institutes in the U.S. One function is that of executive committee for the institute as a whole, which in the U.S. would be composed typically of department heads, the vice-director for research, and the director as chairman, as in the U.S.S.R. The second function is typically carried out by a separate group in many institutes in the U.S. and is called a scientific council made up of scientists who are not regular members of the organization, but who are invited once or twice a year (or perhaps more frequently) to evaluate the scientific work of the institute. This scientific council in most U.S. institutes has no operating functions. It is difficult to know whether the scientific council in the U.S.S.R. institute would appear above the director's box on an organization chart, or whether it would more appropriately be on the same level as that of the director, with a dotted line denoting a primarily advisory function.

Finally, we turn to the position of the research administrator. Every institute visited has such a person and the title is usually a variant of "vice-director for administration" or simply "director of administration." As in most U.S. organizations, he has primary responsibility for finances, supplies, apparatus, equipment, furniture, repairs, maintenance, and other such service activities. The size of his staff tends to vary with the size of the institute as a whole and in some of the larger institutes the administrator may have a staff of over thirty persons working in a number of separate departments. The administrator's position in the organizational hierarchy is also difficult to locate precisely. He reports to the director of the institute but he has very little if anything to do with any other scientists. Although he reports directly to the chief executive, he is not a part of the executive committee nor is he typically considered a member of the executive hierarchy. Interestingly enough, I did not meet him personally at most of the institutes visited, with one or two exceptions, when the director wanted a precise figure or fact I had asked about and he consulted the administrator. With this background, it is now possible to examine the role of the research administrator in more detail.

ROLE OF THE RESEARCH ADMINISTRATOR

As already noted, the administrator reports directly to the director of the research institute and may have a fairly large staff. Furthermore, he is responsible for more or less the same kinds of activities as is his counterpart in the U.S. Some differences begin to appear as we note that the

Soviet administrator is typically trained in what would be the American equivalent of business accounting and business procedures. It is not considered essential, or even desirable (as it is frequently considered here in the U.S.), that he have a scientific background or that he come from the ranks of the scientists. This difference becomes somewhat accentuated when we note his absence in greeting a foreign visitor, where the analogous situation in an American institution would find the administrator one of the more important men present at such a meeting. This is particularly to be expected when that visitor is more interested in problems of organization than in the substantive content of the work of the institute.

It is at first surprising to hear him referred to as the "bookkeeper" and his job described essentially as a bookkeeping one with few if any policy-making responsibilities. This term as used there implies more than simply the keeping of the financial books, referring also to "keeping the books" on maintenance, equipment, and so on. In many respects, we find that he occupies a position sometimes designated in American organizations as that of chief clerk. He has administrative responsibility for the clerks who work under him but has no other decision-making functions.

It is not surprising, therefore, to find that typically he is paid considerably less than most of the research scientists—normally only somewhat above the research technician with no advanced training. While laboratory technicians may earn approximately twelve to fifteen hundred rubles per month and the director of a research institute may have a base salary of at least five to six thousand rubles a month, the chief administrator earns approximately twelve hundred to two thousand rubles per month.[6] The range for the administrator indicates primarily the differences in size of organization and length of experience. A researcher starting out with the first advanced degree probably earns about eighteen hundred to two thousand rubles a month. In short, there can be little doubt that the chief administrator, who is referred to as the bookkeeper and whose duties correspond to those of a chief clerk, is in fact paid as one would expect a bookkeeper or chief clerk to be paid compared with the more technically trained research scientists in the research hierarchy.

For the research administrator there is little or no conflict concerning authority and control over science and scientists. These are exercised by the scientists themselves and not by a lay administrator. The Soviet administrator, when compared to his American counterpart, occupies quite a subordinate position in the research organization, despite the fact that the two have essentially the same titles and many of the same functions in a research organization.

The American research administrator is paid a good deal more than most American scientists in the same research organization and frequently is paid nearly as much as many senior scientists. In the organizational hierarchy, he is always at or near the top of the organizational structure. Although his duties may correspond very closely to those described for the Soviet administrator on a formal basis, the American administrator has many decision-making functions, overtly or covertly.[7] Many of these, incidentally, seem to stem from the unwillingness of the research director to make the decisions himself. The American research director often feels that he has little time for purely administrative decisions, and furthermore, the administrator is often thought to be better equipped to make them. In the U.S. it is often considered desirable for an administrator to have a scientific background, and not infrequently chief administrators in research organizations are recruited from the ranks of scientists. The American administrator is definitely a public figure and in fact serves to save the research director's time in public relations. He frequently exercises authority over scientists with regard to the kind of equipment they can get, space allotment, and adherence to budgets, although much of his authority is exercised indirectly, frequently with the budget or some such impersonal instrument as the indirect mechanism employed.[8]

Finally, when the American administrator is not a scientist, it is not very likely that he can move up much higher in the scientific research institute. This is, of course, similar to the Soviet situation. But frequently the American administrator, even without scientific training, who moves out of the scientific realm whether in the same organization (e.g. an industrial firm or the government) or whether from an organization in one institutional sphere to another, can move up very high in the organizational hierarchy by virtue of his *expertise* as an administrator.

In sum, the American research administrator is better paid, compared with his Soviet counterpart and with scientists in the research organization. He enjoys much higher prestige in America and, of course, he is the source of many more conflicts and problems in a research organization.[9]

This brief description indicates some vital differences in the role of the research administrator in the U.S. and the U.S.S.R. Since we are dealing with essentially similar types of organizational structures and with organizations concerned with roughly similar problems, handled in approximately the same way, and whose over-all size is roughly comparable, we are faced with the question: Why is the role of the Soviet administrator so very different? We are amazed that the Soviet administrator occupies such a subordinate position in the research organization compared with the American administrator. Of course, we could with equal validity ask the question: Why does the American administrator occupy such a superordinate position in the research organization relative to his Soviet counterpart? Asking the question both ways raises interesting subsidiary problems, some of which are considered in the remaining sections of this paper.

POSSIBLE EXPLANATORY FACTORS

As an American commenting on the Soviet scene, it seems to make sense to try to amplify the question in terms of Soviet experiences first. The first obvious question is what happens to all the administrative tasks? Obviously, the American administrator and his staff have much to keep them busy; in fact, they always seem overburdened with a variety of administrative problems. Who takes care of these problems in the Soviet research institute?

This seemingly simple and obvious question turns out, of course, to be fairly complex upon closer examination. For one thing, we must ask whether there is the same "amount" of administrative work and detail in the Soviet and American research institutes. We must also inquire whether the Soviet administrator has approximately the same kinds of duties but simply a lower status, or whether he has lower-status duties and a lower status as a consequence.

We are almost forced to start with the notion that the Soviet scientists, as compared with their American counterparts, tend to view the content and the boundaries of research administration differently. The Soviet view of the research administrator essentially restricts him to a bookkeeping function and in terms of administrative theory, might be labeled the pure execution of policy.[10] The American view of the chief administrator is often much broader. The hypothetical distinction between the execution of policy and the formulation of policy often does not work out in practice. Furthermore, the American scientist's tendency to delegate any problem that he considers essentially nonscientific results in a concept of the chief administrator's role as essentially residual—it becomes in effect all things and all functions which the director or the other top scientists are unwilling, unable, or reluctant to do themselves. In return for the alleged freedom resulting from a broadly conceived view of administration, the American scientist-director must also give up some of the areas of decision making which at the same time he continues to feel are still his prerogative; hence the almost continual underlying conflict between the administrator and the scientists in many American research organizations.

In the U.S.S.R., and for that matter in most of the rest of Europe, the scientist and the director of scientific research organizations appear to be much less reluctant than their American counterparts to assume administrative duties which have a bearing on the conduct of the research.[11] They cheerfully delegate keeping the books and other financial personnel records, and similar bookkeeping-type operations to a chief clerk, who is called a research administrator. But most other so-called nonresearch duties the Soviet scientist, as well as the European scientist generally, seems more willing to do himself. In general, it may be said that scientists at all levels, from the laboratory head to the director, are more willing to involve themselves in the nonclerical aspects of administration—and especially anything which is viewed as connected with the effective conduct of the research itself.

It is not simply as a matter of prestige that the American scientist-director argues in favor of sloughing off administrative duties. Far more important in the eyes of most scientists is the opportunity to concentrate on the conduct of research without being diverted by what seem extraneous organizational and administrative responsibilities. If it is true then, as we have asserted, that the Soviet scientist is far more willing to engage in administrative tasks than his American counterpart, does he in fact spend less time on research, since presumably he has to spend more time on administration?

The answer is paradoxical indeed. Most directors of American research institutes seem to have little, if any, time for their own research. The Soviet director, on the other hand, asserts that he spends most of his day on the conduct of his research and that this is in fact his first duty. When asked for an estimate of how much time a director in the Soviet medical institute had to spend on administrative duties, he typically answered that it was an average of about an hour a day. This increased, of course, at certain times of the year when new budgets had to be in, but generally the time reported spent at the institute not devoted to research was extremely low. We then have the apparent paradox of the Soviet director more willing and more likely to engage in administrative duties than his American counterpart and yet being able to spend considerably more time on his own research than his American counterpart.

Dismissing some of the more tenuous kinds of answers, we can only suggest one rather startling possibility. There is simply less administration. This is exceptional on two counts. We could expect that, given the same type of activity and the same size in comparable organizations, the administrative duties (not counting the purely routine ones, which are handled by clerks in both situations) would be roughly the same in order to meet the requirements of maintaining the organization. It might even be expected by some that, given comparable organizations, the level of administrative duties in the Soviet organization would be considerably higher because of the nature of Soviet society with its greater emphasis on centralization and its general bureaucratic tendencies.[12] But I must conclude tentatively that there is probably less administrative detail and bureaucratic red tape in the Soviet medical institute.

WHY DOES THIS DIFFERENCE EXIST?

It might be concluded that less administration and red tape would be possible in the U.S.S.R. because of the relative simplicity of the financial support structure. One of the obvious reasons for a complicated and large administrative staff in many U.S. medical institutes is the complexity of the financial structure and the necessity to keep track of the dozens and sometimes hundreds of different grants from different agencies with differing termination dates, differing rules concerning permissible practices, differing requirements for progress reports, renewal procedures, and so on. In the Soviet medical institutes, which are under the jurisdiction of the Academy of Medical Sciences, the budget stems from that single source.

Is it very likely that this difference in the financial support structure accounts for differing administrative loads in similar medical institutes in the two societies? The answer, perhaps strangely, is that this explanation is not very likely because when we examine organizations in the U.S. essentially similar to the Soviet ones with respect to financing, we do not find this to be the case. One of the best examples of such a comparison would be one of the National Institutes of Health, which also has a single source for its budget, namely, the Department of Health, Education, and Welfare and ultimately Congress. Despite this single source, or perhaps because of the characteristic federal accounting and auditing regulations, the reporting procedures and the administrative load generally are probably not very different from that found in most other U.S. medical organizations of similar size and scope. In fact, it may be suspected that the administrative load is at least as heavy, if not heavier, in such an organization.[13] It is probably the case then that the particular kind of financial structure is not of central significance in this context, although as pointed out in the previous paper, it may influence an already existing level of administration.[14]

If it is probably not the relative simplicity of the financial structure, what other possible factors might account for the hypothesized lower level of administrative activity in the Soviet medical research institute? Perhaps the Soviet government is willing to require fewer formal controls which in turn reduces the amount of administrative activity simply because they tend to trust their scientists more than we do. This is an intriguing hypothesis because it is probably true that the average American feels that the Soviet government trusts none of its citizens at all, while the American government, industrial firm, or scientific research institute under private auspices would seem more likely to trust their scientists. Unfortunately, it was not possible to obtain any data which would either confirm or deny this hypothesis. If, however, there is less administration in otherwise comparable organizations, then a factor such as this may play an important role.

There is relatively little disagreement that the scientist is accorded considerably more prestige and is relatively better paid and rewarded in the material sense in the U.S.S.R. than he is in the U.S. Conversely, the administrator, at least in the medical research institute, enjoys far less prestige and material reward than the administrator in the U.S. To the extent that the prestige accorded, as well as the material rewards, reflect an evaluation of the relative importance of the activities carried on by scientists and administrators, we have perhaps an additional small bit of evidence in support of the hypothesis that the Soviet scientist is trusted somewhat more.

Another factor of potentially great significance is the nature of higher authority over the organization. The director of the medical institute is responsible to the Academy of Medical Sciences and specifically to the scientists who make up the Presidium of the Academy. He is thus responsible directly to other scientists and not to government administrators or "politicians." His American counterpart is typically responsible to a board composed of laymen who are not often very familiar with the nature of science. Being unfamiliar, they are much more likely to require reports, statistics, and data, which they can understand and which in turn require the services of, and enhance the importance of, the administrator.

Returning to an earlier injunction that the question must necessarily be asked both ways, namely, why does the Soviet institute seem to have "less" administration and why does the American institute have "more" administration, we are led to inquire into some of the consequences of administrative decisions and programs. Administration, in the American sense of the term as defined here, is necessary in order to accomplish a minimum of co-ordination, communication, and control in an organization. But presumably these should be the same in the U.S.S.R. and the U.S. given similarity in organization and its activity. Part of the problem, however, is co-ordination, control, communication for whom and for what purposes? At its simplest, these are necessary for the director; he must be able to exercise control functions and may need help for this. But it becomes more complicated when the director is in turn responsible to other authorities and must provide certain information to them, primarily for purposes of control. As already noted, the complex multiplicity of research budgets in many American institutes may require the exercise of control to meet the differing criteria of a large number of organizations, all of which have provided funds for part of the larger program.

The American research director's having to account for the activities and expenditures of his research organization to a board of trustees or directors—to laymen in general (at least with respect to the intricacies of scientific research)—tends to force the director to provide certain types of nontechnical reports and information. Since these board members may have little technical knowledge of the substance of the research, and since they tend to have a great deal of knowledge concerning the operation of large organizations, both they and the director of organizations responsible to them feel that certain types of reports are most desirable to indicate proper control and reasonable progress although they may have little intrinsic value for the conduct of the research. All of these inevitably increase the administrative load and, in fact, make it very difficult for the director to spend much time on co-ordinating the research itself, let alone doing any of his own.

The additional problem of raising funds, not at all unimportant in most American research institutes, also consumes a good deal of the time and energy of a director and administrators to whom such functions can be, and frequently are, delegated. In the U.S.S.R., on the other hand, whatever the problems concerning the amounts and scope of the financial support, it is a single body of *scientists* to whom the director must go for his financial support for the following year. The men of the Academy of Medical Sciences are presumed to have a fairly intimate knowledge of the scientific character of the work and are less likely to require reports which we might consider normal for boards of trustees here.

Finally, two other far more speculative factors which may affect research administration in the two countries should be mentioned. The first has to do with the contemporary origins of the large-scale research institute in the two societies. In the U.S.S.R., it is apparently the case that the university institute, following the old European tradition, was expanded into a large-scale organization under the Academy. In the process, the high prestige and the relative autonomy of the scientist (with some notable exceptions of political incursions) was maintained. In the U.S., on the other hand, there was little tradition for the relatively atutonomous institute, whether attached to the university or not, and the scientist in general enjoyed relatively little prestige or autonomy. By the time research in the U.S. was expanded in the university and outside, and the complexity of the research organization grew with this expansion, the organizational model which many felt worth imitating was the successful big business enterprise. Moreover, the business organization model was borrowed at a time when the organizational specialist—the administrator—was becoming increasingly important.[15]

The other major factor has to do with the diversity of not only our financial support structure but also the occupational structure for scientists in the U.S. Titles vary from institution to institution, salary scales vary from one institutional sphere to another (industry versus government etc.), and in general there is diversity with respect to most aspects of the employment, supervision, and evaluation of the scientist. This necessitates the collection of a good deal of information to provide some basis for the evaluation of scientists and institutes.

In the U.S.S.R., on the other hand, there is a single system, with many subdivisions to be sure, defining salary scales in different types of institutes, employment grades related primarily to educational attainment, and other more or less fixed criteria. Thus large areas open to discretion in the U.S. are fixed in the U.S.S.R. and require relatively little administrative action.[16] There are, of course, numerous formal and informal ways of by-passing this otherwise inflexible structure which need not be considered in any detail here. The point to be stressed is that having this centralized and generalized system of promotion policies, grades of employment, salary schedules, etc., may actually reduce the administrative load as well as the amount of discretion that can be exercised in any specific institute. Whether the perceived disadvantages of this centralization outweigh this particular advantage is yet another question.

In closing this section, it must be emphasized again that we are primarily concerned with exploring several significant aspects of the administration of research institutes. Obviously, neither the short period of time spent in the U.S.S.R., nor the preliminary nature of my inquiries permit anything other than a very tentative analysis. It should also be obvious that the various possibilities, theoretical and otherwise, which may account for the apparently sharp differences in the administration of medical research encountered in the U.S.S.R. and in the U.S. have hardly been exhausted. In subsequent studies of this problem, these are among the hypotheses deserving of further exploration. In the final section which follows, I turn to an examination of some of the implications of my observations and the hypotheses just noted.

SUMMARY AND CONCLUSIONS

The observations, that the character of research administration and the role of the research administrator in roughly similar types of medical research organizations in the U.S.S.R. and the U.S. are different, call for an explanation. We want to know why this is so and how these differences operate, as well as how this affects the conduct of medical research.

How this is accomplished is possibly easier to describe, and the main points previously made can be summarized briefly here. The primary difference revolves around the definition of the chief administrator. In the medical research institutes of the U.S.S.R., he is defined primarily as a chief clerk. In the U.S. there is no single clear-cut definition, but in general he tends to be defined as something much more than a chief clerk, varying from general business manager to a general manager of an organization. In the U.S., the chief administrator normally has some decision-making functions while in the U.S.S.R. he appears to have practically none. This difference in definition leads to obvious differences in recruitment patterns as well as in the rewards involved in the job.

For the Russians, there is little or no problem concerning the type of person to be recruited for this job. He does not require any advanced education. He must be a competent keeper of books and records (financial and others), and, to be a chief administrator in a fairly large institute, he must be able to supervise the activities of a number of subordinate clerks. For the American research organization, on the other hand, the character of the desirable recruit for chief administrator tends to vary. Some believe the best sort of person for this position is a man who knows how to run and manage an organization. An underlying assumption is that most large organizations, irrespective of their particular activities, are essentially alike with respect to organizational problems, and consequently the best type of man for this position is a specialist in administration who is, with respect to organizations, a generalist. That is, he can move fairly easily from running a research organization to running a soap factory. Another school of thought, however, believes that there is something fairly unique about the management of a scientific research organization and tends to favor a former scientist or at least a man with scientific background who has administrative experience or at least displays a flair for administration. Involved in such a flair is the ability to deal with people and to talk with scientists, in particular to understand their problems as well as their general antipathy toward large bureaucratic organizations.

Given the Soviet requirements and definition, the man recruited need not be paid a very high salary relative to others in the research organization. He is, in effect, a fairly low-level, white-collar worker among considerably better trained and more advanced personnel in the various scientific fields. In the U.S., on the other hand, the man recruited must be paid a fairly high salary relative to other scientists because he too has advanced training, and what is most important, his market includes other types of large organizations where he commands a high salary.

We should certainly expect that the differences built into these two conceptions of the chief administrator should manifest themselves in other ways in the research organization. As already noted, we can enter- tain one of the two major possibilities: Either the amount and character of administration (management control, etc.) is roughly equivalent in the Soviet and American medical institute of the same size and charac- ter, in which case we should expect that the functions of the administra- tor in the U.S. setting are carried out by one or more functional substitutes in the organization; or, it is possible that the amount and character of general administration is quantitatively and qualitatively different in the Soviet institutions and hence few, if any, functional substitutes may be necessary. Our tentative analysis appears to favor the latter possibility although some questions and modifications must be considered.

First, it has been suggested that there is in fact "less" administration in the Soviet institutions and that, furthermore, the scientist himself, and in particular, laboratory and department heads as well as the scien- tific director of the institute, appear to be more willing to carry on some so-called administrative duties, which tend to be shunned by their American counterparts. Most important, these Soviet scientists report that such duties do not infringe on their research time and, in fact, are far more likely to report that they do their research. This suggests the hypothesis that given a reduction of administrative requirements, and an adequate division of labor with respect to the remaining require- ments among the scientists, it is possible to have a more effective orga- nization in which the primary goal of the pursuit of scientific research is not diminished significantly.

In fact, it might be argued that the apparent saving of time in delegat- ing many management activities *bearing directly* on research is in the long run a myth. The structure becomes far more cumbersome, cleav- ages and antipathy may arise between the research people and the administrative people, and the administrator is forced to make decisions in situations where scientific competence and intimate knowledge of the scientific research is necessary. This results in additional mechanisms in the organization to reduce cleavage and to communicate information, which may be far more cumbersome than an ordinary division of labor among the scientists themselves. If the scientist is willing to accept some minimum amount of administrative duty as part of his job, and as part of the price he must pay for the benefits derived from working in a large complex organization, then the net results in terms of what he can accomplish scientifically may be far greater than if he delegates many of these management functions to specialists in management. Such a step would be extremely difficult in many American research organizations because, among other things, it would necessitate the reduction in status, prestige, and monetary rewards of the chief administrator as he is now defined.

It is unfortunately not possible to discuss relative differences in the effectiveness of the conduct of research in the U.S.S.R., and the U.S. medical research institutions.[17] This is so for many obvious reasons, including our lack of adequate measures, but also because of differences in emphases, relative time devoted to the attack on different sorts of problems, and a host of ordinary but complicated problems of assessing the effectiveness of any kind of organization. One point which has some implication for general organizational theory must, however, be stressed. In general, our observation and analysis force us to ask how much administration is necessary in a complex organization. We have tended to assume, perhaps without sufficient evidence, that the level of administrative activity in research organizations (as well as in others) is at, or very near, the minimum necessary for co-ordination, control, and communication considered adequate to maintain the organization. The findings tend to throw some doubt on the validity of this assumption, at least for medical research institutes, and in a very speculative way possibly for most other types of complex organizations as well.

In summary, it seems highly possible that the Russians really do use much less formal administration in scientific organizations than we have thought possible. I have tried to suggest some of the factors that may contribute to this and, in particular, would stress the strategic role of the larger society as well as differences in approach toward large-scale complex organizations. The nature of the financial structure, the kinds of controls exercised by higher authority external to any given organization, as well as the general prestige level of scientists relative to administrators and others seem to affect the situation. It is hoped that additional empirical research can be conducted inside the U.S.S.R., as well as further comparative research in other countries and in other types of organizations in the U.S., to test some of the assumptions and hypotheses suggested here as well as to move closer toward a theory of complex organizations.

NOTES

1. Norman Kaplan, The Role of the Research Administrator, *Administrative Science Quarterly*, 4 (1959), 20-42.

2. Revision and extension of a paper read at the 127th annual meeting of the American Association for the Advancement of Science, December, 1960. Some of the ideas were initially developed in a lecture on "Comparative Research Organization," delivered at the Fifth Institute on Research and Development Administration, American University, Washington, D.C., April, 1960. This investigation is part of a larger series of studies on the organization of scientific research. Grateful acknowledgment is made for the support of these studies by a Public Health Service research grant (RG 5289), from the National Institutes of Health, Division of Research Grants, U.S. Public Health Service.

3. Most of the Soviet institutes visited were selected prior to my arrival in the U.S.S.R. on the basis of available knowledge here concerning their focus on medical research generally, and on cancer problems in particular. I am particularly grateful for the advice and suggestions offered by the late Dr. C. P. Rhoads, director of the Sloan-Kettering Institute, and Dr. John R. Heller, then director of the National Cancer Institute, and now president of the Memorial Sloan-Kettering Cancer Center. The selection of institutes, as well as initial contact with their directors prior to my arrival in the U.S.S.R., was greatly facilitated by the availability of an excellent document compiled by David P. Gelfand, *A Directory of Medical and Biological Research Institutes of the U.S.S.R.* (U.S. Public Health Service Publication No. 587; Washington, 1958). Finally, Mrs. Galina V. Zarechnak, of the National Library of Medicine, very kindly made available a prepublication draft of her study of the history and organization of the Soviet Academy of Medical Sciences, which provided valuable background information helpful in the selection procedure as well as in the subsequent interviews with Soviet medical scientists.

4. I am pleased to record my gratitude to the institute directors, vice-directors, and other Soviet scientists who helped me to explore some of these problems of research organization. I am especially grateful to Professor S. A. Sarkisov, a member of the Presidium of the Academy of Medical Sciences, and Professor N. N. Blokhin, the director of the Institute of Experimental Pathology and Therapy of Cancer in Moscow (Dr. Blokhin has since become the President of the Academy of Medical Sciences), for their help in facilitating my visits and interviews, and in general, for enhancing my welcome at the various medical institutes in the U.S.S.R.

5. For a general, and somewhat critical, review of the history and organization of the Academy of Medical Sciences based primarily on Soviet documentary sources, see: Galina V. Zarechnak, *Academy of Medical Sciences of the U.S.S.R.; History and Organization,* 1944–1949 (Public Health Monograph No. 63; Washington, 1960). See especially her charts and descriptions of Soviet research institutes, pp. 12 ff.

6. These are 1959 rubles. It is difficult to translate these earnings into terms which permit suitable comparisons with the U.S. Furthermore, it is unnecessary to do so for our purposes here since the object is to show that the administrator's salary tends to be much closer to that of technician or beginning scientist, and not, as in the U.S., closer to that of the senior scientists, associate directors, or even department heads.

7. The observations on the role and status of the American research administrator are drawn largely from an earlier paper. Cf. Norman Kaplan, *op. cit.*

8. Kaplan, *ibid.,* p. 33.

9. *Ibid.,* for other evidence see E. Orowan, "Our Universities and Scientific Creativity," *Bulletin of Atomic Scientists,* 15 (1959), 237–238; L. Kowarski, "Psychology and Structure of Large Scale Physical Research," *ibid.,* 5 (1949); A. M. Brues, "The New Emotionalism in Research," *ibid.,* 11 (1955).

10. See, for example, the discussion by Herbert Simon, where he questions the distinction (attributed to Frank J. Goodnow) between policy and administrative processes, "Recent Advances in Organization Theory," in *Research Frontiers in Politics and Government,* (Washington, 1955), esp. pp. 24–26. This kind of distinction has been emphasized by many political scientists commenting on the alleged stability and resilience of the civil service apparatus in Great Britain, France, and other nations in the face of marked changes in the political leadership of the state. This thesis is explicitly challenged in a brilliant analysis of the Nazi Germany case by Frederic S. Burin, "Bureaucracy and National

Socialism: A Reconsideration of Weberian Theory," in Robert K. Merton *et al.*, eds., *Reader in Bureaucracy* (Glencoe, 1952), pp. 33-47.

To our knowledge, the distinction between the formulation and the (mere) execution of policy has been confined almost exclusively to the political sphere. It has not been studied adequately in other kinds of large non-governmental organizations. Is there, for example, a "neutral" apparatus in large corporations which remains essentially intact in the face of sharp changes in the leadership and control of the company? Our analysis here points to the possibility that the Soviets effectively avoid the problems which may arise if the distinction is recognized insofar as the scientists keep administrative policy-making functions for themselves rather than delegating these and by downgrading the administrator to the level of a chief clerk.

11. Published evidence for this statement is admittedly scanty. However, it was strongly supported by my own observations and interview data. In some German laboratories, for example, the Director explicitly provides "on the job training" in administrative duties for his young postdoctoral research assistants. Usually, the young man is given responsibility for "helping" with purchasing activities for a six-month period, and then may be shifted to equipment maintenance for a similar period, and so on. This is viewed, in part, as a continuation of the traditional apprenticeship pattern to ensure that the young man will have gained the experience necessary to qualify him for a more senior post ultimately. Another consequence, of course, is that the director's own total administrative load is lightened considerably by being shared with subordinates. But, significantly, the director delegates some administrative responsibility to other *scientists,* and not to professional full-time administrators. To some extent, I suspect that this pattern is less a deliberately considered policy and more an extension of the traditional patterns of the small research institutes to the much larger organization which is becoming more prevalent today. The absence of this kind of strong tradition in the U.S. is perhaps partly responsible for the greater reliance on professional administrators here.

12. The stereotype of excessive red tape and bureaucracy in the U.S.S.R. is widely supported in the literature and is generally shared by most foreigners visiting the Soviet Union. How much of this stereotype can be attributed to the facts of the case, and how much to preconceived ideas coupled with inadequate comparative analyses is difficult to determine. To our knowledge there have been no studies of bureaucratic tendencies and administrative proliferation in the research institutes of the U.S.S.R. However, medical scientists have commented on the "medical bureaucracy" in the Soviet clinical practices and in the hospitals. Cf. the comments in the U.S. Public Health Service, *The Report of the United States Public Health Mission to the Union of Soviet Socialist Republics* (Public Health Service Publication No. 649: Washington, 1959), especially p. 25.

Much has been written on the bureaucratic facets of Soviet industrial organization, but even here, this notion has been sharply criticized. See, for example, David Granick, *Management of the Industrial Firm in the U.S.S.R.* (New York, 1954), especially the concluding chapter in which Granick makes an explicit attempt to compare the extent of bureaucratization in Soviet and non-Soviet industrial organization. He notes, for example, "It appears an open question whether Soviet industry is not . . . less bureaucratic than are most giant firms in. capitalist society" (p. 262). Granick attributes the fact that many Western observers see so much bureaucracy in the U.S.S.R. to their treatment of planned and centralized control over the economy as being synonymous with "bureaucracy." It should also be noted that there has been increasing concern with the growing bureaucratization of private business organization in the U.S. A study by Seymour Melman of this problem over a fifty-year period in the U.S. cites an increase of 87 per cent among productive workers compared with a 244 per cent in administrative personnel in American manufacturing industries in the period 1900–1940 ("The Rise of Administrative Overhead in the Manufacturing Industries of the United States 1899–1947," *Oxford Economic Papers,* 3, N.S. [Feb. 1951], 62).

13. It must be emphasized that this comparative evaluation is purely impressionistic. It is based largely on available documentary sources and talks with scientists and administrators at the National Institutes of Health. It would certainly be desirable and worth while to check this further in a more precise quantitative fashion.

It should also be emphasized that our impressionistic comparison is between the seven institutes of the National Institutes of Health and their intramural research organization with somewhat similar types of institutes in the U.S.S.R. under the central administration of the Academy of Medical Sciences. This comparison is not intended as a reflection of the effectiveness or policies of the National Institutes of Health administration structure or its administrators. In fact, its administration, as a whole and at the institute level, seems to be highly regarded by the National Institutes of Health as well as by other scientists and research officials who have any familiarity with it. In my own experience, these institutes, when compared with *others of the same size and scope in the United States,* are consistently highly rated in this regard. The comparison with institutes of the Academy of Medical Science, however, highlights the importance of the external environment and the demands stemming from it, which may affect administrative requirements within the organization.

14. Kaplan, *op. cit.,* p. 32.

15. Some confirmation of the importance of administrative personnel in American industry may be found in Melman, *op. cit.*

16. For a comparative study of scientific personnel systems, see: Edward McCrensky, *Scientific Manpower in Europe* (New York, 1958), Chapter vii is particularly relevant inasmuch as it contains a discussion of Soviet practices compared with others.

17. In recent years there have been numerous reports evaluating the "quality" and other characteristics of medical research in the U.S.S.R. by American and other Western medical scientists who have visited the Soviet Union. It would obviously be presumptuous of me, a layman wth respect to the medical sciences, to give my own evaluation. However, my impression from reading many of these reports and from talking with some of the medical scientists who have been there, is that Soviet medical research is generally viewed as competent, and in particular subfields, as quite outstanding. The growing program of translation of Soviet medical and scientific journals must also be viewed as evidence of the importance attached to Soviet research.

An extremely useful selected and annotated list of references has been compiled by Elizabeth Koenig of the National Institutes of Health Library: *Medical Research in the U.S.S.R.,* (Public Health Service Publication No. 710; Washington, 1960). Among the most relevant reports in terms of the institutes I visited are the following: J. R. Paul, "American Medical Mission to the Soviet Union," *Scientific Monthly,* 85 (1957), 150-156; M. B. Shimkin, "Oncology in the Soviet Union," in *Year Book of Cancer,* 1957–58 (Chicago, 1958), pp. 506-510; M. B. Shimkin, and R. E. Shope, "Some Observations on Cancer Research in the Soviet Union," *Cancer Research,* 16 (1956), 915-917; J. Turkevich, "Soviet Science in the Post-Stalin Era," *Annals American Academy Political Social Sciences,* 303 (1956), 139-151; H. Hamperl, Pathologie in USSR (Pathology in the U.S.S.R.), *Deutsche Medizinische Wochenschrift,* 82 (1957), 416-419; C. W. Scull, M. Nance, F. Grant, and G. F. Roll, "Some General Observations on Medical and Pharmaceutical Research in the Soviet Union," *Journal American Medical Association,* 167 (1958), 2120-2123; *The Report of the United States Public Health Mission to the Union of Soviet Socialist Republics, Including Impressions of Medicine and Public Health in Several Soviet Republics* (Public Health Service Publication No. 649; Washington, 1959); *U.S. Public Health Service, United States-U.S.S.R. Medical Exchange Missions, 1956; Microbiology and Epidemiology* (Public Health Service Publication No. 536; Public Health Monograph No. 50; Washington, 1957).

PART III

Science and Politics

An important stimulus to interest in a sociology of science has come from concerns about the relations between "Science and Politics," the theme of part three. Interest has centered on questions of autonomy in science, the social function of science, and the social responsibility of scientists. It is not surprising that the literature in this field is better characterized in terms of rhetoric, speculation, and political ideology than empirical research and theoretical interpretation. The significance of this literature is that it has generated ideas and hypotheses which sociologists have yet to fully exploit. The "state of the field" is illustrated by the work of Don K. Price. Price conceives science and politics to be at opposite ends of a "spectrum of knowledge," the limits of which are defined by truth (science) and power (politics). The spectrum is empirically translated into the progressively greater concern with "the responsible use of power in the application of value judgments to questions which science is never quite able to answer" (Price 1965, p. 118). Price's faith in American constitutional politics defines his perspective on science and government. His work reflects his experience and ability as an analyst of social structure, and adds a systematic quality to his work which, while it does not make what he does sociology, makes it accessible to sociologists. His analyses of the structural responses of the American constitutional system to the "scientific revolution," the nature of the scientific community, and the social psychology of the relations between scientists and politicians are a source of interesting working hypotheses.

Price notes, for example, that changes in the structure of "federalism" based on the government contract give scientists the opportunity to develop policy programs without having many restrictions imposed on them by party doctrine. Price views science as one of four estates (scientific, professional, administrative, and political) corresponding to the four functions of government and public affairs. The estates, according to Price, seem to be a feature of all scientifically advanced countries. But the mutual independence of the estates, Price hypothesizes, varies. This mutual independence is not determined by the nature of science and technology or the mode of production; it is determined by the way men conceptualize the relationships between political power and truth. In the United States, this relationship is commonly viewed as follows: the closer the estate is to the "truth" end of the truth-power spectrum, the more entitled it is to govern itself; the closer the estate is to the "power" end of the spectrum, the more accountable it is to the public. The basic flaw in Price's analysis is his unabashed commitment to the American constitutional system. The consequences of such a commitment for scientific analysis are not different from those associated with the Marxist-socialist commitments of a J. D. Bernal—significant insights are plentiful, but too many unquestioned and unquestionable assumptions interfere with the development of a systematic theory. Commitments to "systems" are consistent with scientific analysis only where such commitments are themselves products of and subject to scientific inquiry.

Shils, in this section's lead article, comments idealistically on the atomic scientists' movement in a "free society." The essay is particularly important in light of the emergence of radical scientific movements in the United States during recent years. Shils outlines the conditions for freedom in science and in society: the free society is characterized by "a plurality of autonomous spheres bound together by a sense of affinity and by the collaboration of equals;" and, "intellectual freedom can be guaranteed only by the mutual esteem which grows out of the sense of fundamental affinity and of the equal value of diverse activities in a pluralistic society." If these conditions were, in fact, characteristic of our society, there would be some foundation for Shils' argument that the scientists should affirm not only "the intellectual sphere and its claim to autonomy," but equally affirm the values of other functions and spheres, such as politics and economics. When he called for this affirmation in the 1950s, Shils believed that the continuing activity of the atomic scientists' movement was "imperative if the free society is to withstand the deforming pressures of the age of nuclear weapons." But "deforming pressures" have affected scientists, politicians, and everyone else. And some form of class analysis seems to be imperative for understanding the social and political activities of contemporary scientists. "Bulletin of the Atomic Scientists" is now the subsidiary title of *Science and Public Affairs*. The Federation of American Scientists is the orthodox wing of contemporary science for society organizations. The myth of pluralism in America is, if not yet dead, moribund. Yet Shils' idealism does not entirely obscure his sociological imagination. He presents elements of a model of free science in a free society which should be kept in mind as we seek to humanize the conditions of our existence.

In a paper which owes much to the contributions of Don K. Price, Lakoff attempts to achieve a "balanced" perspective on science and government in the United States: we cannot, he urges, have confidence in Marxist utopias, but neither can we be "paralyzed by the prophets of doom who depict the advance of knowledge as a Frankenstein monster about to turn on its creator;" we live in a balance of terror that should inhibit overconfidence, but we cannot, "as Raymond Aron has said, 'forget the duty of hope.'" The result is an optimistic interpretation of the United States' experience with the problems of responding to and directing science and technology in the years since 1945. The United States has, Lakoff argues, had considerable success in expanding its basic knowledge and applying it. This optimism is tempered somewhat by Lakoff's recognition that American success has not always been a product of "pristine idealism;" desires for security, power, and prestige have prevailed "most of the time."

Lakoff is concerned with the problem of "how to balance our desire for knowledge and technical innovations with our desire for a good life and a good society." That the United States has "sought to achieve just such a balance" as Lakoff contends is highly debatable. But we endorse his conclusion that the need to achieve such a balance requires carefully scrutinizing the promotion of science by government. Since Lakoff discusses the structure and function of the federal government's science advisory apparatus in some detail, readers should be aware of the Nixon administration's plans for revamping the federal science establishment announced in early 1973. These included elimination of the President's Science Advisory Committee and the Office of Science and Technology. The director of the National Science Foundation was slated to take on the role of federal science advisor, reporting to one of the president's "special assistants."

Schilling discusses problems in the relations between scientists and high-level policy makers. These problems, from a policy-making perspective, involve (1) conflicting advice, (2) bias, (3) predisposition to certain policy perspectives (ranging from "naive utopianism and naive belligerency" to "science serves mankind"), and (4) organization and politics. Schilling concludes that the particular positions of scientists in the policy process are not as important as the purposes and political theories of statesmen in determining the contributions science and technology will bring to international politics.

In a chapter selected from his *France in the Age of the Scientific State* (1968), Robert Gilpin identifies three stages in the West's industrial revolution. The first involved application of the scientific method to mechanical invention and was associated with the political ascendancy of Great Britain in the eighteenth century. The second stage originated in nineteenth century Germany where scientific theory was applied in industry and technology. The third, the contemporary stage, is characterized by the institutionalization of basic science and its integration with other institutions which utilize that science. This stage is rooted in the rapidly decreasing lead time between scientific discovery and technological application and is associated with the political ascendancy of the United States and the Soviet Union. The French fear that the scientific and technological gap between France and the United States, and the dependence of Europe on American technology, form a basis for increasing American political influence on the Continent. The ultimate threat is the integration of American political influence with American economic dominance. From the French point of view, the political problem is how to generate a common European science and technology policy and a common foreign policy in relations with the United States. Only common policies can prevent Europe from becoming increasingly irrelevant as an independent power in a world whose affairs are dominated by scientific nation-states of continental dimensions. In his book, Gilpin concludes that the French goal of competitive participation in the international system of scientific states is not likely to be achieved. He argues that the solution of generating common European policies is inconsistent with the French goal of national political and economic independence and autonomy. Gilpin's orientation to a pluralistic system of competing scientific states is not consistent with the vision of free science in a free society, and suffers from the same empirical and theoretical flaws as pluralistic "theories" of power within individual nation-states. His study is noteworthy for placing problems of science and power in an international perspective.

Chen's "Science, Scientists and Politics" returns us to Shils' concerns about free science and free society, this time with reference to the indoctrination and "thought reform" of scientists in Communist China. Chen attempts to describe (1) the Chinese Communist viewpoint on science, scientists, and politics, and (2) the status of scientists and scientific research in Communist China (Cf. Kwok 1965). Chen's theme and his bias are summarized in the closing lines of this selection: under the influence of years of political indoctrination and "thought reform," China's scientists have "bowed low, but their backs are not broken, and the day may yet come when they will stand erect and make their contributions as free men in an atmosphere of intellectual freedom." Chen's article raises again the question of what is "objective" science? Is it possible to "do science" in an apolitical or non-political context? Is science, by virtue of being a social activity in which men compete for and utilize scarce societal resources, political by definition? Is it, in fact, possible for scientists to pursue "science for its own sake?" Do societal or organizational factors "bias" the "objectivity" of a "scientific community" in a manner analagous to the way in which psychological factors bias individual judgments and conclusions?

The questions of purity, objectivity, and disinterestedness in science are discussed in Restivo's article, "The Ideology of Basic Science." Drawing on Maslow's psychology of science, and the concept of science as a social activity and social process, Restivo argues that the commitment to basic science *can* havè an ideological function. The "commitment to basic science," as part of a web of faith and optimism in the scientific world-view and as a rationale for not relating scientific work to the social world, is also discussed.

REFERENCES

Gilpin, Robert. 1968. *France in the Age of the Scientific Estate*. Princeton, New Jersey: Princeton University Press.

Kwok, D. W. Y. 1965. *Scientism in Chinese Thought, 1900–1950*. New Haven: Yale University Press.

Price, Don K. 1965. *The Scientific Estate*. Cambridge, Mass.: Harvard University Press.

EDWARD SHILS

The Scientists' Movement in the United States*

In 1944 there began a new current of thought and action among American scientists. It originated within the Manhattan project and arose from the depths of a troubled concern about the application of their scientific work. It raised no moral question about the rightness of their own actions in the realization of the atomic bomb, but it insisted that their will be consulted about its applications. For about two or three years there pulsated with an intensity, which varied with institutions, age, and the state of international relations, a current of anxiety, political alertness, and the desire for original and courageous action to prevent the harmful use of the achievements of science.

Edward Shils, "Freedom and Influence: Observations on the Scientists' Movement in the United States," *Bulletin of the Atomic Scientists* 13 (1957): 13-18. Reprinted by permission of Science and Public Affairs, the Bulletin of the Atomic Scientists. Copyright © 1970 by the Educational Foundation for Nuclear Science.
*An article based on a paper presented to the International Study Group on Academic Freedom sponsored by the Committee on Science and Freedom (Congress of Cultural Freedom, Paris, August 1956).

Its novelty lay not only in the considerable scale on which it touched the life of American scientists but also in its content. The earlier interest of American scientists in the social repercussions of their work had been not only isolated and scattered, but it was part of a radical, more or less Stalinist, criticism of American society; it was less interested in the integrity of science than in the derogation of the existing social order. The American Association of Scientific Workers which embodied this tendency never found an echo in the sentiments of American scientists. The new movement, organized in the Federation of American Scientists and expressing its views in the *Bulletin of the Atomic Scientists* though it swept or drew into its ranks many who had once shared the Leninist-Stalinist view and some who still did, did not concern itself with a radical criticism of America. Whereas the A.A.Sc.W. had complained of the suppression of science by capitalism, the new movement was impelled rather by the fear of the ways in which science had been and might be applied. Most of their leaders had had no contact with public life previously, and their desire to call to the attention of the public and its constituted authorities consequences which might arise from the presence of nuclear weapons was not the product of any doctrinal prepossession. The gradual realization of the bomb, and then the two detonations in Japan had shaken them into a worried conviction that they alone possessed an awful knowledge which, for the common good, they must share with their fellow countrymen and, above all, with their political leaders. Free from the technocratic fringe of left-wing scientism, the movement which emerged never had a program which would make rulers out of scientists. It accepted the general structure of government and of scientists in it and sought only in an informal way, and amateurishly, to transmit to legislators, administrators, publicists, and civic leaders the awareness which their scientific experience had given them.

Beginning with the campaign for the development of a feasible scheme for international control of the uses of atomic energy and the elimination of its military application, and at the same time the civilian control of atomic energy, the scope of their interest gradually broadened as new issues arose. What was first conceived as an emergency, in which a few specific problems required resolution, established itself as a chronic condition. Loyalty and security policies, the genetic consequences of atomic bomb and hydrogen bomb tests, and the political desirability of such tests, the estimation of the destructive power of nuclear weapons and the possibilities of civil defense, the nature of nuclear warfare and the possibilities of maintaining peace in the age of nuclear weapons, the economic development of backward areas of the world and the contributions of science to this, and, increasingly, the discovery of the optimal relations between scientists, scientific work and institutions, and the rest of the community, provided a natural agenda for the movement.

Problems had to be pursued into areas of social life which might have appeared earlier to be unconnected with the interests of responsible scientists. The interest in the international control of atomic energy, which was at first their exclusive concern, and which, in fact, is the original contribution of the movement to public discussion, led to the study of the more general features of disarmament and then to the whole range of foreign policy. The study of the significance of atomic bombs has led the scientists' movement to attend to matters of military technology and military strategy and planning; the concern with civil defense has led into the problems of industrial location and the psychology of the family; the development of atomic power has forced the scientists' movement to consider the advantages of private as compared with public enterprise, of monopoly and competition; the desire to protect the integrity of science and the status of scientists has required reflection about the status of intellectuals and intellectual activities of all sorts in modern societies, liberal and totalitarian. So, the simple and urgent problems of ten years ago have become knottily involved with nearly every aspect of society. The decision of the *Bulletin of the Atomic Scientists* to add to its title "A Magazine of Science and Public Affairs," and even to consider changing its title to *Science and Public Affairs,* accurately recorded the broadened scope and permanence of the emergency once considered to be no more than a transient and grave distraction.

The clamorous flood of problems and the limited resources of the scientists to deal with them gave little time for thought about fundamentals or a general philosophy. There is no trace of a political affiliation in the American scientists' movement although there is probably a disposition in the direction of the Democratic party. The movement has scarcely had an ideology—such ideology as it has, has nothing to do with any prevailing current of political ideology. Here and there one or another of its leaders might have expressed the belief that the scientific mode of thought provided a specially valuable preparation for public life and for judgment on questions of public policy and that therefore scientists had more than their distressing knowledge to contribute to American politics. The scientists' movement has, however, been remarkably free from the delusions of "scientism," from the scientific variant of the idea of the "philosopher-king." Every issue has had to be confronted on its merits, and the humane and flexible viewpoint, at once liberal and realistic, which emerges from the pages and deeds of the scientists' movement is the precipitate of numerous discrete actions. No principles to govern the relations of science and society have been promulgated, nor are they likely to be promulgated in the near future; instead, a more differentiated judgment has been schooled, and a more realistic understanding of the obduracy of the facts of social and political life has been developed.

II

The Federation of American Scientists, with its headquarters in Washington and its local branches and affiliates at a few universities and national laboratories, and the *Bulletin of the Atomic Scientists* are the products and generators of this movement. They have lived from hand-to-mouth for a decade. They have been run by amateurs who have remained scientists and scholars, on scant time, grudgingly torn from their own scientific and academic work. Unlike the trade unions of scientists, they have never interested themselves in questions of salaries, hours or tenure, or conditions of work. When they have spoken on behalf of the rights of the scientific profession, it has always been for its right to pursue the truth, and to be free from irrelevant intrusions in doing so.

The small groups or cluster of groups which constitute the scientists' movement in the United States are bound together only in a loose and ill-coordinated organization. The two major organs have no formal connection. The real link which binds the two organs together and which binds the small network of the more intensively and actively interested with the sympathetic matrix of the scientific profession is an informal consensus. The movement is neither a party nor a sect, nor has there ever been any serious effort to turn it into such. It is the more articulate expression of a widespread mood of scientists who are unable to give literary form to their sentiments or who grudge the time required or who lack self-confidence for the public representation of their views on questions of policy relating to science. Those at the center of the movement know that they are the objects of the hope, the projection of the aspirations of many with whom they have no contact, and this sustains them.

The center, the actual life and work of the movement, draws on a very small number of persons with very small funds and very little time and energy at their disposal. Active collaboration in writing, editing, organizing, making representations, raising funds, investigating, etc., is probably the product of the efforts of little more than one hundred persons, with active but less intensive support from not many more. Despite its epoch-making novelty and its eminently meritorious achievements, the movement has not succeeded in enlisting the vigorous participation or even the explicit attachment of the scientific profession in the United States.

III

What have been its accomplishments? The movement has definitely made a mark on American life. The Federation of American Scientists and, even more, the *Bulletin of the Atomic Scientists*—even though they can claim only a few specific victories (and no complete victory on any major problem)—can justly declare that they have installed themselves into the conscience and intelligence of the upper levels of American public life, and that, through the latter, their influence has radiated outward toward the whole politically interested population. The most important publicists of press, radio, and television who write and speak on matters connected with science heed with some measure of respect what the scientists say through their two organs. Many Senators and Representatives, especially those on the relevant congressional committees, turn an interested and sometimes even hospitable ear to what the scientists associated with or sympathetic to this movement have to say. The Executive Branch of the government, although it seldom obeys, listens with discomfort and respect to what the organs of the scientists' movement say, and it feels the need to reply and to adapt its conduct to render it less vulnerable to the scientists' criticism. When the Gray Committee decided that Dr. Oppenheimer was a security risk, it also felt it had to defend itself against the kinds of criticism which would come forth from the scientists' movement. When, during the 1956 Presidential election, the Republican administration wished to support its present position on hydrogen bomb tests, it recognized that it had to appear to have the support of leading figures in the scientific community in order to undo the impression created by other members of the scientific community. In vital matters of national policy, the support of scientists was sought by both of the contending parties.

In brief, the scientists' movement in the United States—and the stirrings in the great professional scientific associations which these two groups, by their incessant and dignified activity, occasionally engender—has introduced a new element into American public life. On almost every issue which has aroused the interest of our scientists' movement, something arising from the efforts of the scientists has stuck and deflected the course of political or administrative action toward what was almost always a more reasonable course—through prodding, reminding, pointing out, through the embodiment of an outlook or state of mind which reasserted the values of detachment and generosity of judgment, of freedom from tyrannous passion, and of the desirability of objective inquiry and of calm reflection.

To the educated public, the *Bulletin of the Atomic Scientists* has brought assurance and support. The small circulation—15,000 including foreign subscriptions and gifts—has sent its ripples far beyond the zone of its regular readership. The resonance which it has found in the American and foreign press has helped to imprint in the American public an awareness of its independence, its detachment, its reasonableness. Many who would otherwise have been more reconciled to iniquities, have been stiffened in their inner resistance by the feeling that there was someone who was thinking on these problems; some few were even encouraged to act with more courage and to express their views in public places. Backbones have been stiffened and hearts have been cheered in many quarters by the scientists' movement, and many minds have been made more thoughtful on all the issues of public policy in which science is involved.

IV

This very cursory survey of the American scientists' movement produces some observations concerning the place of science in the system of public life.

The first observation emerges from the fact that this movement is not only unique in the history of the United States, it is unique in the contemporary world. No other country has witnessed such a phenomenon. Perhaps none could. There were other countries where during the past decade intellectual liberties were placed under the shadows, but none of these gave rise to a scientists' movement. In the totalitarian countries where liberty was not merely threatened but actually subverted, there were no such movements, nor were there any such movements in any of the countries where the encroachments on the autonomy of the scientific community were slighter than they were in the United States.

It is a paradox that the past decade during which the uproar of anti-intellectualism and of distrust for scientists was louder than it has ever been in America, was also the decade of the greatly enhanced influence of scientists within public bodies and of a moderate but nonetheless unprecedented effectiveness of scientists outside the government seeking to influence opinion and policy. Both facts testify to the increased esteem in which scientists have come to be held in the country —and the hostility toward scientists which appears to contradict this assertion can indeed be reasonably interpreted as the Parthian shots of an adversary in retreat. It is of a piece with the paradox that when intellectual freedom was subject to threats and menaces, the scientists' movement showed the presence of freedom by the courageous vigor with which it used that freedom to criticize wrong policies and to combat encroachments on the freedom of science.

The scientists' movement in America is a product of freedom and was rendered possible by the fundamental confidence of those who participated in it that their actions had some chance of success. Both from a conviction of personal potency and from the belief that the political and social systems respond to the action of responsible individuals, the movement drew momentum. Where intellectuals lack self-confidence, or deem their politicians to be brutes and their fellow citizens contemptible, no scientists' movement could come to birth.

This indicates to us some of the limits of what can be achieved by such a movement. It can work for liberty only where a good measure of liberty already exists and where the main agents of the movement feel reasonably secure about their capacity to survive an altercation with the reigning authorities. A scientists' movement could not have been effective in the emergent Stalinist Soviet Union or in the incipient regime of National Socialism in 1933. Even if there had been an aspiration to form such a movement in the Soviet Union or in National Socialist Germany, it would have been brutally crushed at once. There is still not the slightest sign of such a movement in the Soviet Union despite the "thaw," and this silence certainly cannot be accounted for by the absence of problems to comment on, or by "harmonious" relations between scientists and politicians and administrators in the Soviet Union. The absence of a scientists' movement in the Soviet Union is a function of the incapacity of the ruling group in the Soviet Union to tolerate an independent organization which would criticize according to its own lights, and to the mixture of fear and habituation which among Soviet scientists leads them to assent to any policy promulgated by their political authorities.

On the other hand, where, as in Great Britain, the situation of intellectual freedom is satisfactory, there is not sufficient incentive for the creation of a scientists' movement even though there are ample tasks for such a movement. It is in regimes which are strongly entrenched in liberty but in which there are chronic anti-liberal storms that movements like the American scientists' movement have a great task, and likewise in regimes like Poland, occasionally Jugoslavia, and—with unhappy brevity—Hungary, which are feeling their way in the effort to untie themselves from totalitarianism. In the latter, if the authorities are sincere but uncertain, the independent action of the intellectuals in going forward to meet them, stressing the need for freedom while permitting their feelings of underlying affinity with the rulers to be evident, can do much to aid in the establishment of a fruitful intellectual liberty.

Italy is perhaps another country where a scientists' movement could arise and work beneficially. There, constitutionally provided freedom and a great tradition of free science encounter institutional obstacles which can be reduced by public disclosure and persuasion. Likewise in free countries like India which have no tradition of free science, a scientists' movement could help to instruct the opinion of the ruling strata. But the absence of such a movement in India shows how important the traditions of free science and of civility are for the emergence of a scientists' movement. During the recent embroilments with McCarthy and his fellow-travellers only small numbers in the membership of the scientific profession were willing to exert themselves to speak or write in public or to make representations to administrative and political authorities. The scientific profession was one of those most affected by the loyalty-security mania and the distrust of intellectuals, but their response in overt behavior was not massive. Most scientists preferred to stick to their work, to carry on with their ordinary routines and to confine their disapproval, which was nearly universal, to embittered comment among their colleagues. In this respect they were not very different from their colleagues in countries where there is no scientists' movement and where there are problems aplenty for such a movement. Efforts to stir these American scientists into writing their opinions or experiences for the *Bulletin* were frequently unsuccessful. On ceremonial occasions, or when some part of the government behaved with especial crudity, a few of the more eminent occasionally spoke out in strong tones. For the most part, it was in the younger generation of scientists that the earnest and disinterested sentiments which supported the scientists' movement were found.

It is noteworthy that the American scientists' movement has found little response among a vast sector of an academic profession which prides itself on being scientific—namely, the social scientists. Practically every problem dealt with by the scientists' movement calls for knowledge of social and political institutions, of economies and law, but, despite some outstanding exceptions, social scientists have remained aloof, indifferent or disdainful of the lay social science of the leaders of the scientists' movement. This was almost as true for the action of social scientists concerning loyalty and security problems as it was for the other problems of policy in which the scientists' movement became engaged. The same could be said of the rest of the American intellectual community.

Why do crises of intellectual freedom find the academic profession, in its majority, so supine and complaisant? Why are the great battles for academic freedom fought—and when won—won by a heroic handful?

The position of academic freedom is almost always delicately poised. One of the sources for this instability is the indifference of the academic profession itself to its own freedom as a community within the larger society.[1] Only if their own freedom or that of their immediate colleagues is infringed, would most scientists or scholars in most countries bestir themselves, or even pay attention to issues of academic freedom, and then they would do so in a narrow perspective.

There are good and reasonable reasons for this. Scientists do not enter science and could not do good scientific work if their concerns lie elsewhere than in science itself. A true passion for science or scholarship does not necessarily or easily combine with a concern for justice and liberty.

Furthermore, scientists respect scientists. They do not respect busybodies and gadabouts and those who spend their best hours working on nonscientific subjects. The young scientist who spends his time on such matters is not likely to advance in his career, not primarily because his seniors dislike such activities but because they think he is showing insufficient scientific zeal. Nor for that matter can the young scientist really spend much time on extra-scientific affairs and still do outstanding work in his subject.

Then, there is sheer cowardice, the desire for safety which can be given an honorable face. No one who has witnessed his own behavior or that of his colleagues in situations threatening to academic freedom will contest the contention that this phenomenon is very common. When the force of cowardice coincides with the pressure of professional specialization and the scientists' love of his subject and his distrust of the dilettante who meddles in realms beyond its boundaries, the total enfeeblement of the defenses of intellectual freedom is considerable.

V

Academic freedom and academic influence rest ultimately on opinion and on that particular opinion called respect or esteem. There are many institutional safeguards, but none is significant if there is not a substantial body of opinion which tolerates or supports academic freedom because it respects intellectual work and those who do it.

In the United States too many of us in the struggle through which we have just passed have overstressed the utilitarian argument for intellectual freedom—the argument that the harassments which the enthusiasts of loyalty and security sought to impose on the intellectual community were injurious to military defense and economic welfare. We did not stress sufficiently that these impositions wounded the dignity of one of the most important spheres of human action. We failed to stress the inherent value of our sphere of life, not because we did not believe in the intrinsic importance of it, but in part because of our lack of confidence in the will to understand among the sections of the population to whom we addressed our arguments. We felt a great cleavage between ourselves and the others, and we felt too little common ground to believe that an affirmation of the values we espoused would carry any weight with those we sought to persuade.

This uncertainty stems from an insufficient corporate self-esteem and from distrust of authority. (It also comes from the phantasy of the harshness and toughness of political and economic authority, in juxtaposition with the image of intellectuals as weaklings.)

In the last analysis, academic freedom and academic influence rests not on university constitutions or on financial resources or on laws or formal advisory bodies but on the esteem in which intellectual life is held in any particular society. The worst arrangements for loading a university council or Board of Governors with businessmen, lawyers, civil servants, and politicians are compatible with the utmost academic freedom, if the relevant sectors of that society highly esteem intellectual activity and academic institutions. And even the best constitution guaranteeing academic self-government is hopeless if the political and economic powers of a society despise intellectual activity and deny the value of those who carry on these activities. And the esteem in which intellectual activities are held by those outside the intellectual community is to a significant extent a function of the esteem in which those within the intellectual community hold themselves.

VI

The self-esteem of the intellectuals does not, however, foster academic freedom when it expresses an extreme belief that the only proper regime is that of the "philosopher-king." The fundamental nature of the free society—a plurality of autonomous spheres bound together by a sense of affinity and by the collaboration of equals—is infringed on and harm done to all sides when scientists and scholars esteem themselves to the point where they regard the elite of the political and economic spheres as unworthy, incompetent, and repugnant. The partial successes of the scientists' movement in the United States in the agitation for civilian control of atomic energy and international control, came because, in the very first weeks of the campaign in Washington in the autumn of 1945, the scientists who went to Washington were able to persuade certain legislators that they came with respect for the politician's function—and that they came neither as aspirants to rule nor as experts usable as instruments, but as collaborators with a basic identity of concern and outlook and with special knowledge. In the flurry of antagonism against intellectuals amongst the more irreconcilably distrustful of American politicians, the uproar was softened and held in check by the legislators who had become the good friends of some of the scientists who had gone to Washington to explain the necessity for civilian and international control.

Intellectual freedom can be guaranteed only by the mutual esteem which grows out of the sense of fundamental affinity and of the equal value of diverse activities in a pluralistic society. If this is denied by scientists and scholars, they do so at their own peril. If it is denied by politicians, administrators, soldiers, and businessmen, they injure the balance necessary for a good society, they render their own authority more precarious, and they wound the sacred values of man's existence. The agenda of the scientists' movement requires not just the affirmation of the value of the intellectual sphere and its claim to autonomy but the equal affirmation of the value of other functions and spheres. From these affirmations flow the magnanimity and sympathy which are essential to the free society and the free intelligence.

VII

Freedom, which the scientists' movement has done much to protect, is only a precondition of influence. It is not identical with it, and the limited though genuine successes of the movement in influencing popular and governmental opinion occasion somber thoughts about the prospects of the movement in this new terrain. The chief instruments of influence available to the scientists' movement are enlightenment and rational persuasion. But what can enlightenment hope to achieve when its nutriment is denied to it? When the government refuses to disclose important data about radioactive fall-out or about ground-to-air missiles, etc., what can the scientists' movement conclude and counsel on such problems as were raised during the recent discussion of the continuance of hydrogen bomb tests? The split in the camp of the scientists was evidence of the immaturity of the movement, of its limited success in campaigning against excessive classification, and of the ultimate limits of a movement which rests its hopes on public enlightenment in an international political situation in which a certain and not easily determinable minimum of information security is an unchallengeable necessity. What can a scientists' movement in one country accomplish when its success requires the modification of profoundly rooted prejudices in countries to which it has no access and in which their fellow scientists are not inclined or allowed to undertake a corresponding mark of enlightenment and rational persuasion. The greatest success of the scientists' movement in America was the creation of the idea of the international control of atomic energy and the persuasion of the responsible leaders of public life of the validity of the idea. The failure of this idea was its greatest failure: it was failure in a cause which was at first its chief justification. Yet it was a failure almost predestined by the nature of the Soviet Union.

All these obstacles to the scientists' movement—the political apathy and scientific preoccupation of the scientific profession, the woodenness of bureaucrats, the distraction and distrust of politicians, the disregard of the larger public and, above all, the terrible and constant danger of a nuclear war with the Soviet Union—are ample explanations for the only moderate success of this unique movement which has just celebrated its tenth birthday. Since, however, the movement is not doctrinaire, since it faces each problem with its best abilities and without any commitment to a paralyzing doctrine or a set of unrealistic principles, and since it acts in the medium of a free society which gives it room to adapt itself, to learn and to grow, these obstacles give no ground for ultimate discouragement. Rather, they only make the continuing activity of the movement all the more imperative if the free society is to withstand the deforming pressures of the age of nuclear weapons.

NOTES

1. While obviously everyone worth his salt wishes to be free to teach what he thinks true, to do the research which interests him, and to report his results as he finds them, there are quite a few academic persons who are more than willing to do what is proposed to them because they wish to please or because they have no ideas of their own and no independent intellectual curiosity.

SANFORD A. LAKOFF

The Scientific Establishment and American Pluralism

I

In the process of coping with the technological challenges of the last two decades the United States has indeed created what Don K. Price has called "the scientific establishment." As Price has shrewdly observed, the establishment of science in the United States can in certain respects be profitably compared to the establishment of religion elsewhere. "The plain fact is," Price has written, "that science has become the major Establishment in the American political system; the only set of institutions for which tax funds are appropriated almost on faith, and under concordats which protect the autonomy, if not the cloistered calm, of the laboratory."[1] And yet, in one crucial respect, this establishment is a very peculiar one, in that, as Price has also pointed out, it is far from the cohesive and even monolithic structure that we usually think of when we speak of a state church.[2] If anything, the scientific establishment is about as fragmented and pluralized as anything could be, even in a country distinguished for its pluralism in so many respects.

In general our political pluralism takes two forms: first, a system of society in which public government has a preeminent but not exclusive place—a system in which there are also private governments, with more limited power, but with a vital role and a real share in the exercise of power—private governments like those of corporations, labor unions, universities, professional associations, political parties, and even churches; secondly a system of public government in which there is a deliberate effort to encourage fragmentation and decentralization, whether through the separation of powers, through federalism, or through the dispersal of executive and legislative powers among a variety of relatively autonomous agencies and committees.

These two forms of pluralism are very much in evidence in what we may call the scientific establishment. This is clear if we examine any of the statistics concerning the diffusion of scientific activity. In fiscal 1964, when total federal expenditures for research and development reached a record high of $14.9 billion, most of these funds paid for work actually performed outside government itself. Based on the 1962 figures, which are roughly comparable, almost two-thirds (63 percent) of these federal government disbursements were received by private industrial contractors; 17 percent went to universities, university-affiliated research centres, and not-for-profit corporations; and only 20 percent paid for work done in government laboratories.[3]

The distribution of scientists and engineers according to employers tends to follow the distribution of expenditures. One survey indicates that 81 percent of the country's engineers are employed by private industry, 13 percent by all levels of government, and 3 percent by colleges and universities. According to the same source, 43 percent of our scientists are employed by industry, 30 percent by colleges and universities, and 17 percent by government, particularly the federal government.[4] Of all scientists now engaged in research and development it is also estimated that 60 percent work on projects directly or indirectly supported by government funds.[5]

Indeed, even if we confine our examination to scientific activities within the public sector, which is to say to the government agencies that undertake or supervise work in this general category, it is apparent that they do not fit into some neatly integrated administrative structure. For the most part the agencies that are heavily involved in scientific and technical work are housed in separate entities and jurisdictions, sometimes for reasons of historical accident rather than logical fitness. Scientific activities in the federal government range over ten departments and involve twenty-seven independent agencies.[6] The Department of Commerce houses the Weather Bureau, the National Bureau of Standards, the Census Bureau, and the Coast and Geodetic Survey. The Geological Survey, however, is in the Department of the Interior, as are the Fish and Wildlife Service, the Office of Saline Water, and the Bureau of Mines. The Agricultural Research Service is of course in the Department of Agriculture. The Department of Health, Education, and Welfare includes the Public Health Service, which, in turn, oversees the research and disbursements of the National Institutes of Health. The Department of Defense maintains its own supervisory agencies, the Office of Defense Research and Engineering and the Advanced Research Projects Agency, as well as advisory committees and R & D organizations run by each of the armed services. And even the Post Office contains an Office of Research and Engineering.

In all these instances, scientific activities are components of the more comprehensive missions of cabinet-level departments of the executive branch. In addition, however, there are also in the executive branch agencies outside the departments which exist solely for the sake of undertaking, supporting, and supervising scientific activities in particular areas. All of recent creation, they are the Atomic Energy Commission, the National Science Foundation, and the National Aeronautics and Space Administration. In other independent line (or operating) agencies, scientific activities are incidentally important to specific responsibilities, as in the case of the Agency for International Development and the Arms Control and Disarmament Agency. The regulatory agencies—the Interstate Commerce Commission, Federal Power Commission, Federal Trade Commission, Federal Communications Commission, and Federal Aviation Agency—also perform functions involving scientific research and technical evaluation. And to add to the confusion, or perhaps merely profusion, there are the variously public corporations—the TVA, which is a mixture of public and private, and the non-profits, such as Rand, which are privately managed but publicly supported.

The pluralism of the scientific establishment in the executive branch is one example of the general situation. Scientific activities also claim the attention of Congress, for the most part through the committee system. The programs conducted within most of the executive agencies are scrutinized and authorized by the regular standing committees charged with oversight of the general areas in which the agencies function. In addition, several new committees have been established to deal with the programs of the newer autonomous agencies: the Joint Committee on Atomic Energy and the House and Senate committees on space. The Legislative Reference Service has only recently acquired a new division of science and technology. Congress maintains longstanding relationships with two quasi-governmental bodies, the National Academy of Science and the Smithsonian Institution, both chartered by act of Congress.

In short, if we consider only this one sector of the scientific establishment, the public sector, we see a structure made up largely of horizontally parallel rather than vertically integrated segments. The scientific establishment in the government is hardly monolithic or regimented. It too is distinctly pluralistic. If we were to add to this picture the structure of the nongovernmental sectors, we should have a vision of a pluralistic universe capable of impressing even a modern astronomer. Fortunately for the political scientist, it is not necessary to intuit, deduce, or leave aside the question of how this universe is regulated and by whom. The rulers of governments, and the rules by which they proceed, may be difficult to describe with perfect accuracy, but they are not shrouded in impenetrable mystery.

II

In considering the coordination of science in the present day, it may be well to remind ourselves of Henri Saint-Simon's project for a Supreme Council of Newton which was to propagate the new religion of science and its chief dogma, the law of universal attraction. Saint Simon even went so far as to suggest that to guide its affairs the council elect a secretariat which would be considered the "Pope and clergy of the physical scientists."[7] As audacious or merely amusing as this proposal now appears, we had better not dismiss it as altogether out of the question. Too many of the even more fantastic prophecies of Aldous Huxley's *Brave New World* have come close to realization for us to be smugly confident of the impossibility of a kind of global sacro-technocracy. For the time being, however, the world has not yet come to this pass, and in the United States, at any rate, we are still addicted to a more limited notion of what science stands for and a more pragmatic and freewheeling attitude toward the way it ought to be managed. In fact, the present system by which American science policy is coordinated, as it has so far evolved, is almost as pluralistic as the medium in which it tries to operate.

The evolution began with the experience of World War II. The Office of Scientific Research and Development, under Vannevar Bush and James Bryant Conant, had served the War Department well as a means of establishing and maintaining contact between the military and the private institutions upon whose facilities the government had to draw for much of its military R & D. After the war the Defense Department experimented with a number of efforts designed to perpetuate the essential characteristics of OSRD. During the Korean War, when it became plain that Cold War tensions required a stepping up of weapons research, the DOD established a Science Advisory Committee within the Office of Defense Mobilization. By 1957, however, pressures had built up which all but demanded that scientific advice be available at a higher level than the ODM. Accordingly, President Eisenhower issued an executive order which in effect transferred the Science Advisory Committee from the Defense Department to the Executive Office of the President, where it was reconstituted, and continues its existence, as the President's Science Advisory Committee (PSAC). At the same time, President Eisenhower also appointed a Special Assistant for Science and Technology who was elected chairman of PSAC, a custom which continues to be followed.

From the point of view of the President, the creation of PSAC and the office of the Special Assistant were necessary steps because through them he would have access to a source of advice on technical questions completely free of commitments to the operating agencies under his direction.[8] PSAC was to be composed of qualified scientists from outside the government who would be brought together to serve as a kind of board of review for controversial proposals which might emanate from the agencies. From the point of view of the scientific community, these steps offered an opportunity to restore the links between nongovernmental scientists and the presidency that had been badly damaged, if not altogether severed, by the change from a Democratic to a Republican administration—relations which were both symbolized and exacerbated by the widely resented withdrawal of J. Robert Oppenheimer's security clearance.

In time the functions of PSAC and the Special Assistant expanded to include not only questions in which scientific matters were important ingredients, as in the case of weapons systems, but also questions pertaining to government policy toward science and scientists in general—questions such as those arising in connection with the training of scientists, university-government relations, and the identification of national needs in science. PSAC has drawn help from two satellite bodies, a group of consultants-at-large, composed of former PSAC members, and another group composed of consultants from government agencies, as well as from panels able to call on a roster of over 300 scientists in the universities and industry. In cooperation with the National Science Foundation, the staff of the Special Assistant began to collect information designed to provide a basis not only for the review of existing programs but for the initiation of new efforts where they might prove to be advisable. When the NSF was created in 1950, some of those instrumental in bringing it about entertained the hope that it would become the focus of concern for the overall objectives of the science programs and also that to some extent it would oversee the operation of at least the major programs. But for various reasons, among them the controversy over the definition of its objectives and its relatively low position on the administrative totem pole, NSF has chosen to confine itself to the more modest role of supporting and advocating basic research.

PSAC and the Special Assistant have taken full advantage of their relatively high position and their presidential mandate to review and initiate programs along a broad front. PSAC's recommendations contributed to the decision to cancel the B-70 program and proposals for nuclear-powered aircraft. Its advice was also heeded in the establishment of the Office of Defense Research and Engineering in the DOD. And PSAC has also directed a good many studies designed to suggest improvements in government science policy. Since 1957 no less than eighty panels and subpanels have been convened by PSAC from its roster. PSAC itself has been composed of from fifteen to eighteen scientists, until recently almost all from the physical sciences, who have come to Washington for monthly two-day meetings and have, in addition, individually chaired and participated in the work of the panels.

Through PSAC the Special Assistant has sought to fulfill the President's need for qualified and·unbiased advice. If this were the only direction in which the Special Assistant could move to aid the President, however, he would not have much leverage with which to work for the President's objectives within the executive branch. As Richard Neustadt acutely observes, in our system of shared powers and under modern conditions "federal operations spill across dividing lines or organization charts; almost every policy entangles many agencies; almost every program calls for interagency collaboration." Although the President is formally in command of the executive branch and although he is vitally interested in every aspect of its operation, the agency administrators are responsible not only to him but also to "Congress, to their clients, to their staffs, and to themselves."[9] Clearly, if the Special Assistant was to be at all successful in implementing the advice of PSAC and in helping to coordinate ongoing programs, he needed another arm—an arm that would not be directed outward, as PSAC is, toward the university and industrial research laboratories, but inward toward the executive agencies.

The special assistant received such a second arm with the creation in 1959 of the Federal Council for Science and Technology. The FCST, which was also set up under an executive order, brings together the highest officers with policy rank and scientific responsibilities from each of the nine departments and agencies having the largest scientific programs. As in the case of PSAC the Special Assistant serves as chairman, again by custom rather than prescription. And again, as in the case of PSAC, the FCST appoints committees to study and report on specific problems. In two respects, however, there are important differences between the operation of these complementary bodies. When the Federal Council adopts a report of one of its committees, or when, in a less formal way, it comes to certain conclusions about what ought to be done in any given area, its decision amounts to a kind of treaty among the major agencies. In addition, in several instances the FCST has been instrumental in setting up interagency coordinating committees, notably the Federal Radiation Council, the Interagency Committee on Oceanography, and the Interagency Committee on Atmospheric Sciences. It is still too early to assert with confidence that these interagency committees will succeed in overcoming the gaps and duplications in federal programs they are designed to avoid or in securing the more intensive cooperation they are designed to promote. There is no doubt, however, that the Federal Council is significantly successful in providing a forum for the exchange of ideas among agency personnel on a number of levels which might never take place otherwise. The council has also been clearly successful in enabling the Special Assistant to keep a running check on the operations of the major agencies and to offer his services, as the President's broker, in smoothing out any interagency tangles and disagreements.

Through the establishment of PSAC and the FCST the powers and responsibilities of the Special Assistant were considerably strengthened and extended. But this very enlargement helped to make his job more difficult to perform. It was in order to alleviate this difficulty that in 1962 Congress transformed the staff activities which the special assistant had begun to accumulate into a permanent statutory agency, the Office of Science and Technology. OST was formed along lines first proposed by Senator Henry M. Jackson's subcommittee of the Senate Committee on Government Operations. This proposal won support because the staff of the Special Assistant, the members of PSAC, and the responsible officials of the Budget Bureau all agreed that a change was necessary and all preferred that it take the form of a moderate rather than a radical reorganization. A number of considerations weighed heavily with these scientists and administrators:

1) The budget for the various activities, including PSAC, under the direction of the Special Assistant, had to be drawn from the funds available to the President for all White House activities. This budget had grown to a figure in the neighborhood of $750,000. So as not to

impose too great a strain on the general White House budget, it was considered advisable to ask Congress to allocate separate support for the work of the Special Assistant. This would enable him to expand his staff and his activities to whatever extent became necessary, subject to congressional approval.

2) The Jackson subcommittee had recommended the creation of an office of Science and Technology partly to improve the ability of Congress to inform itself about the government's science programs. Understandably, responsible congressmen wanted easier access both to information about the science programs and to the opinions of those in a position to review them. Inasmuch as Presidential assistants are customarily considered responsible to the President but not to Congress, the obvious solution was to establish a formal channel through which the legislators could reach the one man in government with a broad understanding of what was afoot. It was also felt by congressmen that just because the Special Assistant is a key contributor to the making of policy involving science, he ought to be "more visible," or, in other words, available to Congress, as he would be if he were performing an analagous function as a political appointee in the State Department or the DOD.

3) There was a possibility that if this relatively moderate reorganization proposal were not accepted, pressure might build up in Congress for a more radical change, along the lines previously indicated in the various proposals for a Department of Science, proposals which have met with all but universal disfavor within the executive.

4) Events had demonstrated convincingly that science and technology would continue to be a vital area of policy concern for the forseeable future. It was therefore advisable to put the operations of the Special Assistant and his staff on a permanent basis. If matters remained on an *ad hoc* basis, staff members would find themselves at the mercy of political considerations every four years and the removal of valuable documents as personal papers could not be prevented. No harm had yet come to the staff or their functions due to the impermanence of their mandate, but without a regular statutory charter there was no guarantee against future damage.

Since there seemed to be several advantages and no obvious drawbacks to the proposal, the reorganization plan was submitted to Congress and OST was established. With this last touch effected, it soon occurred to the Members of Congress to take advantage of their new access to information. In 1963 a Select Committee on Government Research was appointed under the chairmanship of Representative Carl Elliott of Alabama and began a broad series of inquiries into the techniques of coordinating and supporting science. At the same time, a subcommittee of the House Committee on Science and Astronautics, which had until then been preoccupied with the space program, was set up under the chairmanship of Representative Emilio Daddario of Connecticut with a roughly similar, if somewhat less ambitious, purpose.

III

If these congressional investigations have resulted in no major overhauling, it is because neither committee discovered much to find fault with in the system of advice and coordination which has so far been evolved. All those who have had important experience in this system, including all three past special assistants, indicated in their testimony before both committees that on the whole they were well pleased with the system now in operation.

This widespread satisfaction no doubt reflects an appreciation of the improvements that have been made in recent years rather than any unshakable conviction that the changes which have been set in motion will necessarily prove adequate. Indeed, such is the system that the very strengths it exhibits could become the sources of serious weakness.

Perhaps the greatest of these strengths is that in its pluralistic structure it shares the chief characteristic of the scientific establishment itself. As a result it is able to represent and mobilize all of the components of the constituency with which it deals without becoming the special preserve of any one of them. Through its relations with the National Research Council of the National Academy of Sciences, it secures highly valued estimates of long range needs. Through the Federal Council, which Jerome Wiesner has referred to as a "subcabinet for science and technology,"[10] and the informal relations which develop between the Special Assistant and the agency heads, it oversees programs in being, adjusts differences, and promotes cooperation. PSAC, meanwhile, stands somewhere in the middle, trying to integrate long-term estimates and on-going programs in the form of practical proposals. The Special Assistant acts as the President's lieutenant in matters where settled policy needs to be enforced, operating both by persuasion and by invoking the implicit threat of presidential sanction. At the same time he serves the President by bringing to the attention of the Chief Executive matters of importance which reach him through the various channels he now has open to the agencies, to university and industrial laboratories, to the professional associations, and to Congress; and by providing technical assistance and clarification whenever the President needs such help. The staff of OST, few in number but high in quality, can be used to assist wherever their help is most needed, and can also be relied upon to identify issues that might otherwise escape the attention of one or another of the elements in the network. Meanwhile, of course, the operating agency and bureau chiefs are free to run their programs without irresponsible interference and with the tacit assurance that, provided they maintain good relations with the Special Assistant, they are not likely to find any of their programs suddenly canceled or cut back because their viewpoints are unknown or misrepresented in the White House.

Another very considerable strength of the present system is that it provides welcome assistance to the Bureau of the Budget. In some respects it might have been expressly designed to help the Budget Bureau perform its important functions more efficiently. With the help of the OST staff, the PSAC panels and FCST committees all produce information and evaluations which make it considerably easier for the Bureau to do an effective job of reviewing appropriation requests. In addition, the Special Assistant and the FCST act as a buffer between the Budget Bureau and the agencies whenever conflicts arise between overall executive policy and the interests of particular agencies. These conflicts can be ironed out in the forum provided by the Federal Council or through the good offices of the Special Assistant without requiring the Budget Bureau to engage in a direct confrontation of strength with the departments and agencies, which the Bureau understandably prefers to avoid.

These strengths are surely considerable. But so are the potential weaknesses. It is quite possible that the long range projections of the National Academy, just because they are made by a relatively independent body outside the policy-making system, will not be translated into programs. The Federal Council is a subcabinet for science *only* in the sense that it brings operating agency heads together and not in the sense that these agency heads are appointive agents of the presiding officer. It is therefore possible that FCST will not secure the degree of cooperation that may be necessary if competition for appropriations is keen or if serious sacrifices are called for by the recommendations of the NAS and PSAC. As to PSAC, perhaps its chief potential weakness is that just because it need not embody a representative cross section of the scientific community, it may become an overly homogeneous pressure group working on behalf of narrow institutional or disciplinary interests.[11] Since it is in no formal way responsible to the constituency it represents, there can be no assurance that it will be balanced and objective in its judgments.

Nor, finally, is the Special Assistant in an unambiguously strong position. Under his various hats, he carries the key to the effectiveness of the entire system, but the key does not belong to him. His ability to achieve implementation of the advice he solicits and to coordinate the work of the agencies depends to a very great degree on the influence he has with the President. Even when relations are very good, as between President Kennedy and Dr. Wiesner, the Special Assistant can never be perfectly sure of his ground. Although the evidence for a definite judgment is not yet publicly available, it would seem that Dr. Wiesner had considerable influence with the President in connection with the negotiations leading to the nuclear test ban treaty with the Soviet Union. In at least one other case, it is known that he was overruled. Reportedly, Dr. Wiesner and James Webb, the director of NASA, disagreed over the question of whether an earth orbit or a lunar orbit would be the most advisable course for the manned lunar landing project. Dr. Wiesner is said to have favored the earth orbit, in part because of the dividends it might have for other programs, including the military space program. President Kennedy reportedly overruled Dr. Wiesner on the ground that since the space agency had the responsibility of carrying out the assignment, it would be best to allow the agency to decide upon the method, regardless of secondary considerations. Apart from any experience, however, it is clear that without considerable and sustained presidential support, the Special Assistant would find it extremely difficult to carry out any recommendation that might meet with agency objections.

One somewhat subtle consideration which militates against these potential weaknesses should be noted. In only a few years the Special Assistant and PSAC have managed to extend their influence into the operating agencies by persuading the departments to create new high-level positions for scientist-administrators. At the instigation of PSAC, several departments, including State, Agriculture, Commerce, and Interior—and in effect though not in name, Defense—now have assistant secretaries for research and development. All these posts have been filled by people associated with or approved by PSAC. These appointments, together with those in the Office of Defense Research and Engineering, have enabled the Special Assistant and PSAC to establish better ties to the departments and agencies than might otherwise exist. It would be a mistake to think that appointees to these posts become catspaws of the Special Assistant. In each case they have been men of excellent qualifications and proven independence. As members of a department they are bound to develop loyalties to it and feel great concern for its special missions. Nevertheless, their previous association with PSAC undoubtedly makes informal coordination and communication easier than it would be with personnel who have come up through the ranks of the agency or who are appointed without PSAC approval.

IV

In general, then, the American scientific establishment is regulated according to the same pluralistic principles that define its very structure. This pluralism surely distinguishes the modern establishment of science from the traditional establishment of religion. Nevertheless, there are certain ways in which the comparison remains suggestive.

Both types of establishment confer benefits and run certain risks. In each case regular and assured public support frees the practitioners to pursue their callings, releasing them from the need to divert much of their energy to the job of finding sources of financial assistance. Establishment may also be said to encourage dedication, whether it be to the practice of faith or the pursuit of knowledge, by conferring a measure of public approbation upon those willing to undertake the training and the discipline necessary to the work. On the other hand, any institution or set of institutions enjoying such a protected position is vulnerable to corrupting influences, and certainly to constant imputations of corruption.

For just this reason, in ages past, the question of ecclesiastical establishment provoked considerable critical debate. In seventeenth century England, while modern science was still in its swaddling clothes, zealous Puritans denounced the Church of England as corrupt and tyrannical. In reply, the formidable Anglican spokesman, Richard Hooker, only needed to cite the example of Scotland, where the Puritan churches were no less firmly established. The English Puritans, he charged, were fanatic enthusiasts, enemies to all order and to reason, and not even faithful disciples of Calvin. If established churches could become temporarily corrupt, he asked, how much more corrupt would matters be if they were left to the private judgment of everyone who claimed to be directly inspired?[12]

Although the critics of the modern scientific establishment form no united party or sect, they nevertheless do exist, and again the accusation of corruption has arisen. One type of criticism concerns the role of the scientist as an adviser to government. Scientists are accused of suborning their noble pursuit of truth to the inhumane rattle of the war machine, presided over by the "High Brass" in the Pentagon.[13] It is said that their participation in government permits them only to provide technical service while compelling them to acquiesce in policy decisions they can have no part in making. Others contend that scientists can have far too great an influence with statesmen unable to question their technical judgment, and that they may use their proximity to power to intrigue against colleagues with whom they disagree.[14]

Apart from such accusations of political compromise, there are also critics who suggest another kind of corruption—intellectual corruption. The acceptance of government subvention is criticized because it is said to rob the scientist of the freedom to follow his creative instinct wherever it leads him. Instead, by a Machiavellian process mixing carrots and sticks, inducement and coercion, he becomes committed to paths prescribed by others—paths which often lead him to the temptations of power and pelf. Because government spending must be justified in terms of well-defined social goals, a kind of Gresham's Law is said to operate, by means of which applied research drives out basic research, and the development of useful instruments drives out both forms of research. In the process, the solitary and more genuinely creative ways of "little science" allegedly are slighted in favor of "big science,"[15] with its stifling bureaucratic overorganization, its penchant for team research, its disease of "projectitis." Seemingly innocuous university research projects, installed at the behest of government agencies, are said to represent harmful interference both with the academic freedom of faculty members and with the goals of effective and balanced education.

A defender of the scientific establishment would probably answer these critics, much as Hooker answered the Puritans, by arguing that some kind of establishment is better than none. He might well reject most or all of the criticisms as exaggerations of problems the system is well prepared to deal with on the whole, if not in every instance. When scientists give advice they merely contribute one strand to a much larger fabric of decision-making. When they accept funds, it enables them to do research of all kinds, and not merely pragmatic and military work. At the same time, if the defender of the establishment were completely candid, he might well confess to some doubts of a different sort—doubts which assail even those who value government support for science in its present form.

In abstract terms these doubts might be summed up as the difficulty of reconciling the demands of constitutional democracy with the requirements of effective scientific research. Representative government may be said to work best when its decisions embody a broad consensus, when the goal of the decision is clearly defined, and when the method of achieving the goal is appropriate and easily made subject to public scrutiny. In science, however, decisions must be made by those best qualified to make them, results are often uncertain, multiple approaches may be advisable, and processes are so esoteric that no one but a few experts—often the ones at work on the projects—can say whether things are being well managed or not. In more practical terms this difficulty takes a host of forms. Should decisions on such allocations be made by the most qualified scientists, even if they represent a relatively small number of institutions, or should they be made by a more representative body? How much should it cost to learn a scientific truth? Should funds go to institutions where the level of work is relatively high or to those where the level might be improved? How is it possible to compare what may possibly be learned from an investment in research with what could be obtained by the same expenditure in the form of goods and services required by some other socially desirable goal? How, indeed, can anyone make intrascientific comparisons of space exploration with high-energy physics, or of oceanography with weather forecasting? Scientists, moreover, no less than clerics, are subject to human bias and self-interest. When, like churchmen and monastic orders, they disagree over matters of doctrine and compete for control of the benefices, how is any collegial assembly, whether of cardinals or Nobelists, to choose between them fairly?

No amount of academic or congressional investigation will fully answer these criticisms and doubts. Some of them involve questions of value more than matters of fact. Others are inherent as problems in any attempt to govern something as ungovernable as the pursuit of truth. To govern, as Pierre Mendès-France once said, is to choose. To an extraordinary degree, the choices that face modern nations, the United States in particular, involve our response to, and our direction of, science and technology. It is no exaggeration to say that most of the public dilemmas we are compelled to live with are profoundly affected by what is done with and through science and technology. Nor are these forces by any means beyond our control, as the philosophers of technological determinism have tried to persuade us. If we cannot have the same confidence as Marxists claim to have in the happy influence of progress in the "mode of production," nevertheless, we should not be paralyzed by the prophets of doom who depict the advance of knowledge as a Frankenstein monster about to turn on its creator. We cannot assume, it is true, that under modern conditions the freedom of choice may not itself turn out to be fatal. The balance of terror in which we live must constantly inhibit any such overconfidence. But at the very least we can say with Raymond Aron: "We have lost our taste for prophecies; let us not forget the duty of hope."[16]

The experience of the United States, in the years since World War II, does in fact provide some grounds for hope. Throughout this difficult and challenging period we have been trying, with considerable success, to expand our knowledge of nature and to make use of what is learned. It would be naive to say that we have always done so out of pristine idealism. Most of the time our motivation has been a compound of the desire for security, for power, and for prestige. But whatever our immediate motives, the result has been that as a nation we have gradually come to recognize that we can no longer afford the luxury of borrowed genius, the romantic reliance on backyard hits and misses, and the all but open contempt for learning and theorizing that has so often diminished the value of our pragmatic inclinations.

In historical terms, even this is no mean improvement. For the longest period of American history, the support of science was not considered a proper activity of government. The founding fathers rejected the proposal for a national university. Our first federal observatory came into being disguised as a naval storehouse. The Smithsonian Institution was only created when Congress reluctantly agreed to accept the bequest of an Englishman. The atomic bomb came to us with the physicists expelled from Europe. Until very recently, Americans excited by the progress of science could only express embarrassment—as Jefferson, Franklin, and John Quincy Adams did—that this great republic, from which the philosophers of the Enlightenment expected so much, should in fact have done less to foster science than royalty and revolutionary dictatorships across the Atlantic.

Recent experience, however, shows convincingly that with respect to science at least, we are a changed country. Nor is this change, of course, only a matter of our attitude toward science. Ours is a social system far removed from the state of nature in our dependence upon artificial conveniences and upon one another. It is a highly organized society where all of us, whether we are in business or government or in the professions, are to some extent compelled to become organization men. Few would seriously wish to go back to the state of nature, whether it be to Walden Pond or the Everglades. Nor could we go back, even if we wanted to, because the progress of knowledge is irreversible. We can learn, but we cannot unlearn. Our problem is rather how to balance our desire for knowledge and technical innovations with our desire for a good life and a good society. Our problem is to determine how we can enjoy the blessings of growth and enrichment without suffering what Justice Brandeis feared would be "the curse of bigness."

It is for this reason that the involvement of government in the promotion of science demands close attention. For in this effort the United States has sought to achieve just such a balance. On the one hand we have sought to develop an enormous and diversified capacity to produce and to utilize knowledge; and on the other hand we have tried to arrange and manage this far-flung system in a way that respects intellectual initiative and personal responsibility, that promotes cooperation rather than coercion, and that provides as much decentralization of decision-making as is compatible with the demands of our most urgent priorities.

NOTES

1. Don K. Price, "The Scientific Establishment," in Robert Gilpin and Christopher Wright, eds., *Scientists and National Policy-Making,* New York, 1964, p. 20.

2. See particularly his pathbreaking discussion of support for science as "federalism by contract" in *Government and Science,* New York, 1954, Chap. 3.

3. Statement of Dr. Jerome B. Wiesner, *Hearings Before the Select Committee on Government Research,* House of Representatives (88th Cong. 1st Sess., 1963), Washington, D.C. 1964, p. 96.

4. Statement of Leland J. Haworth, *Hearings Before the Subcommittee on Science, Research, and Development of the Committee on Science and Astronautics,* House of Representatives (88th Cong. 1st Sess., 1963), Washington, D.C. 1964, p. 370, Table 3.

5. Testimony of Dr. James R. Killian, Jr., *Hearings Before the Select Committee on Government Research,* Part 2, p. 757. (See Footnote 4 for complete reference.)

6. *Federal Organization for Scientific Activities,* 1962, National Science Foundation (NSF pp. 62-37), Washington, D.C. 1963, p. 3. Four agencies—the Weather Bureau, Coast and Geodetic Survey, Institutes for Environmental Research, and Environmental Data Service —have recently been consolidated into the "Environmental Sciences Services Administration" (ESSA) within the Department of Commerce.

7. Frank E. Manuel, *The New World of Henri Saint-Simon,* Cambridge Mass., 1950, p. 126.

8. For fuller discussions of PSAC see the essays by Harvey Brooks and Robert N. Kreidler in *Scientists and National Policy-Making, op. cit.;* and Carl William Fischer, Jr., "Scientists and Statesmen: A Profile of the Organization and Functions of the President's Science Advisory Committee," [reprinted in Sanford A. Lakoff, *Knowledge and Power* (New York: Free Press, 1966)].

9. Richard E. Neustadt, *Presidential Power: The Politics of Leadership,* New York, 1962, p. 89.

10. Jerome B. Wiesner, *Hearings Before the Select Committee on Government Research, op. cit.,* p. 273.

11. Fischer, in his paper on PSAC, . . . notes that 63 percent of all PSAC members have been from the physical sciences and 77 percent from the physical, mathematical, and engineering sciences; and that 70 percent received their graduate academic training at seven universities.

12. See Richard Hooker, *Laws of Ecclesiastical Polity* (1593), London, 1954 (Everyman's Library Ed.).

13. See Edward Speyer, "The Brave New World for Scientists," *Dissent,* Vol. VIII, No. 2 (Spring 1961), pp. 126-136.

14. See C. P. Snow, *Science and Government,* Cambridge Mass., 1962.

15. See Derek J. de Solla Price, *Little Science, Big Science,* New York, 1963.

16. Raymond Aron, *War and Industrial Society,* London, 1958, p. 60.

WARNER R. SCHILLING

Scientists, Foreign Policy, and Politics

The central problems occasioned by the participation of scientists in the determination of high policy are not nearly so novel as is generally supposed. The scientist has been brought into the councils of government because he possesses specialized skills and information believed relevant to the identification and resolution of particular policy problems. His relationship to the policy process is therefore a familiar one, that of an expert. Just as Sputnik I precipitated the establishment of a Special Assistant to the President for Science and Technology, so the earlier problems of fighting World War II and insuring postwar employment had brought the Joint Chiefs of Staff and the Council of Economic Advisers into the Offices of the President.

The central problems in policy-making posed by the entry of scientists into the policy process are thus formally no different from those

Warner R. Schilling, "Scientists, Foreign Policy, and Politics," in Robert Gilpin and Christopher Wright (eds.), *Scientists and National Policy-Making,* New York: Columbia University Press, 1964; pp. 359-383. This essay is a revised version of an article published in the *American Political Science Review* LVI, No. 2 (June 1962), and is reprinted by permission of the American Political Science Association.

associated with any other experts involved in the determination of national security policy. In particular, four such problems can be noted. (1) Like all experts, scientists will at times disagree, and the nonscientist (be he politician, administrator, or an expert in some other field) will confront the problem of choosing a course of action in the face of conflicting scientific advice. (2) Like all experts, scientists will at times manifest certain predispositions toward the resolution of the policy problems on which their advice is sought, and the nonscientist will confront the problem of identifying the policy predilections peculiar to scientists and being on his guard against them. (3) The nonscientist and scientist will confront one problem in common, and that is how to organize themselves to maximize the contribution that science can make to the government's programs, opportunities, and choices. (4) The scientist will confront a problem common to all experts who participate in the American policy process, and that is how to engage in politics without debasing the coinage of his own expertise.

THE PROBLEM OF CONFLICTING ADVICE

The difficulties the nonscientist confronts in choosing a course of action in the face of conflicting scientific advice seem inherently no more formidable than those a nonexpert would face in deciding what to do in the event of conflicting advice from economists, soldiers, or specialists on Soviet foreign policy. There are at least seven procedures that the nonexpert can follow in such circumstances, singly or in combination, and they appear to have about the same promise, for better or for worse, regardless of the kind of experts involved.[1]

The first step the nonscientist can take is to make certain that it is really conflicting *scientific* advice he is receiving. In the fall of 1949 President Truman asked Secretary Acheson to look into the disputes then current within the Atomic Energy Commission and elsewhere about the consequences of undertaking an intensive effort to make an H-bomb. Upon investigation the Secretary of State concluded that the scientists involved were not really very far apart except on the foreign policy issues that were his and Truman's responsibility to decide.[2]

Procedures two and three are simple: the nonscientist may be guided by quantitative or qualitative features of the division (he can side with the majority, or with that side whose past record is the more confidence-inspiring). Failing these, there is, four, the "principle of least harm," and, five, the "principle of minimal choice." In the former, one chooses that course of action which appears to involve the least cost if the technical premise on which it is based proves to be wrong. Thus in

World War II, given the American belief that the Germans were hard at work on an A-bomb, it seemed more sensible to spend $2 billion on the assumption that the bomb could be made than to do little or nothing on the assumption that it could not. In the case of the "principle of minimal choice," one chooses that course of action which seems to close off the least number of future alternatives. This was the character of President Truman's first decision on the H-bomb. He decided to go ahead in the effort to explore the feasibility of an H-bomb, but nothing was decided about technical steps of a greater political or military consequence (for example, testing a device if one were fabricated, or preparing to produce the materials that would be required for weapons production in the event of a successful test).[3]

In the case of procedure six the nonscientist can make his choice among conflicting scientists on the basis of whichever technical estimate is most in accord with policy on which he was already intent. (In contrast to the first procedure, where the nonscientist endeavors to factor out of the conflict the policy preferences of the scientists, here he is factoring into the conflict his own policy preferences.) In the spring of 1942, the British scientists Henry Tizard and F. A. Lindemann (Lord Cherwell) diverged by a factor of five in their estimates of the destruction that could be accomplished by an all-out campaign to bomb the homes of German civilians and also in their judgments about the consequences that even the lesser amount of destruction would have for the military course of the war. (Lindemann thought that it would be "catastrophic," Tizard that it would be "most damaging" but not "decisive.") The importance of the issue lay in the fact the Naval Staff was pressing Churchill to allocate some of the bombers that the Air Staff planned to use in the campaign to the Naval Staff's own anti-submarine effort. The Air Staff, which had long been persuaded of the efficacy of strategic bombing, found Lindemann's calculations "simple, clear and convincing." The Naval Staff was similarly impressed by Tizard's. The final decisions were Churchill's, and he was greatly influenced by Lindemann's estimate—an influence presumably not unrelated to his own interest in presenting the Russians with a dramatically visible contribution to the war against Germany.[4]

In procedure seven the nonscientist is guided by his own sense for the scientific and technical problems involved. In the 1949 H-bomb debate, some of the politicians involved were little deterred by the fact that the scientists were by no means confident that they could make such a weapon and by the possibility that an all-out but failing effort might entail very high costs for the A-bomb program. These politicians were willing to press ahead in part because of their belief that the scientists were not really aware of their own potential. Similarly, when the German soldiers, scientists, and engineers engaged in the development of the V-2 divided on the question of whether it should be launched from mobile or fixed batteries, Hitler's own technical enthusiasm for large, hardened bunkers led him, unwisely as it turned out, to decide on behalf of the latter.[5]

In concluding this survey of the problem of conflicting advice, it should be noted that one of the more likely outcomes is that the actions of the contending scientists may prove much more influential than the procedures followed by the nonscientist. Divided experts will not always be equal in their physical or personal access to the decision-maker, in the persistence with which they state their case, or in the force and clarity of their arguments. Thus, in the H-bomb debate, there were instances where equally qualified scientists differed greatly in the time and energy they spent circulating their views of the technical (and political) prospects, and such differences were by no means without consequence for the judgments of others.[6]

THE PROBLEM OF BIAS

Discussion of the policy predispositions displayed by scientists must be entered with considerable caution. The major theoretical premise involved is that all experts will evidence certain predilections with regard to policy and policy-making which are the result of the character of their expertise: their skills, knowledge, and experience. Since experts differ in the skills, knowledge, and experience they command (or in the responsibilities with which they are charged), they will differ in the biases they characteristically exhibit. Thus scientists, soldiers, and diplomats jointly concerned with a policy problem are likely to approach the question of how and in what manner it should be resolved with rather dissimilar predispositions.

These points, however, are easier stated than demonstrated. To begin with, it should be clear that insofar as policy is concerned "the scientific mind" is as much a chimera as "the military mind." Scientists, like soldiers and the rest of us, differ greatly in the ideas they have about the political world and the things that will (or ought to) happen in it, and their views on foreign policy matters are far more likely to be reflective of these differences than conditioned by their common professional skills and interests. Moreover, even if differences in expertise or responsibility were the only factors determining the views of policy-makers (and they certainly are not), one would still have to take account of the fact that scientists are as varied in their professional skills and pursuits as soldiers. The perspectives of a theoretical physicist engaged in basic research are no more to be equated with those of an organic chemist engaged in applying extant knowledge to the improvement of an industrial product than is the outlook of a staff officer in Washington drafting a war plan to be considered identical with that of a general in charge of a theater of operations.

In addition to these difficulties, analysis must also contend with the fact that it is directed toward a moving target. The policy perspectives that a physicist may have developed as a result of two decades in a university laboratory are unlikely to endure without change after a few years on a Washington advisory committee. Many American scientists are well along the same route that transformed the policy perspectives of large numbers of the American military profession during the war and immediate postwar years. As a result of new problems and new responsibilities, these soldiers acquired new skills, knowledge, and experience. In consequence, with regard to their approach to foreign policy, some are, for all practical purposes, interchangeable between the Pentagon and the State Department, and one could wish that there were more diplomats equally well equipped to work on both sides of the Potomac.

With these reservations in mind, six policy perspectives will be presented here which seem moderately characteristic of many scientists, most of them physicists, who have participated in national security policy in recent times. Most of these predispositions were first evidenced during their work with the military during World War II, and the extent and manner in which they have been later operative in reference to larger foreign policy issues is not always easy to document, since most of the sources are still classified. Needless to say, in outlining these predispositions, one is presenting a cross between a caricature and a Weberian ideal type, not describing real people. In discussing these predispositions, this writer does not mean to convey the impression that they are either "good" or "bad" from the point of view of policy or policy-making, or that one or another of these predispositions may not also be evidenced by groups other than scientists. The point to this discussion is that if certain orders of scientists are indeed prone to these or other policy predispositions, the nonscientist will be wise to be alert to them, even if on occasion he should conclude that they are all for the good.

NÄIVE UTOPIANISM OR NÄIVE BELLIGERENCY

C. P. Snow has described the scientist as an impatient optimist in his approach to social wrongs; he is quick to search for something to do and inclined to expect favorable results.[7] Certainly, the scientist's profession inclines him to look at problems in terms of searching for a solution to them. When this perspective is turned to problems of international politics, however, the scientist's approach often appears open to the characterization of "näive utopianism or näive belligerency."[8] His approach to international relations appears simplistic and mechanistic. It is almost as if he conceives of policy being made primarily by forward-looking, solution-oriented, rational-thinking types like himself.

In these perspectives the scientist is likely to find little in common with the diplomat (who is inclined to believe that most of his problems have no solution, and who is in any event too busy with the crises of the day to plan for tomorrow), or with the politician (whose approach to problems is so spasmodic as to seem neither analytical nor rational, and whose policy positions are anyway soon blurred by his efforts to accommodate to the positions of others), or with the professional student of international politics (who, when the opportunity permits, lectures the scientist on the elegant complexity of the political process, but who never seems, to the scientist at least, to have any really good ideas about what to do). It is perhaps these differences in perspective that lead the scientist on occasion to seem "intellectually arrogant"; it is as if he concludes that those who have no promising solutions or are not seeking them cannot be very bright. In his predisposition toward action and solutions, the scientist comes closest to sharing the predilection of the soldier for decision, which may be one reason why their partnership has been so spectacularly successful.

The "Whole-Problem Approach"

The first grant made by the United States government for experimental research was in 1832 to the Franklin Institute. The scientists were asked to investigate the reasons for explosions in steamboat boilers. They reported back not only with a technical explanation but with a draft bill to provide for federal regulation of steamboats.[9] In this they evidenced the scientist's predilection for the "whole-problem approach." The reluctance of scientists to apply their expertise to mere fragments of the total problem, especially under conditions where those who prescribe the fragments do not reveal the whole of which they are a part, was evident in the work of both British and American scientists during World War II. Military officials initially approached the scientists with requests for the development of particular weapons and devices without revealing the military problems or reasoning responsible for their requests. The scientists objected to this procedure, and they were eventually able to persuade the soldiers to inform them of the general military problems involved in order that the scientists might reach their own conclusions about the kinds of weapons and devices the military would need to meet those problems.[10]

The predisposition to want to be told and to deal with the whole problem no doubt has its base in the professional experience of scientists (and one of the central credos of science) that good ideas on a problem may come from the most unexpected quarters and that the widest possible dissemination of information about a problem will significantly enhance its chances for an early solution.[11] Still, there are problems and problems; some are open to determinate solutions, and others can be resolved only through the exercise of political power. The point about the "whole-problem approach" . . . is that it not only helps propel the scientists from an advisory to a political role but it serves to make the scientist somewhat blind to the fact that he is so moving. In its most extreme form, the "whole-problem approach" coupled with the "intellectual arrogance" perspective can lead to instances like the following: on one high-level advisory committee concerned with several areas of national security policy, a scientist whose formal claim to participation was a knowledge of infrared-ray phenomena was reportedly quite free with his proposals for what political policies should be adopted with regard to the United Nations.

QUANTUM JUMPS VERSUS IMPROVEMENTS

A number of scientists have advanced the proposition that the military tend to be more interested in improving existing weapons than in developing radically new ones, and they have urged that a separate civilian agency be established to undertake such development. Both scientists and soldiers have explained this difference in their approach to military research and development, "quantum jumps versus improvements," with the hypothesis that the soldier's interest in developing entirely new weapons must always be inhibited by his concern for the possibility that war may come in the near future, since in this event his interests are best served by improving existing weapons. It has also been suggested that military leaders, who must be prepared at any time to ask others to take up the weapons at hand and fight with them, cannot afford to let themselves or others become too impressed with the deficiencies of those weapons as compared with others that might have been developed.[12]

An explanation for this difference, less flattering to the military, is the occasional assertion by scientists that theirs is a profession which stimulates original and creative thought, while that of the military tends to develop minds which accept the existing situation without too much question. As indicated in the discussion of the first predilection, this is a judgment which the scientist may extend to the diplomat and the politician as well.

The difficulty with quantum jumps in foreign policy, however, is that the structure of both the domestic and the international political process is normally such as to make them infeasible. Thus, diplomats and politicians are accustomed to seeing the same old policy problems come around year after year, and they are generally intent on policies which promise only slow and modest change. Scientists, on the other hand, have been demanding and searching for quantum jumps in foreign policy ever since the end of World War II. It is symptomatic that the first proposal developed by the Advisory Committee on Science and Technology to the Democratic National Advisory Council, established in 1959, was for the creation of a new scientific agency, independent of the State and Defense Departments whose function would be "to face all the problems of disarmament."[13]

TECHNOLOGY FOR ITS OWN SWEET SAKE

In the summer of 1945, after the A-bomb had been tested but before the first drop on Japan, the Director of the Los Alamos Laboratory, J. Robert Oppenheimer, suggested to his superior, General Leslie Groves, that if some improvements were made in the design of the bomb it would be more effective. Groves decided against the improvements because he did not want to incur any delay in the use of the bomb, which he expected would end the war with Japan. In the summer of 1943, after the Director of the German V-2 project, General Dornberger, had finally secured a first-class priority for the use of the weapon, those responsible for producing it in quantity were increasingly handicapped by the scientists and engineers who kept improving but changing its design. Dornberger was finally obliged to issue a flat order against any further improvements.[14]

There was nothing irresponsible in these scientists' actions. Charged with the technical development of weapons, they would have been remiss in their responsibilities if they had failed to call attention to the prospects for improvement. The point to the examples is that scientists and engineers, in the pursuit of their own responsibilities and interests, may easily lose sight of those of the policy-maker.

This predisposition, "technology for its own sweet sake," appears to have its roots in two more of science's central credos: the belief in the value of pursuing knowledge for its own sake and the belief that the best motivation for the direction of research is the strength and character of individual curiosities. But the direction and strength of scientific interests and curiosities is not necessarily coincident with the requirements of military or foreign policy. One of the most recent examples of the scientist's capacity to get caught up in a challenging problem (assigned, to be sure, by policy-makers) is afforded by the ingenious ideas scientists have conceived for evading nuclear-test detection systems and for the design of new systems to detect those evasions. In the light of the later course of negotiations, an American statesman who believed there was considerable foreign policy gain in a test-ban treaty and who believed that the Russians were at one time seriously interested in such a treaty might well conclude that the formula developed by the British scientist Watson-Watt for meeting wartime military requirements— "Give them the third best to go with; the second comes too late, the best never comes"—was not without its implications for meeting peacetime foreign policy requirements.[15] This observation is not intended as an argument that the interests of the United States would have been better served by a test-ban treaty with a "third best" detection system than by no treaty at all. The point is rather that the policy-maker must be sensitive to the prospect that because of the constant advance of technology his only real choices may be of this order.

THE SENSE FOR PARADISE LOST

This predisposition is likely to be more characteristic of the scientists who had their graduate training and early professional experience in the years before World War II than of those who have known only war or Cold War conditions.[16] The prewar scientists took it as an article of faith that certain conditions were essential for the progress of science, in particular that scientists be free to select their research problems and that both scientists and scientific information be free to move among as well as within nations.[17] All of these conditions were violated during World War II, and as a result of the Cold War they were never fully reestablished. The nuclear physicists had had perhaps the most highly developed sense of international community. They were relatively few in number, had close personal relationships at home and abroad, and had been experiencing an exciting exchange of discoveries since Rutherford identified the nucleus in 1911. They also lost the most, for theirs was militarily the most sensitive knowledge, and the pages of the *Bulletin of the Atomic Scientists* offer eloquent testimony to their ideological disturbance.

The result is that the senior scientists tend to be especially sensitive to possibilities which hold some promise for restoring the former order. They may usually be found on the side (or in front) of those urging freer exchange of scientific and military information with allied governments, less secrecy in the circulation of scientific (and sometimes military) information, and more extensive cultural, and especially scientific, exchanges with the Soviet Union. Similarly, the major activities of the Foreign Policy Panel of the President's Science Advisory Committee (PSAC) and of the Office of the Science Adviser to the Secretary of State have been in connection with the Science Attaché program, the facilitation of international scientific programs and conferences, and the exchange of scientists with the Soviet Union.[18]

SCIENCE SERVES MANKIND

For at least 300 years the Western scientific tradition has assumed that the unrestricted generation of new knowledge about the world was a social good. Over these years science in its purest form (the discovery of the facts of nature for knowledge's sake alone) became increasingly an autonomous social institution; research scientists were largely disassociated from the practical applications of their discoveries, but they took it for granted that these discoveries would ultimately benefit mankind.[19] The advent of nuclear and bacteriological weapons systems which have the potential of destroying so much of mankind and his works has called this faith sharply into question. It does not take much imagination to wonder if man, in view of his apparent inability to escape from the order of conflicts which have historically resulted in war, would not be better off in a world where the knowledge that has made the new weapons possible did not exist. For some of the senior nuclear physicists this is more than a philosophical question. They are unable to avoid a sense of personal responsibility; they reason from the premise that they were few, and if they had acted differently weapons development might not have taken the turn it did.

In the immediate postwar years, the apparent contradiction between the good of science and the evil of war was resolved by the expectation that the very destructiveness of the new weapons would lead man to renounce at last the folly of war. The course of foreign policy in later years has weakened these expectations but not destroyed them, as the recent flurry of arms-control proposals premised on the rational self-interest of both sides in avoiding mutual destruction testifies.

The need to preserve their sense of service to mankind led some American scientists to refuse to work on weapons. Similarly, there are reports that several Russian scientists were imprisoned, exiled, or placed under surveillance for refusing to participate in weapons work between 1945 and 1953, and a number of Germany's elite physicists announced in 1957 that they would have no part in nuclear weapons work.[20] Such cases are dramatic, but nowhere have they prevented the development of weapons on which governments were determined. The more consequential resolutions have been those in which scientists have simply identified the good of mankind with the strength of their nation or have endeavored to develop new weapons systems which would be as effective as the old in promoting national policy but which would result in less slaughter if used. This was part of the rationale behind the recommendation made by a group of American scientists in 1951 that the government undertake the development and production of a large number of A-bombs for tactical use in the ground defense of Western Europe. Their hope was that such an innovation would relieve the United States of the burden of having to rely solely on the threat of strategic bombing to contain the Red Army.[21]

The failure of the United States to orbit a satellite before the Soviet Union did was the result of the State Department's insensitivity to the political implications of the event and the decision of the President and the Secretary of Defense not to let a satellite program interfere with military missile programs. A small part of the story, however, is to be found in the reluctance of some of the American scientists involved in the programming of the International Geophysical Year to see an American IGY satellite propelled by an operational military weapon. Their preference for the less-developed but non-military Vanguard over the Army's Redstone appears to have reflected a combination of the "sense for paradise lost" and the "science serves mankind" predispositions, in this case an interest in showing the world the peaceful side of science and in demonstrating that the scientists of the world could cooperate in the interests of knowledge as well as compete in the interests of nations.[22]

THE PROBLEMS OF ORGANIZATION AND POLITICS

With regard to the two remaining problems to be discussed—how to organize relations between science and government and how the scientist can participate in policy-making and still keep his expert standing —four points seem deserving of special emphasis: (1) the problem of organization, especially in the area of foreign policy, is still very much in the research and development stage, and so it may long remain, considering the precedent set by the problem of how to relate military experts and foreign policy; (2) in many areas of policy it will never be possible to specify what constitutes "the best" organization; the way in which policy-makers are organized is not without influence on the kind of policies they will produce, and so long as there are differences over policy there will be no agreement about organization; (3) in the American political system, at least, the science expert at the high-policy level has no real hope of keeping out of politics, his only choice is in the character of his political style; and, finally, (4) it should not be forgotten that organization and policy-making are not the same as policy; successful instances of foreign policy capitalizing on or guiding developments in science and technology will not automatically follow just because scientists have been liberally injected into the policy-making process.

ORGANIZATION

Current American organization in the area of science and foreign policy still reflects the emergency responses to the Russian ICBM and Sputnik I. One effect of these events was that scientists were rushed to the most important single center of power, the Office of the President, by means of the creation of the Special Assistant to the President for Science and Technology and the President's Science Advisory Committee.

The President certainly needs men around him who are sensitive to the areas of interaction between science and foreign policy. But a case can be made for the proposition that the center of gravity for the input of scientific advice into the policy-making process should be at a lower level than the White House. The President's political interests lie in keeping the staff about him small and generalized. Well-developed plans and programs will have a better chance of maturing in the larger and more diversified facilities that departments and agencies can provide. Secondly, as C. P. Snow concludes in his account of the differences between Tizard and Lindemann, there are risks in having a single science adviser sitting next to the center of political power. Although it should be noted that Churchill fared better with a single science adviser than Hitler did with none ("The Führer has dreamed," Dornberger was told, "that no [V-2] will ever reach England"), Snow's point has merit and it holds for institutions as well as for individuals.[23] The President will generally find his choices facilitated by the existence of multiple and independent sources of scientific advice.

This is a condition that already prevails in the case of many of the departments and agencies whose actions have significant foreign policy consequences, especially in the use of scientists by the Department of Defense, the Atomic Energy Commission, and the National Aeronautics and Space Administration. It is, however, a condition notably absent in the case of the Department of State. As it now stands, the President has more scientists to advise him on the scientific and technical aspects of various foreign policy issues, particularly in the national security field, than has the Secretary of State.[24]

Established in February, 1951, the Office of the Science Adviser in the Department of State has yet to become a point of vantage in the determination of high departmental policy. Deprived in its original charter of any jurisdiction in the atomic energy field, the Office was moribund within four years of its birth. The number of overseas science attachés administered by the Office went from a peak of eleven in 1952 to zero at the end of 1955 and the Office itself languished without a scientist from February, 1954, to January, 1958. Resurrected in the aftermath of Sputnik I, the Office—as of February, 1962—consisted of some fourteen science attachés overseas and a Washington staff of six, of whom only three, including the Director, were professional scientists. Nor were these three scientists supplemented by technical personnel elsewhere in the Department. There were no scientists, full or part-time, in the Department's offices for space and atomic energy, political-military affairs, or policy planning. Of the Department's line offices, only the Bureau of Intelligence and Research maintained a technical staff.

As might be inferred from these arrangements, most of the policy-makers concerned believed that their needs for scientific advice were adequately met through formal and informal communication with scientists employed in the operating agencies and departments and with the President's own Science Advisory Committee. The Department's Science Adviser, as a participant in the activities of both the President's Committee and the Federal Council on Science and Technology, stood available to facilitate such communication. Otherwise, both the demands placed upon the Office and its own interests served to limit its activity, as previously noted, to a relatively narrow range of foreign policy problems.

In the summer of 1962, spurred by a recommendation from the President's Scientific Advisory Committee, the Department reorganized the Office into the Office of International Scientific Affairs with a provision for expanded personnel (perhaps a total of eighteen, including nine scientists) and a charter designed to lead it into more active participation in the policy or line offices. Moreover, as a result of the Department's decision in the spring to disband the office for space and atomic energy and to divide its responsibilities among a variety of bureaus and offices, the new Office now shares with several other bureaus a mandate over the "peaceful" uses of space and atomic energy. Although this writing is too close to these events to assay the results of the reorganization, the Office would appear to have little prospect of becoming for the Secretary of State the functional equivalent of the President's scientific advisory apparatus, at least in the areas where science and technology impinge on national security policy. Responsibility for the military applications of space and atomic energy were distributed to the European Bureau and the Office for Political-Military Affairs, and here the use of scientists and the Department's science adviser remains as before.[25]

Whether the Department of State would be better served by an "in-house" scientific competence in these fields is a question that an outside observer cannot easily answer. Much depends on the validity of the expectations that the Department can rely on the scientists of the operating agencies and the President's Committee to alert it to developments and information relevant to foreign policy. Even more depends on how determined the Department is to play an active and influential part in shaping the scientific and technical activities of the government to conform to *its* conception of national needs and priorities. (The two conditions are, of course, not unrelated. The more influence the Department exercises in determining the goals and programs of other agencies, the more confident it can be that scientists in those agencies will call the Department's attention to goals and programs which they believe to be receiving too much or too little attention.) In the final analysis, the question of the Department's organizational needs can only be answered in terms of the strength and content of its policy interests. In the field of national security policy, the Department has yet to define its responsibilities and identify its interests in such a manner as to point to the need to expand its modest political staff, much less to create a science advisory body to help this staff monitor and direct the course of science and technology as they affect such policy.

ORGANIZATION AND PURPOSE

Since administrative organizations exist for the purpose of serving policy goals and implementing policy programs, it is to be expected that those who differ on the goals and programs of policy will differ about the proper design of administrative organizations. The desire of many scientists in 1945 to see atomic energy used for peaceful rather than military purposes was one of the reasons for their political campaign to place the post-war atomic energy program in the hands of a civilian commission instead of the War Department. Similarly, more recent differences about how to organize the government's space effort reflect, in part, policy differences about whether space will or should be an area for major military operations.

The same point can be seen in the proposal to create a Department of Science and Technology which would include the variety of "little" science programs now scattered throughout the Executive structure (for example, those of the Weather Bureau, National Bureau of Standards, and the Antarctic Office), but would exclude those of the Department of Defense, the Atomic Energy Commission, and the Space Administration. The hope behind this proposal is that, combined together, the "little" programs would be able to compete more effectively in the struggle for government dollars with the "big" science programs of the military, atomic energy, and space organizations.[26]

The question of the "best" science organization is thus inescapably tied to the question of what is the "best" science policy. But who can demonstrate whether science and foreign policy would be better served by allocating dollars to a program to control weather or to a program to explore Mars? There are no determinate solutions to problems of this order. Neither, for that matter, is there any one "right" amount of the nation's scientific resources that should be allocated to basic as compared to applied research. Differences on policy questions such as these are unavoidable among scientists and nonscientists alike, and they can be resolved in but one manner; through the interplay of power and interest in a political arena.

This condition, plus the increasing dependence of scientific programs and research on government funds, plus the increasing consequences of the choices the government makes in allocating those funds, all promise to put the politicians and the scientists under increasing pressure. As the opportunities for further development in each of a thousand different scientific fields mushroom with the acceleration of scientific knowledge, whatever the government decides to support, it will be deciding *not* to support more. Indeed, it is not too difficult to see the scientists becoming practiced advocates and lobbyists for the government's support of their cherished fields and projects, or to imagine the day when politicians start to complain about "interscience rivalry" and begin to fancy that, if only there were a single Chief of Science, competition and duplication could be ended and the nation could have an integrated science policy.

SCIENTISTS IN POLITICS

The American political system is not one that insulates its experts from the politics of choice.[27] The scientist involved in high-policy matters is likely to find himself propelled into the political arena, either by a push from behind or by his own interest in seeing that the "right" choices are made. Some of the incentives the scientist may have to follow up his advice with an effort to see that it is accepted (and to take a hand in a few other matters while he is at it) were outlined and illustrated in the preceding section. It is equally important to recognize that the scientist may find himself on the political firing line, placed there by a politician interested in using the scientist's prestige as an "expert" to disarm the critics of his (the politician's) choices.

Thus, prior to the moratorium on nuclear tests, the Eisenhower administration appeared to be using scientists and their scientific facts on fall-out as a means of justifying and defending a policy that was obviously compounded of a variety of considerations besides that of the radiological hazard. The comparison with Truman's use of the prestige of the Joint Chiefs of Staff to defend his choices in the Korean War comes easily to mind. So, too, do the statements of various Republican leaders that they had lost confidence in the Joint Chiefs and their determination, when they came to power, to get rid of the "Democratic" Chiefs and to appoint Chiefs in sympathy with Republican policies.

The scientist, in short, is not likely to orbit the centers of political power emitting upon request "beeps" of purely technical information. He will inevitably be pulled into the political arena. If his participation there is to be either productive or personally satisfying, both the scientist and the nonscientist need to be highly conscious of the character of their activity and the problems involved. The scientist (and many a nonscientist) must learn that the making of foreign policy is not a quest for the "right" answers to the problems of our time. There are only hard choices, the consequences of which will be uncertain and the making of which will often seem interminable in time and irrational in procedure.

The debate and disagreement over these choices will be heated and confused under the best of circumstances, but emotion and misunderstanding can be eased if scientists and nonscientists are both alert to the limits as well as the potential of the scientist's contribution. On the scientist's part, there is the obvious need to exercise the utmost care in making clear to himself and to others the areas where he speaks as a concerned citizen and those where he speaks as a professional expert. More difficult will be the task of learning how and to whom to address himself in each of these capacities when he is dissatisfied with the outcome of a policy decision in which he has participated. There is, as Don K. Price has pointed out, no clear code in Washington to govern the conduct of dissenting experts, only a "flexible" set of possible relationships with one's immediate superiors and those whose authority competes with or exceeds that of one's superiors. In contrast to the soldier, who can find some although not complete guidance in the doctrine of "civilian control," the very nature of the scientist's intellectual habits and many of his policy predispositions may make especially difficult his task in determining the limits to which he can stretch his dissent.[28]

On their part, the nonscientists need to recognize that scientists can hardly be expected to remain politically indifferent or inactive about the policy issues with which they are involved (especially when no one else in Washington practices such restraint). It was the naïveté of this expectation that was so appalling in the conclusion of the Gray Board that Oppenheimer was a security risk because (among other reasons) "he may have departed his role as scientific adviser to exercise highly persuasive influence in matters in which his convictions were not necessarily a reflection of technical judgment, and also not necessarily related to the protection of the strongest offensive military interests of the country."[29]

It is unlikely that civil-scientist relations will ever get any worse than this. With time and experience one can expect many of these problems to be eased, but it would be unrealistic to expect them to disappear. Military experts have participated in the making of foreign policy far longer than scientists, and the question of how they can best do so is still the subject of more than a little disagreement.

POLICY PROCESSES AND POLICY

In closing this discussion of scientists and the problems of their organizational and political relationships to others engaged in the determination of foreign policy, it is important to remember that the policy process can bring minds together but it cannot make them think. It is worth noting that in the political and administrative structure of the Soviet Union no scientist is as institutionally close to the Premier as is the Special Assistant for Science and Technology to the President of the United States and that there is no equivalent of the Science Advisory Office in the Russian Ministry of Foreign Affairs.[30] Yet one would not say that the foreign policy of the Soviet Union has appeared either ineffectual or insensitive in its response to developments in science and technology.

The circumstances attendant on the development of radar by the British from 1935 to 1940 provide a useful insight into both the potential and the limits of effective organization. Essential, obviously, were the scientific and technical ideas that Watson-Watt and his colleagues had in mind in 1935, ideas which in turn were the result of the earlier years of research they had been free to conduct in the facilities of a government laboratory. Certainly, it was important that there were administrative scientists in the Air Ministry who were so alert to the military problems of the Air Force that they could see on their own initiative the need to establish a special scientific committee for the study of air defense (the Tizard Committee) and who were so alert to the work of the scientific community that they made their first request for information to Watson-Watt.[31] Of consequence, too, was the fact that the personal and political relations of the members of the Tizard

Committee with the members of the military, administrative, and political hierarchies whose interest and cooperation were vital for the subsequent progress of the research and development program were relations characterized by mutual ease, respect, and understanding.

But these conditions would not have led from the formation of the Tizard Committee in 1935 to a chain of operational radar stations by 1940 and a Fighter Command practiced in their use if it had not been for the military ideas of members of the Royal Air Force. It was they who first thought of the formation of a committee to look specifically into the problem of detection, they who recommended more funds than those first proposed by the Tizard Committee for the development of an electromagnetic detention system, and they who were responsible for the decision to start constructing the stations and training the personnel while the equipment was still under development.[32] The explanation for this interest and support is to be found in their theories about the next World War. They believed the Germans were planning to engage in the strategic bombing of Great Britain, and they wished to be prepared for it.[33]

The point is obvious but important. British scientists and science organization were in the final measure only ready tools. They were good tools, but the use to which they were put was the result of the kind of ideas the military men had about war. The same will hold in the other areas in which science may affect foreign policy. The contributions that science and technology will bring to international politics will largely turn, not so much on the particular arrangements of scientists in the policy-making process, but on the purposes of statesmen and the theories they have about the political world in which they live.

NOTES

1. *Cf.* the implication in the following remarks of Glenn T. Seaborg, the Chairman of the Atomic Energy Commission: "Scientists don't necessarily have to make the final political decisions, but it might be easier to let a capable scientist learn political reality than to teach a politician science." Quoted in the *Bulletin of the Atomic Scientists*, XVII, No. 2 (February, 1961), 79.

2. In this and subsequent undocumented references this writer has drawn upon personal interviews during 1956-58 with participants in the H-bomb decision.

3. For the "principle of least harm," see Bernard Brodie, "Strategy as a Science," *World Politics*, I, No. 4 (July, 1949), 479n. On the H-bomb choice, see Warner R. Schilling, "The H-Bomb Decision: How to Decide Without Actually Choosing," *Political Science Quarterly*, LXXVI, No. 1 (March, 1961), 37-38.

4. See Sir Charles Webster and Noble Franklin, *The Strategic Air Offensive Against Germany, 1939–1945* (London, H. M. Stationery Office, 1961), Vol. 1, chap. 6, especially 331-36, 340-43, 371, and Winston S. Churchill, *The Second World War: The Hinge of Fate*

(Boston, Houghton Mifflin, 1950), pp. 121, 328, 333-34. For more spirited accounts, see C. P. Snow, *Science and Government* (Cambridge, Mass., Harvard University Press, 1961), pp. 47-51, the review of this book by P. M. S. Blackett in *Scientific American,* CCIV, No. 4 (April, 1961), 192-94; the Earl of Birkenhead, *The Professor and the Prime Minister* (Boston, Houghton Mifflin, 1962), pp. 257-67, and C. P. Snow, *Appendix to Science and Government* (Cambridge, Mass., Harvard University Press, 1962), pp. 23-30.

5. Maj. Gen. Walter Dornberger, *V-2* (New York, Ballantine Books, 1954), pp. 97, 158-160, and Lt. Gen. James M. Gavin, *War and Peace in the Space Age* (New York, Harper and Bros., 1958), pp. 76-77.

6. Note should also be taken of the problem the policy-maker faces when all his experts *agree.* This writer is unable to suggest a useful procedure here (other than variations on procedures five, six, and seven above); but that the problem is a real one can be seen in the conclusion of the German physicists that it would be infeasible for any Power to develop an atomic bomb during World War II. Some of the German scientists later stated that political considerations were partly responsible for their advice and for the fact that they made so little progress themselves on an A-bomb (*cf.* procedure one).

The German work on the A-bomb during World War II is described in Samuel A. Goudsmit, *Alsos* (New York, Henry Schuman, 1947). For various appraisals of the influence exercised by political considerations, see Robert Jungk, *Brighter Than a Thousand Suns* (New York, Harcourt, Brace and Co., 1958), pp. 88-104, Hans A. Bethe, "Review of *Brighter Than a Thousand Suns,*" *Bulletin of the Atomic Scientists,* XIV, No. 10 (December, 1958), 427, and William L. Laurence, *Men and Atoms* (New York, Simon and Schuster, 1959), pp. 90-93.

7. C. P. Snow, *The Two Cultures and the Scientific Revolution* (New York, Cambridge University Press, 1959), pp. 9-11.

8. The author is indebted to Hans Speier for the phrasing of this point.

9. Don K. Price, *Government and Science* (New York, New York University Press, 1954), pp. 10-11.

10. This persuasion was largely accomplished through demonstrations of the military utility of the scientists' taking such an approach, although in the early history of the M.I.T. Radiation Laboratory a certain amount of polite bargaining was apparently practiced. One scientist involved, whenever told that the reason for a request was a problem for Washington, not him, to worry about, adopted the practice of working on something else until he was given a description of the problem involved. For a brief summary of the British experience, see Alexander Haddow, "The Scientist as Citizen," *Bulletin of the Atomic Scientists,* XII, No. 7 (September, 1956), 247.

11. General Leslie Groves, who directed the Manhattan Project, was especially sensitive to the scientists' tendency to take on the whole problem. (Some even advised him on how the garbage should be collected at Los Alamos, an act which may possibly have reflected self- rather than scientific interest.) One reason for his effort to compartmentalize the work scientists were doing was his fear that "if I brought them into the whole project, they would never do their own job. There was just too much of scientific interest, and they would just be frittering from one thing to another." *Oppenheimer Transcript,* p. 164.

12. See, for example, Lloyd V. Berkner, "Science and National Strength," *Bulletin of the Atomic Scientists,* IX, No. 5 (June, 1953), 155, 180.

13. See the *Bulletin of the Atomic Scientists,* XV, No. 10 (December, 1959), 412.

14. *Oppenheimer Transcript,* p. 33, and Dornberger, *V-2,* pp. 134-37.

15. Sir Robert Watson-Watt, *Three Steps to Victory* (London, Odhams, 1957), p. 74.

16. In 1955 slightly more than half of the active research physicists in the United States were under forty years of age and had received their doctorates after December 7, 1941. DuBridge, "The American Scientist: 1955," p. 1.

17. These assumptions are excellently set forth in Margaret Smith Stahl, "Splits and Schisms: Nuclear and Social," unpublished doctoral dissertation, University of Wisconsin, 1946, chap. 4.

18. For the activities of the Panel and the Office, see James R. Killian, Jr., "Science and Public Policy," address to the American Association for the Advancement of Science, December 29, 1958, as printed in U.S. Congress, Report of the Senate Committee on Government Operations, *Science Program—86th Congress,* 86th Congress, 1st Session (Washington, D.C., USGPO, 1959), pp. 12-13, and *The Science Adviser of the Department of State* (Washington, D.C., USGPO, 1960), State Department Publ. No. 7056.

19. See Stahl, "Splits and Schisms," chap. 4.

20. See Arnold Kramish, *Atomic Energy in the Soviet Union* (Stanford, Calif., Stanford University Press, 1959), p. 105. Kramish states that it is not certain whether the objections of the Russian scientists were technical or political. For the declaration of the German physicists, see the *Bulletin of the Atomic Scientists,* XIII, No. 6 (June, 1957), 228.

21. *Oppenheimer Transcript,* pp. 584, 594-95, 891-94.

22. See Walter Sullivan, *Assault on the Unknown* (New York, McGraw-Hill, 1961).

23. Snow, *Science and Government,* pp. 66-68, and Dornberger, *V-2,* p. 87.

24. There are eighteen scientists on the PSAC; its working panels also contain participants from outside the Committee. In December, 1958, the Committee and the Office of the Special Assistant for Science and Technology had together some 75 scientists and engineers serving part-time. See Killian, "Science and Public Policy," p. 8. The work of the Committee and the Office are additionally described and appraised in U.S. Congress, Staff Study of the Subcommittee on National Policy Machinery Senate Committee on Government Operations, *Science Organization and the President's Office,* 87th Congress, 1st Session (Washington, D.C., USGPO, 1961).

25. The projected size of the new Office is not out of line with that of others in the State Department. The Office for Political-Military Affairs, for example, numbers some fourteen people, of whom one has the formal responsibility for monitoring the military applications of space and atomic energy.

For information on the Office of the Science Adviser, see State Department, *The Science Adviser of the Department of State,* the New York *Times,* July 2, 1962, and September 15, 1962, and Graham DuShane, "Full Circle," *Science,* 129, No. 3291 (January 24, 1958), 175. Additional information was secured from interviews with Department officials in February of 1962 and 1963. The description and interpretation made above are, of course, entirely this writer's responsibility.

26. See Lloyd V. Berkner, "National Science Policy and the Future," address at Johns Hopkins University, December 16, 1958, as printed in *Science Program—86th Congress,* pp. 116-18.

27. This point, especially as it relates to science experts, is discussed in Price, *Government and Science,* pp. 61-62, and in Herman Finer, "Government and the Expert," *Bulletin of the Atomic Scientists,* XII, No. 9 (November, 1956), 331-32.

28. See the discussion in Price, *Government and Science,* pp. 131, 133, 138-42. The point about the scientists' lacking a tradition of civilian control was suggested by William T. R. Fox.

29. AEC, *In the Matter of J. Robert Oppenheimer, Texts of Principal Documents and Letters* (Washington, D.C., USGPO, 1954), pp. 19-20. Note the policy predisposition in the phrase "strongest offensive military interests."

It should not be comfortable for an American to reflect on the career of Peter Kapitsa, a Soviet physicist who was a student of Rutherford and who worked in England from 1922 to 1934 and then returned to the Soviet Union. Kapitsa was placed under house arrest in 1947 and remained there until after Stalin's death. Kapitsa has told Western scientists and newsmen that his arrest was the result of his refusal to work on nuclear energy for military purposes. Kramish believes that his arrest was due to the government's dissatisfaction with his advice on certain technical approaches to weapons development. In either event, it is noteworthy that Kapitsa is believed to have since become, on an informal basis, one of Khrushchev's main science advisers. On the matter of his arrest, see the report by Harrison Salisbury in the New York *Times,* July 11, 1956, the *Bulletin of the Atomic*

Scientists, XIII, No. 1 (January, 1957), 38, and Kramish, *Atomic Energy in the Soviet Union,* pp. 109-110. The information on his recent activity was supplied by the staff of the Subcommittee on National Policy Machinery, Senate Committee on Government Operations.

30. On Soviet government and science organization, see U.S. Congress, Report of the Subcommittee on National Policy Machinery, Senate Committee on Government Operations, *National Policy Machinery in the Soviet Union,* 86th Congress, 2d Session (Washington, D.C., USGPO, 1949), pp. 24-35, 59-62, and Nicholas DeWitt, "Reorganization of Science and Research in the U.S.S.R." *Science,* CXXXIII, No. 3469 (June 23, 1961), 1981-91. The points made above were additionally confirmed by the staff of the Subcommittee on National Policy Machinery.

31. The circumstances provide an interesting variation of the "whole-problem approach." The Tizard Committee was initially interested in techniques for destroying aircraft or their crews, and Watson-Watt was asked in 1935 to investigate the possibility of using electromagnetic radiation for this purpose. He reported that such a use was apparently infeasible. In any event, he went on to note, the aircraft would first have to be located, and, if anyone was interested, electromagnetic radiation might be useful for this. Watson-Watt, *Three Steps to Victory,* pp. 81-83.

32. For the development of radar, see *ibid.,* pp. 108-9, Snow, *Science and Government,* pp. 24-38, 60-61, 74-75, P. M. S. Blackett, "Tizard and the Science of War," *Nature,* CXXXCV, No. 4714 (March 5, 1960), 648-49, and Basil Collier, *The Defense of the United Kingdom* (London, H. M. Stationery Office, 1957), pp. 33, 36-39.

33. Ironically, the British were mistaken in their theory. The German Air Force had no such strategy in mind, and in 1940, when it tried to improvise a strategic bombing campaign, it had neither the equipment nor the doctrine with which to conduct the campaign effectively. See Herbert Dinerstein, "The Impact of Air Power on the International Scene: 1933–1940," *Military Affairs,* XIX, No. 2 (Summer, 1955), 65-7;. Telford Taylor, *The March of Conquest* (New York, Simon and Schuster, 1958), pp. 24-30, and Adolf Galland, *The First and the Last* (New York, Ballantine Books, 1954), chaps. 2-5.

ROBERT GILPIN

The Atlantic Imbalance in Science and Technology

Whatever one chooses to label it—the "science gap," the "technology gap," the "educational gap," or as Americans tend to prefer, the "managerial gap"—the discrepancy between American and European capabilities in science and technology has become a major preoccupation of the French and other Europeans. They are particularly distressed over the U.S. superiority in those "strategic" technologies such as computers, aerospace, and atomic energy that are being spun off by scientific advance and are basic to industrial-military power in the contemporary world. These technologies not only constitute the growth sectors of the modern industrial economy and the foremost expressions of military power, but are altering the very dimensions of social life. "The computer," Christopher Layton has written, "is comparable in its significance for the age of automation to steam-power in the nineteenth century, a force that changes the whole way economic life is run. Yet

"The Atlantic Imbalance in Science and Technology," in Robert Gilpin, *France in the Age of the Scientific State* (copyright © 1968 by Princeton University Press). Reprinted by permission of Princeton University Press.

the know-how is esoteric and easily kept within the magic circle of the firms involved."[1] Having already alluded to the fact that Europe is increasingly dependent on International Business Machines (IBM) and other American companies for its computers and automation know-how, Layton asked a question which the French and other Europeans have asked themselves about the need for a strong indigenous computer industry: "Can Europe afford to do without it? Ultimately, the decision is political. If it is held, politically, that Europe should be a power with a voice of its own in world affairs, one can argue with force that it should also have within its borders a center of power in this key industry that is transforming society."[2]

In practically all areas of advanced scientific technology the French and other Europeans believe the United States possesses a predominant position which it is seeking to preserve and exploit.[3] IBM controls over two-thirds of the world computer market and dominates the multibillion-dollar European market which is growing at the rate of 25 percent a year.[4] At a time when nuclear power generation is in rapid expansion the European atomic energy industry (with the exception of Great Britain and France) is based largely on American nuclear reactors. In aviation the position of the European industry has become extremely shaky and its future is uncertain, even though the aircraft market is expanding at the rate of 13 percent a year.[5] With respect to space technology, the futility of the European situation is evidenced by the fact that European-built satellites of any consequence must ride into space on American rocket boosters. Of much greater importance, the multi-billion dollar, and rapidly growing, field of global communications by satellite will undoubtedly be dominated by the American Communications Satellite Corporation *(Comsat),* the manager of the International Communications Satellite Corporation *(Intelsat).* And, of course, in the military exploitation of these advanced technologies the U.S. is far advanced over the other nations of the West.

What disturbs the Europeans, however, is not merely the present imbalance of technological power between the United States and western Europe, but the long-term consequences of America's over-all and commanding lead in those basic and applied sciences which are constantly producing new technologies of economic and military importance. France, in particular, has been sensitive to this situation. Faced with the American scientific-technological colossus France as late as 1963 found herself on the seventh rung in per capita expenditure of funds for scientific research.[6] In an era when technological innovation is increasingly the fruit of scientific research the French fear that this "science gap" will lead ultimately to the permanent dependence of Europe on the United States for the technologies required for economic growth, public health, and national security. To appreciate this fear, one must understand the changed relationship between science and technology in the contemporary world.

THE NEW RELATIONSHIP BETWEEN SCIENCE AND TECHNOLOGY

Western society is presently entering the third phase of the industrial revolution. The first phase was associated with application of the *method* of science to mechanical invention in the eighteenth century and with the political rise of Great Britain. The second phase originated in late nineteenth-century Germany with the application of scientific *theory* to industrial processes and technological invention. The present phase is due to the narrowing lead time between a scientific discovery and its technological application; this period is characterized by the integration of science as an *institution* for the discovery of knowledge with those other institutions which utilize this knowledge. This latest phase is associated with the political rise of the United States and, to a lesser extent, the Soviet Union.

The essence of the first phase of the industrial revolution lay in the application of the new experimental method of science to an understanding and improvement of industrial processes and techniques. As Mantoux, Ashton, and Gillispie, among others, have shown, abstract and theoretical concepts played little part in the changes in industrial practice.[7] It was rather the spirit and example of their scientific acquaintances which guided the English entrepreneurs who began the first industrial revolution. The theoretical advances in natural history, chemistry, and mechanics of the eighteenth century were not important in themselves for the vast changes which began to take place in the textile, chemical, and power industries.[8] The evidence suggests, for example, that James Watt's pivotal invention of the separate-condenser steam engine in 1769 owed little to work on thermodynamic theory such as that by Watt's friend, Joseph Black, on the concept of latent heat.[9] Black's work did, however, infuse Watt with the "new spirit of inquiry."[10] In Watt's own words, Black, by revealing the "correct modes of reasoning and of making experiments of which he set me the example, certainly conduced very much to facilitate the progress of my inventions."[11] It was only much later in the nineteenth century that a true scientific understanding was obtained of just exactly why Watt's steam engine actually worked.

In the latter part of the nineteenth century the gap between scientific understanding and technological innovation began to close. Prior to this time scientific theory and its elaboration had simply not progressed sufficiently to be of much utility in the solution of industrial problems. A great deal of "normal" science,[12] to use Thomas Kuhn's phrase, had had to transpire between the great theoretical discoveries of the late eighteenth and early nineteenth centuries in chemistry and electricity before these theories could become the foundation of the modern chemical and electrical industries.[13] Furthermore, scientific theory could not be of assistance to technology until the latter had reached the high level of development which it did attain in the latter part of the nineteenth century. In short, science and technology had developed in separate and relatively autonomous spheres until around 1870 when they reached a point in their respective evolutions where scientific theory and technological innovation could become closely intertwined.

This critically important marriage between science and technology took place in Germany. Though industrial research had existed previously, the Germans were the first to undertake the systematic application of the theoretical results of research to technological invention.[14] The achievement essential to this development was the transfer to industry of the laboratory organization and exact techniques of scientific research developed in the German university.[15] This "invention of the method of invention," which Alfred North Whitehead called the greatest invention of the nineteenth century,[16] took place in the German electrical and synthetic organic chemical industries.[17] While the older methods of empirical innovation did not cease to be important, scientific concepts and scientists themselves for the first time began to play a much larger role in technological innovation.

Prior to the second phase of the industrial revolution science had little to do with the political and economic fortunes of nations. The past hegemonies of Spain, France, and England were entirely unrelated to the strength or weakness of their respective sciences, though technology, of course, played an increasingly important part, especially in the political rise of Great Britain after 1560. But the technologies of gunpowder, iron, and ship-building, which provided a base for national power, had little to do with scientific understanding. On the contrary, as Thomas Kuhn has suggested, until the latter part of the nineteenth century science and technology were not only separate enterprises but "the cultural matrix that has supported a flourishing scientific enterprise has not usually supported a progressive technology and vice versa."[18]

Germany was the first nation whose meteoric political rise is attributable in large measure directly to science. Before the twentieth century it was "the only nation that . . . achieved simultaneous eminence in both science and technology . . . from about 1860 to 1930."[19] Especially through the systematic advancement and exploitation of chemical theory the Germans were able, in a relatively short period of time, to transform a backward society into the technically most progressive nation on the continent—a change of no mean consequence for the European balance of power.

Unfortunately we lack any term which denotes this new alliance between science and technology. Yet, the need for some appropriate expression is obvious and such awkward phrases as "science and technology" or "research and development" are employed to cover a vast and continuous spectrum of activities which runs from fundamental scientific research at one end through applied research to technological innovation and prototype development at the other. The boundaries separating one activity from another are rarely distinct in terms of either the motives of the individuals involved or the consequences that flow from their work. Perhaps, however, the term "scientific technology," in contrast to the "empirical technologies" of the past, conveys best this notion of the increasing dependence of technological innovation on scientific theory today.

Currently the Western world is passing through a third phase of the industrial revolution because of the rapidly decreasing lead time between scientific discovery and its technological application (see Table 1). Scientific knowledge is not only advancing at an exponential rate—with a doubling time of less than 10 years—but scientific and technical advance increasingly go hand-in-hand. The discovery of the laser and the transistor, for example, were scientific as well as technological achievements, for which both Nobel Prizes and patents were awarded. As a consequence, man's power over nature and over his fellow man is advancing in abrupt jumps rather than through the accumulation of many slow incremental advances.

TABLE I.[a] Decreasing Lead Time Between Discovery and Application

Telephone	56 years (1820-76)
Radio	35 years (1867-1902)
Television	14 years (1922-36)
Radar	14 years (1926-40)
Atom Bomb	6 years (1939-45)
Transistor	5 years (1948-53)
Laser	5 years (1956-61)

[a]Pierre Lelong, "L'Evolution de la Science et la Planification de la Recherche," *Revue Economique*, No. 1, Jan. 1964, p. 19.

If one inspects the spectrum of technology running from the most empirical to the most scientific, he discovers that in the contemporary world the technologies of greatest importance for national prestige and military power—electronics, aerospace, and atomic energy—are found at the scientific end of the spectrum where the lead time between discovery and application is frequently very short. It is essentially for this reason that basic scientific research has become an important factor of national power. This fact alone sets apart our age.

The essential aspect of the third phase of the industrial revolution is the transformation of scientific, industrial, and political institutions in order to accelerate the advancement of basic scientific knowledge and the exploitation of this knowledge for the innovation of novel and powerful technologies. This development began in Great Britain just before World War Two when natural scientists were taken into government to work on such technical problems as radar. But its full manifestations were not to be seen until during the war itself when the United States mobilized its entire scientific population and institutions for the successful exploitation of science for the military effort. The result was a change in the quality of scientific organization and in the role of science in society.

Today Great Power status accrues only to those nations which are leaders in all phases of basic research and which possess the financial and managerial means to convert new knowledge into advanced technologies. In the case of the two superpowers, eminence in science and in technology go hand-in-hand, and it appears most unlikely that any nation or group of nations can ever again aspire to a dominant role in international politics without possessing a strong, indigenous scientific and technological capability. International politics has passed from the era of traditional, industrial nation-states to one dominated by the scientific nation-states.

In an address to the European Parliament on the need for greater European scientific-technical cooperation, Robert Marjolin, then Vice President of the Commission of the European Economic Community, underscored the significance of the transformation that is taking place:

> The capacity for invention and its corollary, the capacity to exploit invention, now play a part similar to the possession of mineral deposits and sources of energy in the past. To be in the front rank of nations, it is not enough now to possess the equipment for mass production; you have to be able to remold your production and techniques at a pace which there is every reason to believe will grow faster in the future.[20]

THE SCIENCE GAP

If one accepts the validity of Marjolin's statement on the importance of scientific research and development for the modern economy, then the Europeans have genuine cause for concern when they compare themselves with the United States. For example, in 1962 the U.S. spent about four times as much, at official exchange rates, as did western Europe for research and development. Because of differences in research costs the effective expenditure differential is smaller but still striking. Utilizing a "research exchange rate" the United States in 1962 spent about two and a half times as much as western Europe, and about 20 percent more than the Soviet Union. Of greatest political significance is the fact that about 60 percent of this American expenditure was for military and space research, compared with about 33 percent in western Europe for the same purpose. As a consequence, the U.S. leads western Europe by a factor of four at the "research exchange rate" in the militarily significant areas such as atomic energy, aerospace, and electronics, although in nonmilitary research (and politically less significant areas) the American lead is only 1½ to 1 at the "research exchange rate."[21]

Whereas the United States has been expending three percent of its gross national product (GNP) on research and development, the Europeans have been devoting only two percent of a much smaller base. In absolute terms the U.S. is spending about $20 billion per year on research and development, as opposed to about $5 billion for all of Europe (see Table 2). This is an immense gap if one compares it to the period of the early 1930s when expenditures on both sides of the Atlantic were approximately equal, between $50 and $80 million.[22]

TABLE 2.   Estimated Gross Expenditure on R and D (GERD) and Gross National Product (GNP), 1962[a]

	Millions U.S. dollars, off. ex. rate	pop. (millions)	R & D expenditure per capita, U.S. dollars, off. ex. rate	GERD as percent GNP at market price
United States	17,531	187	93.7	3.1
Western Europe[b]	4,360	176	24.8	
Belgium	133	9	14.8	1.0
France	1,108	47	23.6	1.5
West Germany	1,105	55	20.1	1.3
Netherlands	239	12	20.3	1.8
United Kingdom	1,775	53	33.5	2.2

[a]Adapted from Freeman and Young, *The R and D Effort in Western Europe, North America, and the Soviet Union*, Paris, Organization for Economic Cooperation and Development (OECD), 1965, p. 71. This is the most authoritative study of the subject.
[b]Belgium, France, West Germany, the Netherlands, United Kingdom.

Perhaps the most reliable index of scientific-technical power is the number of engineers and scientists employed in research and development, because in the long run the supply of skilled manpower is the primary factor limiting the growth of research and development activities. While definitions of "scientists and engineers" differ, there were in 1962 approximately 436,000 "full-time equivalent" scientists and engineers in the United States, 415,000 in the Soviet Union, and only 148,-000 in all of western Europe (see Table 3).[23] Thus, even though the total population of the United States and western Europe is approximately the same, the former has almost three times as many scientists and engineers.[24]

Of even greater significance, this discrepancy will not only continue to exist for the indefinite future but it will undoubtedly increase because of the greater output of the American system of higher education. "Even if Western European countries' ambitious programmes for the expansion of higher education facilities are all realised, the output of scientists and engineers in the United States in 1970 will probably still be much larger, and the relative size of her total stock will be greater still."[25] Even in qualitative terms it is feared that the Europeans will be farther behind in 1970 than in 1950.[26]

In addition, a significant proportion of this European production of scientists and engineers will migrate to the United States. Between 1956 and 1963 some 1,500 scientists and engineers (5½ percent of the European annual output) migrated *per year* from western Europe to the U.S.; and these migrants tend to be the better scientists and engineers, or at least of higher than average quality.[27] There is no reason at the moment to believe that this "brain drain" will cease (see Table 4).

Raw statistics on research expenditures and scientific-technical manpower, however, are not indicative of scientific achievement. Of this there are no satisfactory quantitative measurements, but there are nevertheless some crude indices one can employ. Perhaps the nearest thing to an international scoreboard for science is provided by the Nobel Prizes for science in the areas of chemistry, physics, and medicine (Table 5). These figures reveal a steady rise in American scientific achievement and a relative decline in western European science. France, especially, has fallen behind with respect to its past performance and in relation to that of other major European countries. Between 1935 and 1965 France, for example, received no Nobel Prizes for science. In 1965 the award for medicine was shared by three French biologists at the Pasteur Institute, André Lwoff, François Jacob, and Jacques Monod; in 1966 the award in physics went to Alfred Kastler at the *École Normale Supérieure.*

In mathematics, on the other hand, France has been more successful in maintaining its traditionally strong position. In 1950, 1954, and 1958 French mathematicians won the coveted Fields Medal, the equivalent of the Nobel Prize in mathematics.[28] But even here there is growing evi-

TABLE 3.[a]   Estimated Manpower Engaged in R and D, 1962

	Scientists & Engineers engaged in R & D (thousands full-time equivalent)	Other personnel engaged in R & D (thousands full-time equivalent)	Tot. personnel engaged in R & D (thousands full-time equivalent)	Tot. pop. (millions)	Tot. working pop. (age 15-64) (millions)	R & D personnel per 1,000 pop.	R & D personnel per 1,000 working pop.
U.S.	435.6	723.9	1,159.5	186.6	111.2	6.2	10.4
Western Europe[b]	147.5	370.8	518.3	176.1	113.9	2.9	4.6
Belgium	8.1	12.9	21.0	9.2	6.0	2.3	3.5
France	28.0	83.2	111.2	47.0	29.1	2.4	3.8
West Germany	40.1	102.1	142.2	54.7	36.7	2.6	3.9
Netherlands	12.6	20.2	32.8	11.8	7.3	2.8	4.5
U.K.	58.7	152.4	211.1	53.4	34.8	4.0	6.1
U.S.S.R.	416.0[e]	623.0	1,039.0	220.0	142.0	4.7	7.3
	487.0[d]	985.0	1,472.0			6.7	10.4

[a] Adapted from Freeman and Young, *R and D Effort*, p. 72.
[b] Belgium, France, West Germany, the Netherlands, United Kingdom.
[c] "Conservative" estimates.
[d] "Project" assumptions.

297

TABLE 4.[a] Migration of Scientists and Engineers to the U.S.

Last permanent residence	IMMIGRATION INTO U.S. (ANN. AV. 1956-61)			IMMIGRANTS AS RATIO OF 1959 OUTPUT OF SCIENCE AND ENGINEERING GRADUATES (IN PERCENT)		
	Scientists	Engineers	Scientists and Engineers	Scientists	Engineers	Scientists and Engineers
France	26	56	82	0.5	1.2	0.9
Germany	124	301	425	6.0	9.8	8.2
Netherlands	34	102	136	7.9	21.8	15.1
U.K.	155	507	661	2.6	17.2	7.4
Total "western Europe"	339	966	1,304	2.5	8.7	5.4
Austria	23	43	67		10.9	7.0[b]
Greece	14	50	64	3.6	20.7	10.2
Ireland	13	32	45	4.7	15.4	9.3
Italy	29	42	71	0.9	1.7	1.3
Norway	6	72	78	3.4	23.8	16.2
Sweden	8	97	106	1.3	16.3	8.8
Switzerland	38	96	134	10.6	22.4	17.0
All Europe (including others)	549	1,684	2,233			
Canada	212	1,027	1,240	12.5	48.0	32.3
All countries	1,114	3,755	4,868			

[a]Adapted from Freeman and Young, R and D Effort, p. 76.

[b]Sources: *Scientific Manpower from Abroad*, NSF 62-24, Washington, D.C.; and *Resources of Scientific and Technical Personnel in the OECD Area*, OECD, 3rd International Survey, Paris, 1963.

dence that the French preeminence in mathematics is now being seriously challenged by Americans.[29]

Turning to another index of scientific productivity—the publication of scientific papers—here again the U.S. enjoys a significant lead over all other nations. As Derek de Solla Price has been able to show, although the number of scientific workers in the United States and the Soviet Union is approximately the same, American scientific productivity in physics and chemistry is twice that of the Soviet Union (Table 6). In fact, American scientists produce about one-third of the world's scientific output, in terms of research papers published, while their nearest competitors—the Russians—account for only one-sixth.

TABLE 5. Nobel Prize Awards for Science, 1901–63

	France	U.K.	West Germany	U.S.	Russia	Italy
Physics						
1901-11[a]	4	2	3	1		1
1911-21	1	3	4			
1921-31	2	2	3	2		
1931-51		3	1	4		1
1941-51		3		3		
1951-63		3	3	15	4	
Chemistry						
1901-11	1	2	5			
1911-21	3		3	1		
1921-31		3	4			
1931-41	2	1	4	3		
1941-51		1	3	4		
1951-63		6	2	5	1	1
Medicine and Physiology						
1901-11	1	1	4		2	1
1911-21	1			1		
1921-31	1	2	1			
1931-41		3	3	5		
1941-51		3		8		
1951-63		6	1	14		

[a]The years are those for which the prize was awarded, not necessarily those in which the prizes were announced. The awards for certain countries such as Japan have not been included.

THE TECHNOLOGY GAP

But the Atlantic imbalance is not merely a matter of scientific research. If it were, the French and many other Europeans would not be so

TABLE 6.   World Production of Scientific Papers[a]

| | PERCENT OF WORLD TOTAL | | | |
Country	Share of GNP 1964	Share of phys. abstr. 1961	Share of chem. abstr. 1965	Share of pop. 1964
U.S.	32.8	31.6	28.5	5.9
U.S.S.R.	15.6	15.6	20.7	7.0
W. Germany	5.2	6.2	6.3	1.8
E. Germany	0.8		2.2	0.5
U.K.	4.8	13.6[b]	6.7	1.6
France	4.5	6.3	4.5	1.4
Japan	3.6	7.8	7.3	2.9
Italy	2.6	3.4	2.7	1.5
Canada	2.2	1.1	2.0	0.6
India	2.2	1.8	2.2	14.4
Poland	1.6	1.5	2.9	0.9
Australia	1.1	0.5	1.2	0.3
Rumania	1.0	0.6	0.9	0.5
Spain	0.9	0.2	0.4	1.0
Sweden	0.9	0.7	0.9	0.2
Netherlands	0.9	5.2	0.8	0.4
Belgium	0.8	0.3	0.6	0.3
Czechoslovakia	0.7	0.9	1.6	0.4
Switzerland	0.7	1.0	1.0	0.2
Hungary	0.5	0.5	1.0	0.3
Austria	0.4	0.2	0.5	0.2
Bulgaria	0.4	0.2	0.5	0.2
All other countries	15.8	0.8	4.6	57.5

[a]Private communication. The reader's attention is invited to Price's interesting statistical studies of science. See especially his *Little Science, Big Science,* Columbia University Press, 1963.
[b]Note: Data known to be swollen because of one or more large international journals published from this nation.

concerned. In addition to this scientific lead, and in large measure be-cause of it, the United States enjoys a commanding lead over western Europe in the innovation of advanced scientific technologies. As in the case of scientific research, there are no adequate quantitative measures of relative technological position, yet rough and meaningful compari-sons are possible. One measure advanced by the Europeans (much to the distaste of American officials and economists) to gauge a nation's tech-nological standing is the quasi-mercantile notion of the "technological balance of payments."[30]

A nation's technological balance of payments compares its payments abroad for technical know-how, i.e., the purchase of licenses, with its receipts from the sale of licenses. Using this measure, most knowledge-able Europeans accept as valid the conclusion that there can be no serious doubt that the United States has a very large and growing "favourable balance," while the principal western European countries have a large "unfavourable balance."[31] Most significant of all, this unfavorable balance (Table 7) arises mainly from the heavy and growing deficit of these European countries to the U.S.; two-thirds of the total French deficit and half that of Germany is with the United States.[32]

TABLE 7.   Estimated "Technological Balance of Payments"[a] (in millions of dollars at off. ex. rates)

	Receipts	Payments	Balance	Ratio of payments to receipts
Transactions with all countries, all industries				
U.S., 1961[b]	577	63	+ 514	0.1
France, 1962	40	107	− 67	2.7
West Germany, 1963	50	135	− 85	2.7
Transactions with U.S. only, all industries				
France 1962	11	53	− 42	4.8
West Germany, 1963	10	52	− 42	5.2
U.K., 1961	17	86	− 69	5.1
Western Europe (including others), 1961	45	251	−206	5.6
Transactions of particular industries with all countries				
West Germany (1963)				
chemicals[e]	19.3	33.8	−14.5	1.7
electrical mach.	10.7	29.0	−18.3	2.7
steel, mach., vehicles	14.2	45.2	−31.0	3.2
France (1960)				
chemicals[e]	10.3	14.0	− 3.7	1.4
electrical mach.	1.7	12.6	−10.9	7.4
mach.	0.2	4.1	− 3.9	17.2
U.S. (1956)				
chemicals[e]	34.1	10.7	+ 23.4	0.31
electrical mach.	21.0	0.7	+ 20.3	0.03
mach.	28.2	1.3	+ 26.9	0.05
vehicles	16.6	2.3	+ 14.3	0.14
Transactions of particular industries with U.S. only				
West Germany (1963)				
chemicals[e]	7.5	13.5	− 6.0	1.8
electrical mach.	0.9	13.5	−12.6	14.9
steel, mach., vehicles	2.5	16.2	−14.1	7.1

[a]Adapted from Freeman and Young, *R and D Effort*, p. 74. It should be remembered that only a part of the total flow of technical knowhow can be measured.

[b]The unadjusted 1961 figures were $707 million (receipts) and $81 million (payments). These figures include some nontechnical payments. On the unadjusted basis the 1962 figures were $807 million (receipts) and $104 million (payments).

[e]Including petroleum products.

In addition to the technological balance of payments situation, Europeans lag behind the United States in a second measure of inventive activity—patents. This is of course a much less reliable measure, because while the purchasing of licenses gives some indications of value, it is very difficult to assess the significance of patents, which can differ greatly in quality. Nevertheless, patent-filing does constitute another rough measure of technological innovation.

The American share of total European patents has averaged about 17 percent from 1957 to 1963 and is much higher than the combined western European share of total American patents (about 10 percent).[33] In the specific case of France, while French patents in the U.S. amounted to only two percent of the number of domestic patents in 1961, U.S. patents in France were equivalent to 45 percent of the number of domestic patents.[34]

But what has been most disconcerting to the French about the state of affairs is not merely the magnitude of their own deficit, with respect to patents and the technological balance of payments, but the rate of growth. In 1930 France was second in the world after the United States in the export of patents. By 1938 she had dropped to fourth place, trailing after the U.S., Great Britain, and Germany. In the postwar world she was no longer an exporter but had become an importer of new technologies.[35]

The deterioration of the French technological situation became cause for alarm after 1959. Between 1952 and 1958 the number of American patents taken out in France increased by 80 percent.[36] Between 1959 and 1962 the French deficit in technological payments doubled to 335 million francs and the number of foreign patents per hundred national patents had increased several times that of any other European countries (Table 8).[37] By 1963 the French were spending abroad 380 million francs more for licenses than they earned in return, and "none of the major industries in France were able to balance their license accounts in 1963."[38] (see Table 9).

TABLE 8.   Number of Foreign Patents Per Hundred National Patents[a]

	1951	1960
West Germany	21	57
U.S.	22	26
France	44	170
U.K.	67	67

[a]Adapted from OECD, *Reviews of National Science Policy—France*, p. 80.

In 1964 the French deficit in technological payments had risen to 400-500 million francs per year and the indications were that this amount *would continue to grow.*[39] As one report pointed out, the available figures on the current state of Europe's technological balance of payments reflect the United States' research lead that existed in the 1950s. The continuing American research and development lead, the Europeans fear, will be reflected in patent and "technological payments" mainly in the latter part of the 1960s and the 1970s.[40]

TABLE 9. French Balance of Licenses and Patents by Industrial Sector, 1963[a] (in millions of francs)

	Income	Expenditure
Oil and fuels	1.2	11.5
Foundry, heavy engineering	6.3	42.2
Automobiles	16.7	27.5
Elec. engineering and electronics	9.5	118.8
Precision engineering, clock and watchmaking, optics	0.9	52.6
Glass	1.5	4.4
Chemicals	80.8	123.9
Dairy	0.6	25.2

[a]Adapted from OECD, *Reviews of National Science Policy—France,* p. 80.

In surveying the Atlantic imbalance in science and technology in terms of its dimensions and dynamics, Frenchmen and other Europeans see little prospect that it can be overcome in the immediate or even distant future. A similar pessimism was expressed of course about the "dollar gap" of the early postwar years, and that soon developed into a "dollar glut."[41] But Europeans, and especially the French, believe the present situation to be one of an entirely new and different order. The problem posed for Europe is not just that of the present size of the gap, but the fact that the American advantages create their own momentum, forcing Europe into a vicious cycle from which it cannot easily escape.

Frenchmen fear that the consequence of the U.S.-western Europe imbalance in science and technology could be the permanent economic subjugation and political domination of Europe by the United States. The rapid American advance in both spheres has added for the French a new dimension to the problem of U.S. hegemony in the Western world. By creating a *force de frappe* and departing from NATO the French have sought to lessen their military dependence on the United States and to enhance their own freedom of action in western Europe. Now, however, they find themselves faced by a new and—from their perspective—even more ominous challenge, for the technology gap has added a new dimension to the problem of American power. The technological superiority of the U.S. and the over-dependence of Europe on American technology, the French fear, increases American political influence in Europe and threatens Europe with United States economic domination.

NOTES

1. Christopher Layton, *Trans-Atlantic Investment,* The Atlantic Institute, 1966, p. 99.

2. *Ibid.*

3. W. W. Kulski, *De Gaulle and the World,* Syracuse University Press, 1966, p. 137.

4. Layton, *Trans-Atlantic Investment,* p. 98.

5. *Ibid.*

6. Assemblée Nationale, Séance du 29 Octobre 1960, *Journal Officiel,* p. 6,016.

7. Paul Mantoux, *The Industrial Revolution in the Eighteenth Century,* Harper and Row, 1961, p. 475; T. S. Ashton, *The Industrial Revolution, 1760-1830,* Oxford University Press, 1961; Charles Gillispie, "The Natural History of Industry," *Isis,* Vol. 48, Part 1, Mar. 1957, 398-407; and Charles Gillispie, *The Edge of Objectivity,* Princeton University Press, 1960, pp. 173-74.

8. For the case of the chemical industry see Charles Gillispie, "The Discovery of the Le Blanc Process," *Isis,* Vol. 48, Part 2, No. 152, June 1957, 152-70.

9. Sir Eric Ashby, for example, among many others, makes this point. *Technology and the Academics,* Macmillan, 1958, p. 5.

10. Carl Condit, "Comment: Stages in the Relationships between Science and Technology," *Technology and Culture,* VI, No. 4, Fall 1965, 590.

11. Quoted in Donald Fleming, "Latent Heat and the Invention of the Watt Engine," *Isis,* Vol. 41, Part 1, No. 131, Apr. 1952, 5.

12. Thomas Kuhn, *The Structure of Scientific Revolutions,* University of Chicago Press, 1962.

13. John Beer, "The Historical Relations of Science and Technology," *Technology and Culture,* VI, No. 4, Fall 1965, 549.

14. A. C. Crombie, ed., *Scientific Change,* Basic Books, 1963, p. 671.

15. *Ibid.,* p. 676.

16. Whitehead, *Science and the Modern World,* Mentor, 1952, p. 98.

17. John Beer, "The Emergence of the German Dye Industry," *Illinois Studies in the Social Sciences,* XLIV, 1959, 70.

18. "Comment on Scientific Discovery and the Rate of Invention," in National Bureau of Economic Research, *The Rate and Direction of Inventive Activity: Economic and Social Factors,* Princeton University Press, 1962, pp. 452-53.

19. *Ibid.,* p. 453.

20. European Economic Community, Press Release, Oct. 19, 1966, p. 1.

21. C. Freeman and A. Young, *The Research and Development Effort in Western Europe, North America, and the Soviet Union,* Paris, Organization for Economic Cooperation and Development (OECD), 1965, p. 11. This is the most authoritative study of the subject.

22. Layton, *Trans-Atlantic Investment,* p. 91.

23. Freeman and Young, *R and D Effort,* p. 12.

24. *Ibid.,* p. 72.

25. OECD, "Resources in Research and Development," Paper prepared for the Ministerial Meeting on Science, Jan. 12-13, 1966, p. 2.

26. This is the conclusion of one French expert on the basis of an exhaustive survey of European education from the elementary to the university level. Raymond Poignant, *L'Enseignement Dans les Pays du Marché Commun,* Paris, Institut Pédagogique National, 1965.

27. OECD, "Resources in Research and Development," p. 4. For a detailed analysis of the "brain drain" see "L'Émigration des Scientifiques et des Ingénieurs vers Les États-Unis," *Le Pogrès Scientifique,* No. 93, Feb. 1966, pp. 38-53.

28. Louis Dollot, *La France Dans le Monde Actuel,* Presses Universitaires de France, 1964, p. 59.

29. *Le Monde,* Sept. 15, 1966.

30. Freeman and Young, *Research and Development Effort,* pp. 51-55.

31. *Ibid.,* p. 51.

32. *Ibid.,* p. 53.

33. *Ibid.,* p. 54.

34. *Ibid.*

35. "55% des Brevets Exploités en France Sont Étrangers," *Entreprise,* No. 185, Mar. 21, 1959, pp. 51-53.

36. "La Recherche Scientifique en France et Dans les Grands Pays Étrangers," *Problèmes Économiques,* No. 809, July 2, 1963, p. 12.

37. OECD, *Reviews of National Science Policy—France,* Paris, 1966, p. 80.

38. *Ibid.*

39. *Le Monde,* Feb. 7-8, 1965.

40. Freeman and Young, *R and D Effort,* p. 54. The implications of American scientific leadership for the future have been nicely spelled out by Antonie Knoppers, *The Role of Science and Technology in Atlantic Economic Relations,* The Atlantic Institute, 1967.

41. Actually the United States continues to enjoy a very favorable balance of trade. The American balance of payments difficulties are due principally to nontrade factors such as American investments abroad, expenditures by American tourists, and the costs of maintaining American troops abroad and fighting the Vietnam war.

THEODORE HSI-EN CHEN

Science, Scientists, and Politics

SCIENTISTS MUST BE REFORMED

The attitude of the Chinese Communists toward scientists is essentially the same as their attitude toward all intellectuals. They need the services of the scientists, but at the same time they distrust them as products of bourgeois society. In order to be of use to the new society, the scientists, like all other intellectuals, must be reformed. They must get rid of all ideas and attitudes incompatible with the spirit of the new age and they must acquire the viewpoint and the ideology of the proletarian-socialist revolution. They must, in other words, undergo thorough "thought reform."

As soon as they came to power in 1949, the Communists launched a nationwide program of political indoctrination. Intellectuals were orga-

Theodore Hsi-en Chen, "Science, Scientists, and Politics," in *Sciences in Communist China*, American Association for the Advancement of Science Publication No. 68, (1961); pp. 59-98. Copyright 1961 by the American Association for the Advancement of Science. Reprinted by permission of the author and the publisher.

nized to engage in political "study" in order to "reform" themselves. Scientists were told that they needed reform as much as other intellectuals and that they must try to overcome the bourgeois attitude of aloofness from politics and to adopt a new philosophy of "serving the people."

One of the functions of the Academy of Sciences, established as early as November 1949, was to organize a thought reform campaign among the scientists. In 1951–1952, when college professors, writers, artists, lawyers, physicians, and other intellectuals were gathered in groups for intensive "ideological remolding," the Academy of Sciences in Peking enlisted several hundred research scientists and technicians for four months of "reformative study." According to Kuo Mo-jo, the president of the Academy, the "remolding" of scientists meant that they should cast off their ideological handicaps, chief among which were liberalism, individualism, and the above-class and above-politics mentality which they had carried over from the past. Scientists, it was stressed in the reform campaign, must identify themselves with the working class, and in order to do so they must first reform themselves with the ideology of the working class.[1]

An important phase of thought reform is the process of criticism and self-criticism in which the scientists are led to examine their own background, to pinpoint their ideological errors and shortcomings, and to make specific confessions and pledges. Many such confessions were made in the thought reform campaign of 1951–1952. Examination of the confessions and pledges shows that they follow a general pattern. A person writing a confession usually begins by relating his family and academic background and then goes on to show how such background has saturated him with bourgeois ideas inimical to the interests of the working class. He confesses concrete and specific failings, which are obviously the targets the Communists have designated for the "ideological struggle." Let us briefly examine some of the most frequently mentioned targets of attack.

CLASS ORIGIN

The Communists believe that a person born of a bourgeois or petty bourgeois family starts life with an ideological handicap, which he must make a conscious effort to overcome. Awareness of this handicap is one of the first steps in ideological reform. Kao Shang-yin, dean of the College of Science of Wuhan University, regretfully stated that he began life in a petty bourgeois family; Ch'en Hua-kuei, head of the Department of Agricultural Chemistry in the same university, said that he was born in a family which for five or six generations had been of the village gentry and landlord class; Ku Teh-jen, of the Department of Electrical Engineering of Sun Yat-sen University, told how he had been influenced in the wrong direction by his father who was an industrialist, and his two uncles, both making a good living in the technical field; Lin Chao-tsung, head of the Chemistry Department of Hunan University, confessed that his was a family of landlords engaged in trade, and this class background had conditioned his thinking and blinded him to the fact that his education had been made possible by the exploitation of peasants.[2] One of the most extended denunciations of family background was made by Liang Ssu-cheng, a well-known architect and a son of the eminent scholar and reformer of the closing years of the Manchu dynasty, Liang Ch'i-ch'ao. He wrote as follows:

> My class origin, my family background, and my education explain the sources of my erroneous thinking. One major source is my father's conservative reformism and his ardent worship of China's old traditions. The other is the education I received at Tsinghua University and in America.[3]

Continuing to criticize his illustrious father, Liang confessed that he had once worshipped his father but had now come to realize that his father's patriotism was actually his loyalty to the feudal and bureaucratic rulers of his day, and that his proposed reforms were no more than an attempt to protect the interests of the landlord gentry class. His great influence on Chinese youth really amounted to a reactionary move to win young people to his conservative way of thinking and to resist the advance of the revolutionary ideology of the proletariat. It is noteworthy that despite this effort to make peace with the new order, Liang was later denounced as an unreformed bourgeois intellectual.

PRO-AMERICANISM

One of the major objectives of thought reform is to eradicate American influence and to stamp out what the Communists call the "pro-American mentality" of Chinese intellectuals. To meet this demand, the scientists in their public statements took pains to denounce the American influence they had received and to pledge determination to overcome this ideological shortcoming. Architect Liang Ssu-cheng's identification of his family background and his American education as the two major sources of his reactionary thinking is expressed in different words in many other confessions. Scientists recalled how they had been dazzled by American materialism and had failed to see how American society was dominated by the exploiting class and how the workers and the Negroes in America were subjected to endless persecution. They confessed that they had been enamored of the American way of life and misled by the so-called American democracy. Biologist Chung Hsin-hsuan said that his pro-Americanism had led him to use English names for his botany specimens and to send Chinese specimens to the United States for identification and for exchange; he had received grants from Harvard University for collecting specimens in China, and he had even done some collecting for the United States Department of Agriculture.[4] Other scientists expressed shame for having received fellowships or United States State Department grants while engaged in study in America. Still others regretted that they had taken part in wartime research in America, thus unwittingly aiding the war-mongering activities of imperialist America. Even sending scholarly articles for publication in American scientific journals was an offense serious enough to deserve specific mention in a number of confessions.

IGNORANCE OR SUSPICION OF USSR

In the thinking of the Chinese Communists, anti-Americanism and pro-Sovietism are two sides of the same coin, and most of China's intellectuals are found wanting in both. Just as the scientists admitted the mistake of admiring the United States, they pleaded guilty of harboring unfavorable thoughts about the USSR. Physicist Chou P'ei-yüan, formerly dean of Tsinghua University and now vice-president of Peking University, testified that he had consistently opposed the Soviet Union, admired the United States, and distrusted the Communist Party. Biologist Kao Shang-yin said that his bourgeois background had prevented him from studying the theories of Michurin. Other biologists hastened to prove their "progressiveness" by declaring their readiness to throw Mendel's law overboard in favor of Michurin's advanced theories. Chemist Ch'en Hua-kuei admitted that he had been misled by Anglo-American propaganda to believe that the Soviet Union oppressed its people at home and pursued exploitative and expansionist policies abroad. Engineering dean Ch'en Yung-ling also confessed that his pro-Americanism had made him suspicious of the Soviet Union so that "when Chairman Mao asked us to lean to one side, I could not bring myself to do so."[5]

ALOOFNESS FROM POLITICS

"Aloofness from politics," "aloofness from class," and "purely technical viewpoint" all refer to the desire of scientists to be allowed to pursue their scientific work without being involved in politics or the class struggle. The Communists argue, however, that class consciousness is the essence of political consciousness, and without political consciousness, a scientist would not know what goals to work for and would easily become the tool of the exploiting class. In bourgeois society, the argument continues, scientists are kept away from politics so that they may unquestioningly serve the exploiting class, but in the proletarian-socialist society, all technical personnel must take an active interest in politics.

The avoidance of politics is also known as the "purely technical viewpoint." A professor of chemistry said that on account of his purely technical viewpoint "he had judged student progress solely in terms of academic achievement without regard to political record or revolutionary fervor."[6] Related to this viewpoint is that of "science for its own sake," which the Communists attack as another expression of bourgeois ideology. In previous years, scientists had tried to encourage young people to study science by using the slogan "national salvation through science." Now they confessed that this was tantamount to discouraging student interest in politics and poisoning the minds of young people. The new viewpoint was that science alone was not enough, it must be guided by correct political ideology. The study of science, therefore, must go hand in hand with political study.

INDIVIDUALISM

In the Communist view, individualism is the besetting sin of all bourgeois intellectuals. Scientists are charged with bourgeois individualism when they pursue scientific study according to their personal plans instead of following state plans, when they prefer to work individually instead of collectively, when they are slow in accepting the guidance and leadership of the Communist Party, and when they seek personal or professional advancement instead of putting the interests of the proletarian cause above all else. To absolve themselves, they painstakingly related how they had been plagued by bourgeois individualism. Biologist Kao Shang-yin[7] expressed regret that he had, out of personal interest, wasted precious years in research on the Paramecium, which was useless to mankind, while he neglected the far more important study of Michurin's theories. Another wrote: "I was not accustomed to the collective way. I feared organization. I feared discipline."[8]

When the new government announced its plan to reorganize higher education by the merging of departments or institutions in 1951–1952, scientists who raised questions were accused of individualism and "particularism" and putting selfish interests above the collective good. One by one, the scientists were brought to terms. The head of the Department of Civil Engineering of Peking University wrote a piece titled "My Bourgeois Thinking Was an Obstruction to the Reform of Higher Education."[9] Architect Liang[10] blamed his individualism and his vainglory for having opposed changing Tsinghua from a comprehensive university to a technological university. Other scientists confessed that they had not only been motivated in their career by a craving for fame and position, but also had encouraged students to consider scientific study as a means of personal advancement. Those who had submitted articles for publication in foreign journals confessed that they had been seekers of personal fame; moreover, in their eagerness to attain international recognition they had allowed themselves to be influenced by "international standards of scholarship" and consequently to become further entrenched in the bourgeois ideology. One said that he had written textbooks solely for fame and profit, while another wrote: "My ambition was to be able to solve purely technical or geological problems, to write a few good articles, and to become a famous scholar.[11]

SELF-DENUNCIATION

The Communists emphasize that thought reform is a very personal affair. Each person must examine his own record to discover his specific personal shortcomings and strive to remedy them. No vague general statements are enough. Pressed to make specific confessions, scientists made public denunciation of their own character and attributed the most unworthy motives to their actions.

To be sure, thought reform is a humiliating and a harrowing experience.[12] It is beyond the scope of this treatise to discuss why the scientists were willing to discard their self-respect and denounce themselves in such a humiliating manner. We merely note in passing that it was the very intention of the Communists that the intellectuals should be shaken out of their complacency and made to humble themselves until they accepted the direction of those whom their past social position might have tempted them to look down on.

SCIENTISTS VOICED COMPLAINTS

That the confessions and pledges did not mean the absense of inner resistance and resentment on the part of the scientists is evidenced by the complaints and protests they voiced in the brief period of 1956–1957 when in the name of the "hundred flowers"[13] the Communist government permitted some criticisms to be made openly and published in newspapers. They expressed resentment over the continuing harassment of thought reform. They objected to the restrictions on scientific study and the loss of intellectual freedom. They pointed out that centralized planning for scientific study was in the hands of politicians who knew nothing about science but had the power to impose their ideas on scientists. Microbiologist Fan Hsin-fang, for example, said that the twelve-year plan for microbiology was formulated without the knowledge of the microbiologists.[14]

Scientists complained that they had to meet so many political demands that they had no time left for scientific pursuits. Besides much time expended in political "study" and thought reform, they had to participate in many political activities, for the Communists insist that thought must be translated into action, and a person undergoing thought reform must prove his ideological awakening by energetic participation in various forms of "revolutionary action." Not only did the Communist Party designate "central tasks" that all "progressive" persons must take part in, but the youth organizations, the trade unions, and other groups also planned activities which demanded the time of the scientists. There was no end of mass campaigns, and failure to participate was sure to invite the charge of "aloofness from politics."

Many scientists resented the domination of the cadres—those ill-educated and sometimes barely literate functionaries who were vested with authority to supervise the scientists. They had the authority to judge whether or not each individual scientist was making due progress on his thought reform. They considered themselves high priests of orthodoxy, ideologically superior to the scientists. They did not hesitate to "teach" the scientists on matters of Party policy and Marxist dogma, for they officially represented the "leadership" of the Party. Some of them were so overbearing that they treated the scientists with contempt and ordered them around as their underlings.

LEARNING FROM THE SOVIET UNION

Numerous criticisms were directed against the decree that scientists should learn from the Soviet Union and avoid relations with the United States and Britain. Mao Yi-sheng said that scientists had put away their American and English books and had come to the point where they did not even dare mention Anglo-American technology for fear of being accused of political backwardness.[15] In 1952, the Academy of Sciences ordered that scientists make an intensive effort to study Soviet science. Botanist Ch'ien Ch'ung-shu testified that at the age of 70, he belatedly found that he had to study the Russian language.[16] The unquestioning wholesale introduction of Soviet science may be illustrated by the report that among the courses offered by the University of Harbin (devoted to technology), 82.2 per cent had almost completely adopted Soviet textbooks and teaching methods and an additional 5 per cent had partially adopted Soviet teaching materials.[17] So much time was spent on translating Soviet books and articles that scientists produced very little writing of their own.[18] The blind adoption of Soviet materials was even extended into the secondary school. A professor of the Peking Normal University reported that a middle school mathematics textbook which was quite adequate was abandoned in favor of a Soviet textbook, which was later found to be sixty years old and admitted by a Russian visitor to be inferior to comparable textbooks in the United States or Britain.[19]

Failure to follow the example of the Soviet Union was sure to invite the criticism of bourgeois thinking. In the Shenyang College of Medicine, a professor of pathology was ordered to teach pathological physiology and pathological anatomy as two separate courses. When he ventured the opinion that in his experience he had found it better to teach them as an integrated course, he was reprimanded for his backward bourgeois viewpoint and told that to teach structure and function as separate subjects was the proletarian way, hence the progressive way.[20] Until 1956, it was not permissible for anyone to question the theories of Lysenko, even though this was already being done in the Soviet Union. A botanist translated two articles by Soviet scientists who were critical of Lysenko and submitted them to the Editing and Testing Bureau of the Academy of Sciences, but the Bureau did not dare publish them.[21] Even Kuo Mo-jo, the doctrinaire president of the Academy of Sciences, was once candid enough to speak against the blind worship of Soviet science. In an interview with a Bulgarian editor during the "hundred flowers" campaign, he admitted that a biologist experimenting with the cultivation of a certain strain of wheat was ordered to destroy what he cultivated because he was not following the Lysenko method.[22]

INTERFERENCE WITH BIOLOGICAL RESEARCH

Scientists met with interference in their work not only on account of the demand for following the Soviet Union, but also for other political and ideological reasons. Numerous examples could be cited from the testimonies given during the months of "blossoming-contending." A few may be cited to show that political interference with the work of scientists was not limited to one or two areas.

Any project not fully in line with the ideas of the authorities or of the cadres representing Party leadership could be attacked as "bourgeois." While discussing biological research, we may cite the experience of an American-trained plant pathologist engaged in research on the cause of a prevalent disease of citrus fruit. Because his hypothesis that the disease was caused by fungi did not agree with the decision of the cadres that the cause was to be found in the water, he was severely criticized for "bourgeois theories and methods" of research. He was denounced in public meetings and subjected to long harassment. In a scientific conference in 1951, he read a paper reporting his research findings, but as soon as he cited American sources, he was asked to stop without finishing his paper. On the other hand, the report of a young Party member who based his discussion on Lysenko was given official approbation.[23]

POLITICS AND MATHEMATICS

Mathematicians found that they had to accept the Party's definition of the role of mathematics. Insisting on the principle of uniting theory with action, the Party leadership had demanded that the study of mathematics must help to solve the practical problems of industry and engineering. This, it is said, is the socialist way: neglect of practical application would indicate the influence of bourgeois ideology. Mathematicians who have not accepted this dictation without question have been attacked as enemies of socialist construction.

In Wuhan University, over 100 meetings of faculty and students of the Mathematics Department were held in July 1958 to debate the question whether the study of mathematics should be integrated with production. Since every thought reform campaign sought to put up a specific target for attack, a professor of mathematics Ch'i Min-yu, was singled out as an example of a mathematician enmeshed in bourgeois ideology. He had maintained that the insistence of immediate application would hamper the development of mathematical theory, and many of his students and colleagues had supported his view. He agreed that theory should originate from practice, but he argued that on more advanced levels it would not be necessary for mathematical theory to return constantly to practice for guidance. To support his view, he gave the example of the airplane, which after leaving the ground could continue to soar to greater heights, and of the bamboo, whose higher joints could be quite removed from the ground. For these views he was attacked as an idealist and a bourgeois scholar. Yielding to pressure, he finally confessed his errors.[24]

This campaign against mathematicians was not limited to only one university. It was part of the officially instigated "struggle between two roads"[25] that was launched in all institutions of learning. In Nankai University in Tientsin, "under the leadership of the Party," students and faculty also waged a struggle against "bourgeois elements and so-called experts" who did not accept the Party policy of uniting theory with practice and who claimed that mathematics was a "special subject" that could not be so directly integrated with production as other sciences. The critics charged that these scholars practiced the bourgeois way of teaching or studying "mathematics for its own sake." They were motivated by individualism and a desire for personal fame and profit rather than by service to the proletarian-socialist cause. It was reported that "under the leadership of the Party," the students themselves, encouraged by the Communist spirit of bold thought and bold action and without the help of the professors, took up the study of actual production problems and applied their mathematical knowledge to these problems.[26]

In the spring of 1960, the Minister of Education announced a plan of educational reform which would reorganize the content of mathematics, the natural sciences, and other subjects taught in the schools and colleges. Much of what is being taught, he said, should be considered out of date because it reflects the mechanical materialism and metaphysics of bygone days. Basic courses in mathematics, physics, and chemistry now taught in the colleges should be taught in the secondary school, while a part of the mathematics now taught in the junior middle school (including the first degree of algebra) should be taught in the elementary school.[27] Euclid's geometry should be replaced by a new system of mathematics based on Marxist-Leninist dialectical materialism.[28] Again, we see politics determining the content of science.

OTHER SCIENCES

We conclude this section with illustrations from architecture and psychology to show how political interference has been pushed into all areas of intellectual endeavor. In architecture, the story may be succinctly told by a few excerpts from a statement by the noted architect Liang Ssu-ch'eng, who played a major role in designing the architectural renovation of Peking after 1949.

> I am a scientific worker who has committed grave errors in my work . . . In the last 20 years I wrote numerous reports on Chinese architecture as well as articles and books on Chinese architectural history and city planning. My writing consistently reflected the viewpoints of idealism and metaphysics. The theories I proposed were formalistic and restorationistic.[29]
> After the liberation of Peking, the Party placed the greatest confidence in me. In May 1945, I was allowed to participate in the planning of the capital city . . . But I consistently opposed the Party and persisted in my erroneous theories . . . Their influence spread from the capital to the rest of the country . . . I squandered the capital obtained by the blood and sweat of workers and peasants; at the same time, I obstructed the socialist construction of the fatherland.

Early last year, the construction department of the government convened a national planning conference, in which severe criticism was made of formalism and restorationism in architectural designing. . . . Unfortunately I was confined to the hospital and the sanatorium throughout last year and was unable to attend that conference and to take part in the criticism and study of architectural thought after the conference. After my recovery, which took place under the care of the Party, the Party again made a great effort to help me and organized for me a number of forums and discussion meetings. Thus I was enabled to see in a preliminary way why my architectural thinking had been wrong . . .

I now understand that no technology can go on without the leadership of the Party. Departure from the leadership of the Party is bound to result in mistakes . . . [30]

Another field of study in which the Communists have actively interfered is psychology. One reason for so much attention to it is that it is a required course in all teacher-training institutions, and the communists want to be sure that teachers develop a strictly Marxist point of view on all matters. The attack on "bourgeois psychology" began in the Peking Normal University, the most highly esteemed teacher-training institution in the country. Instigated by the "Party leadership," students and young instructors organized numerous forums to assail the bourgeois viewpoint of the senior psychologists of the faculty, among them Chu Chih-hsien, Chang Chih-kuang, P'eng Fei, and Chang Hou-Ts'an. The major charge against the psychologists was that they had neglected the study of the class nature of man. They were guilty of substituting the psychological analysis of human nature for the class analysis. They talked about human nature as if there were common characteristics regardless of social conditions. They were thus teaching a kind of psychology incompatible with the age of socialism.

The writings of the psychologists were scrutinized for their erroneous bourgeois viewpoint. The discussion of individual differences in a syllabus was found to contain this statement: "Some people are brave, others are cowardly; some are altruistic, others are selfish; some loyally fight for Communism, others hanker after the decadent capitalist system." This, said the critics, was a clear example of bourgeois thinking, because the author implied that cowardice and preference for capitalism were natural characteristics of man. Elsewhere, there was found a statement that overfatigue and insufficient sleep were likely to have adverse effects on one's mental state. Such a statement, it was pointed out, reflected the love of ease and aversion to hard work characteristic of the bourgeoisie.

The psychologists had taught that children were not class-conscious. They were thus unable to change their supraclass and suprapolitics mentality. They had stressed the physiological basis of human nature, forgetting that human behavior was a product of social life. In one book, the author discussed the increase of sex interest during adolescence and said that it was natural for young people to think of love and romance. He failed to realize, said the critics, that today under Communist leadership, young boys and girls joined youth organizations and occupied themselves with labor and political study and had no time for the silly behavior of bourgeois youth such as described by the psychologists. With the encouragement of the Party, the students demanded a thorough cleanup of all bourgeois ideas and viewpoints in psychology in order to establish firmly the Marxist point of view, to "pull down the white flag and raise the red flag" in psychology as in all other areas of scholarship.[31]

The ideological attack on psychology and psychologists was not limited to one institution. It was a nationwide campaign. As late as February 1960, a national convention of the Chinese Psychological Society still concerned itself with the critique of the bourgeois viewpoint in psychology.[32] The demand on psychologists is to study the psychology of collective living, the characteristics of Communist behavior, and the practical problems pertinent to socialist construction.[33] Psychology, like all other sciences, must directly and expressly serve the proletarian-socialist cause. Any talk of "objective science" is merely a form of bourgeois propaganda which favors the divorce of science from politics.

A PROGRAM PROPOSED BY SCIENTISTS

How the criticisms made by China's intellectuals in 1957 resulted in a violent reaction by the Communists, who adopted severe measures to muzzle all contrary opinions, is now a familiar story that has been told many times. The critics were denounced as Rightists and subjected to virulent attack in a nationwide anti-Rightist campaign. A specific target of attack in the scientific circles was a proposal submitted by a group of scientists pleading for more favorable conditions for scientific study and research. Taking advantage of the limited opportunity temporarily granted to present their views, the scientists prepared a statement which included such topics as "the protection of the scientists," and "the question of leadership." They proposed that scientists should be freed from onerous political and administrative duties in order to devote more time to study and research, that they be excused from such extraneous activities as entertaining foreign visitors, that they be provided with more assistants and adequate equipment, and that each be permitted to pursue the line of study in which he was most proficient. Mildly implying dissatisfaction with the bureaucratic control of science, they said that leadership in scientific research should be "gradually involved in the actual work" and they advised against the subjective a priori determination of who is to exercise leadership and what body is to serve as the "locomotive" of scientific research.

The proposal contained some sharp words against the Communist attitude toward social science. Opposing the rejection of all bourgeois social science as a shortsighted policy, it maintained that the proper policy for social science should be to reorganize it, not to abolish it. It deplored the fact that many social scientists had abandoned their profession because their studies no longer received recognition and the subjects they used to teach had been thrown out of the university curriculum simply because they were not taught in the Soviet Union. It criticized the practice of treating official policies and government decrees as if they were objective laws, and of turning scholars into propagandists or interpreters of official policies. Finally, the scientists underscored the importance of encouraging and fostering young talents, and proposed that instead of giving major consideration to political qualifications in the selection and promotion of students, equal weight should be given to academic achievement, and that the scientists should have more say in such matters.[34]

In the anti-Rightist campaign following the brief period of free criticism in 1957, this proposal of the scientists became known as the anti-Party, anti-socialist scientific program. Leading the attack on the proposal was Kuo Mo-jo, president of the Academy of Sciences, who, in an address in the National People's Congress, called it "a wanton attack on the Party and the government and an attempt to seize the leadership for scientific work and to lure scientists away from the socialist road."[35] Point by point he took the scientists to task. The most reprehensible part of the scientists' proposal, he said, was their challenge of Party leadership and of the goal of socialism. Their rejection of centralized planning amounted to a rejection of the leadership of the Communist Party, while their emphasis on the individual interests and talents of scientists betrayed their complete disregard of the imperative duty of scientists to serve socialist construction. As to their criticism of overemphasizing the political qualifications of students and young scientists, they were merely expressing the bourgeois viewpoint and trying to turn youth against the Party.

RIGHTISTS AMONG SCIENTISTS

Following the lead of Kuo Mo-jo, scientists hastened to join the anti-Rightist campaign. One after another spoke in denunciation of the scientists' proposal and to demand a complete liquidation of anti-Party and anti-socialist bourgeois ideology in scientific circles. At the same time attack was directed against the scientists who had drafted the proposal. They were denounced as Rightists who had plotted against the proletarian-socialist revolution. Soon a campaign was on foot to ferret out all Rightists from among the scientists. Attack was made against chemist Tseng Chao-lun,[36] radiologist Meng Chao-yin,[37] psychologist Kao Chueh-fu,[38] physician Li Tsung-en,[39] and many others. The most sustained attack was directed against Ch'ien Wei-ch'ang, noted physicist and vice president of Tsinghua University, who was mentioned earlier in this paper.

Ch'ien Wei-ch'ang was one of the most highly esteemed scientists in China up to the time of his activity in the "blossom-contending" season. Recognized as a distinguished authority on mechanics and dynamics, he occupied an enviable position as professor at the renowned Tsinghua University. In recognition of his scientific achievements, he was made dean and later vice-president of the most famous technical university in China. He was a deputy to the National People's Congress and had been a member of a cultural team to visit Burma. He held many political positions, which indicated the confidence the Communists had had in him. He was greatly admired and highly praised by his students.[40] He had in earlier years made public confessions of his bourgeois past, he had renounced his American connections, and had pledged to support the Communists. For his part in drafting the scientists' proposal, however, he incurred official wrath and was denounced as a leading Rightist among scientists.

In the university of which Ch'ien was vice-president, the Party committee, i.e., the "leadership," convened faculty and students in successive meetings to examine his ugly record of anti-Party activities. Colleagues and students now vied to expose his detestable past.[41] They testified that he had once said that the Communist Party does not know technical matters and cannot provide leadership, that such matters should be left in the hands of the scientists. He had opposed the government plan for the reorganization of higher education. His words and his attitudes all amounted to defiance of the Party. He had resisted learning from the Soviet Union. As a matter of fact, he had made slight of Soviet science and considered it inferior to capitalist science. He had once said that there was not much to learn from Soviet technology. His ambition was to build an institution in China similar to the California Institute of Technology. In his whole educational career, he had consistently followed the bourgeois road.[42]

Ch'ien was present in these denunciation meetings in his own university. Yielding to the pressure, he mildly confessed that he had been wrong in taking part in the scientists' anti-socialist proposal. His critics, however, refused to accept his statement that he had been misled by other people; they insisted that he himself had consistently held anti-Party views and they demanded from him a more thorough confession.[43] In successive meetings he tried to appease his critics, but the authorities had not said to let him go, and the attacks continued. Subsequently, representatives from the Academy of Sciences, from the Ministry of Higher Education, and from other schools and universities in Peking joined the meetings to amplify the charges against him. They dug into his past to find evidences of conspiracy against the people's state. They called him an impostor, a reckless intriguer, an enemy of the people. Ch'ien Hsüeh-sheng, noted aerodynamics scientist, formerly a member of the faculty of the California Institute of Technology, now in China, said of his colleague: "He is a liar and a time-server. He has not the least spirit of a scientist. He is a most noxious and virulent Machiavellian."[44] One of his colleagues recalled that he had once complained that the standard of scholarship in the university today was not as high as before, and this statement was made the basis of the charge that he really considered the bourgeois system of education superior to the socialist system.[45] Colleagues who had been close to him took special pains to dissociate themselves from him and to express shame and regret for having been under his influence.

Within a period of six weeks, Ch'ien made six public confessions, but the campaign against him was not ended yet.[46] He had become a symbol, and the Communists wanted the "struggle" to be kept alive as long as "Rightist" thinking was still present among scientists. The crucial question in scientific circles is whether or not scientists should accept the leadership of the Communist Party. Obviously, there are scientists who still question the competence of the Party to decide on purely technical and professional matters. These are the Rightists and bourgeois elements who must be brought to terms. Ch'ien has been made the symbol of this recalcitrance. Even today, one still finds in Communist writings reference to the need of struggle against such elements as "Rightist Ch'ien Wie-ch'ang."[47]

POSTRECTIFICATION PRESSURES ON SCIENTISTS

In the rectification campaign that followed the ill-fated "blossoming-contending" of 1957, the intellectuals of China were subjected to new and intensified pressures. Not only were the critics chastised and their views suppressed, but new pledges were demanded of the intellectuals to make sure that they would follow the Communist line. The previous methods of the thought reform were considered not thorough enough. The intellectuals were now called to "surrender their heart" completely and unreservedly to the Communist Party. Scientists, no less than industrialists, writers, and people in other walks of life, were called upon to organize special meetings for the exchange of personal "experience in heart surrender." Actually, these are essentially further elaborations of the earlier confessions accompanied by renewed pledges of allegiance to the Communist Party. In one such meeting, for example, doctors of the Chinese Medical College confessed that they had been too much interested in becoming experts in their respective fields and had neglected the importance of becoming "Red." They had been ideologically backward and now they were ready to "surrender their white heart" to the Communist Party.[48] Prominent members of the Academy of Sciences made a joint declaration pledging to "pursue our scientific work with the dialectical materialism as our guiding principle and to adhere to the Communist outlook on life."[49] Other scientists also found there was no alternative but to yield to the pressure.

These pledges are not made by Communist Party members, but by scientists whom the Communist Party still considers unworthy of Party membership but who must accept the dictation and the control of the Party. Mao Yi-sheng . . . declared that he "would forever follow the Communist Party."[50] Later, in a meeting of the National People's Congress, he took issue with the Rightists who had opposed the leadership of the Party in science. He candidly stated that the "self-reform" of the intellectuals was a "long and painful process." Nevertheless, he declared: "Let all the scientific and technical workers take up political study and self-reform in earnest, clarify their standpoint, and establish the world view of dialectical materialism. Let them, under the leadership of the Party, develop their positive qualities, completely liquidate the bourgeois idealists, and work for the early realization of socialist industrialization."[51]

Mathematician Hua Lo-keng, who is not unknown in this country, felt the necessity of a positive effort to clear himself because he had been one of the group of scientists whose deliberations resulted in the proposal now known as the anti-Party anti-socialist scientific program. He confessed that his political sense was not keen enough to detect the intrigue of politicians who were making use of the scientists. With two other scientists implicated in the same manner, he declared: "We acknowledge our mistake and want to learn a lesson from this affair. From now on, we resolve to strengthen our study of Marxism-Leninism and gradually establish the Communist world-view."[52] On a later occasion, he made an extended statement disputing the view of Rightists who maintained that the Communist Party did not know enough about science to lead the scientists. He said: "Our Communist Party is the political party of Marxism and Marxism is the universal scientific truth. Consequently, the Party itself is the product of science, and by virtue of its pronounced scientific nature, it can surely lead the people of the whole world toward victory in every part of the struggle."[53]

NEW POINTS OF EMPHASIS

The new points of emphasis in the ideological remolding of scientists after the anti-Rightist campaign of 1957[54] may be briefly summarized. First, the scientists must accept the leadership of the Communist Party. Secondly, there are two roads for science, the socialist road and the capitalist road, and every scientist must clearly distinguish between the two and definitely choose the socialist road. Thirdly, science must be integrated with production and scientists must engage in labor and production and become one with the masses. Fourthly, in order to assure effective leadership and service to socialist construction, scientists must accept centralized planning and cast away their bourgeois individualism. They must learn to work collectively instead of individually. They must lay aside their personal interests and be guided by the needs of socialist construction. Fifthly, every scientist must be a "Red expert," who is ideologically sound and technically competent at the same time.

In refutation of the Rightists, scientists had to declare that not only is the Party competent to lead in science, but there can be no scientific progress without Party leadership. Mathematician Hua Lo-king's statement, just cited, was titled, "The Party Can Lead in Science and Education." The pledge to accept Party leadership was often accompanied by a renewed pledge to learn from the Soviet Union. In a science forum in Tsinghua University, it was stressed that scientists must "strengthen the leadership of the Party over science, intensify the study of the Soviet Union, mobilize the masses to surmount difficulties, and speed up the march on science."[55] On other occasions, scientists carefully cited the achievements of recent years and attributed them to the correct leadership of the Communist Party. This leadership it was stressed was applicable to technical matters as much as administrative matters.[56] In further pursuance of this line of thinking, exponents of Party leadership in science turned their criticism to the statement of some Rightists that scientific scholarship had deteriorated in recent years. They pointed out that under Communist leadership more money had been allotted to technical and industrial expansion within a few years than in one hundred years under the old society. In a forum of the technical personnel of the Ministry of Industry and Ministry of Communications, the participants declared that the progress of recent years would not have been possible without the leadership of the Communist Party which firmly maintained the correct policy of following the Soviet Union as a model for China.[57] In discussing the Big Leap Forward in science in 1959, Kuo Mo-jo again called for more study of the Soviet Union.[58]

Two Roads for Science

Much of the ideological struggle is centered on the issue of the "two roads," namely, the socialist road versus the bourgeois road. Scientists are told that they must know which road they are on, and fight against the danger of slipping into the bourgeois road. In the words of Nieh Yung-chen, Communist leader and chairman of the Scientific Planning Committee of the central government, science itself cannot be an objective and science for its own sake is an absolutely wrong viewpoint. Science, he continued, must serve socialist construction and to do so it must rid itself of the influence of bourgeois ideology. He said:

> There are scientific workers who politically support the socialist system and the leadership of the Communist Party, but, by virtue of their bourgeois education and their working in the bourgeois environment over a period of many years, still have not discarded or are unwilling to discard, the bourgeois philosophy of life and bourgeois methods and viewpoint of scholarship, thus reflecting in their work the conflict of the two roads. Consequently, a struggle among scientists between the proletarian and the bourgeois ideology cannot be avoided. This struggle must be carried to the end until the victory of the proletarian ideology: only then will scientists be able to bring about the Big Leap Forward in science.[59]

Labor and Production

To put themselves in the service of socialist construction, scientists must get out of their ivory tower and cease to indulge in theoretical study and research divorced from production and socialist construction. Bourgeois scientists who had had no experience in labor and production must now go to the farms and factories, not only to try to achieve an integration of theory with practice but also to learn from the peasants and workers. Furthermore, this experience can help to purge the scientists of their false bourgeois pride and instill in them a new respect for the working class, according to Communist reasoning.

To break down further the aloofness of the scientists and to stress the rise of a proletarian intelligentsia,[60] Communist propaganda has gone out of the way to glorify the role of "countrified experts."[61] These are peasants and workers who have had little or no formal schooling but have in their actual work devised methods to improve production. They are now hailed as scientists whose feet are firmly on the ground and who are free from the bourgeois evils of the scholarly scientists. In the new local "universities" under the auspices of the communes and factories, these "countrified experts" now occupy professional chairs teaching such subjects as termite control, cotton production, fertilizers, and planting and grafting methods. Practical ingenuity, rather than theoretical knowledge or understanding, characterizes these so-called scientists of proletarian society.

Scientists from the higher institutions have been urged to go to the "countrified experts" and to learn from them. At the same time, "countrified experts" have been appointed as research scientists to work in the research departments of universities and in various institutes of the Academy of Sciences. A peasant who discovered an effective insecticide for the extermination of insects was appointed professor of entomology at Sun Yat-sen University.[62] Workers have been made engineers, and peasants have been designated research workers.

The scientists have not watched this development without concern. Some of them have in indirect language suggested the importance of a good foundation in academic study as a basis for scientific research.[63] The Communists maintain, however, that such concern is the expression of the bourgeois viewpoint and constitutes one of the attitudes that must be changed in the ideological remolding of the scientists. In a forum of the Peking Union Medical College to wage the struggle against bourgeois ideology, one of the specific targets of attack was too much attention to degrees and qualifications. To destroy the monopoly and the vested interests of bourgeois scholars, deemphasis of degrees and qualifications was declared to be necessary.[64] To counteract the scientists' concern with systematic formal education, the Communists say that there is nothing mysterious about science, that the rulers of bourgeois society deliberately make science difficult and mysterious in order to keep it away from the masses, and that the newly liberated masses of proletarian society have rich experience as a basis of scientific knowledge and are now capable of making important contributions to a science.[65] Science, therefore, should be popularized and the "march on science" should be a popular mass movement. Under the title "Science Is No Mystery," the *Jen Min Jin Pao (People's Daily* in Peking) editorialized that "science and technology originate in social practice and material production: there is nothing occult, mysterious, or unattainable about it." It pointed out that, by and large, inventions have come from the working people rather than from scholars and that most of the important inventors in history have come from the oppressed classes and those in low social positions or unfavorable circumstances. That science is difficult is merely a superstition that could have no place in proletarian thinking.[66]

An interesting trick the Communists have employed to shatter the pride of the scholars is what they call a teaching contest between the scholars and the workmen-scientists, and between the scholars and the students. At the Wuhan Institute of Architecture, for example, a contest was staged to compare the lecture of an experienced professor with that of a bricklayer who had no more than elementary education. It was reported that while the professor's lecture was rich in theory and full of erudite principles, the audience at the end of the lecture still did not know how to start laying the first brick. The professor was followed by the workman, whose lecture was based on experience and was so clear and instructive that he was easily voted the winner. Then followed a contest between a professor and one of his students, both lecturing on the subject of concrete foundations. Again, the professor came out on the short end. Even the professors themselves, it was reported, admitted defeat. They lost because their theory was divorced from practice, because they did not have the proper political orientation, and because they had not engaged in labor and production. The students and the workers, on the other hand, had adhered closely to the leadership of the Communist Party: they had high political consciousness and rich experience in labor and production.[67]

THE BIG LEAP FORWARD IN SCIENCE

By recognizing scientists from among the workers and peasants as well as young students, the Communists claim that their "march on science" can become a big mass movement enlisting the service of millions of people. The manifold contributions of these "scientists" add up to a Big Leap Forward in science which promises to produce results as spectacular as the Big Leap Forward in economic construction. Spurred on by the Party, students and faculty as well as "countrified experts" have formulated numerous projects of "scientific and technical research." Most of these are directly related to industrial and agricultural production. The academic study of science for its own sake is supposed to have been supplanted by a serious effort to solve the practical problems of production.[68] "In order to bring about the Big Leap Forward in science," said Nieh Yung-chen, chairman of the Scientific Planning Committee of the state, "it is necessary to understand the basic policy that science must serve the Big Leap Forward in production."[69] He further expounded:

"The demands created by the expansion of production gives an impetus to the development of science, and scientific development in turn promotes the expansion of production. This is the dialectics of production and scientific development."

As in production, the Big Leap Forward in science utilizes the method of the intensive drive. In the course of one night, the students and faculty of Nankai University were reported to have decided upon more than two thousand research projects.[70] Such activities are what the Communists proudly hail as the "Communist way of bold thinking and bold action," fearing no difficulties and allowing no obstacles to stop them. To achieve this spirit, they say, it is necessary to destroy old superstitions, chief among which is the superstition of overevaluating Anglo-American science and the superstitious belief that only scholars are capable of scientific research. Under the leadership of the Communist Party, it is repeated again and again, it is possible to achieve what would be considered impossible in bourgeois society. As one Communist writer put it, "We must destroy the colonial psychology of worshiping foreign countries and the bourgeois outlook of worshiping the experts and distrusting the masses . . . we must liberate the masses from the mentality of belittling themselves and help them know their historical mission. . . ."[71]

No Place for Individualism

The Big Leap Forward presupposes centralized planning under the leadership of the Communist Party. A serious obstacle to the acceptance of Party leadership is the individualism of the scientists, and so the Communists have launched a continuous campaign against bourgeois individualism. Scientists are criticized for their desire to pursue scientific study on subjects in which they are personally interested, rather than projects which are in line with state planning. Many of them prefer to work individually rather than collectively in teams. Others are preoccupied with publication plans to bring themselves profit or fame. All these bourgeois ways must now be cast away.

Red Experts

Whether or not scientists should be politically and ideologically sound is not an open question. The unequivocal answer is that no expert is any good unless he is "Red." The question which has been discussed to some extent is whether "Redness" is more important than "expertness" and whether in the education of scientists it is advisable to try first to make competent experts and then take care of the political and ideological qualifications. The official position and the only position approved in Communist China today is that political qualifications are even more important than technical qualifications and that "expertness before Redness" is a bourgeois subterfuge that cannot be tolerated. Scientists who express themselves in public have taken care to conform to the official viewpoint.

A series of forums were held in Tsinghua University to discuss the relative importance of "Redness" and "expertness." Students and faculty assailed the tendency of some scientists to take "the middle road" or to be satisfied with "pinkness" instead of "Redness." In summarizing the discussions, the president of the university, Chiang Nan-hsiang, declared that politics is the soul of scientific work and that the main reason why Soviet scientists beat American scientists in the race to launch the first satellite in space is that Soviet scientists are motivated and guided by socialist politics.[72] He stressed the class nature of science and said that Communist science must serve the proletarian class.

Similar forums on the relation between "Redness" and "expertness" have been held in other cities and universities.[73] They point out that "expertness without Redness," "expertness before Redness," and "more expertness than Redness" are all dangerous mistakes and the only correct course to take is to put "expertness" in the service of "Red" politics.[74]

The demand that scientists should be "Red experts" is simply another way of saying that they cannot keep aloof from politics. Moreover, in all they do, they must be guided first and last by political considerations. The slogan is, "Let politics take command." In the selection of personnel and in the advancement of students, academic qualifications are often secondary to political qualifications, for the Communists believe that ideological orthodoxy and political dependability can amply compensate for any deficiency in technical ability. In the ideological remolding of government office personnel, it was proposed that politics should be assigned 70 per cent importance, while professional work should rate only 30 per cent.[75] To let politics take command, said the president of Tsinghua University, is to let politics guide professional work, to obey the leadership of the Party, to respect Marxist materialism, to study dialectics, and to follow the mass line of the Party.[76]

SUMMARY AND CONCLUSIONS

In conclusion, we may briefly summarize the Communist viewpoint in regard to science, scientists, and politics. The following ideas are prominent.

1. Science and politics are inseparable. Scientists must be governed by proletarian politics and must clearly understand and directly serve the political goals set forth by the Communist Party.

2. The leadership of the Communist Party is absolutely essential to science and scientists. To obey the leadership requires a firm pledge by every scientist that he surrenders his heart unquestioningly and unreservedly to the Communist Party.

3. Scientists must study Marxism-Leninism and become versed in the methods of dialecticism. In their education, the study of politics and ideology is fully as important as the study of science.

4. The study of politics and ideology is not enough by itself. Knowledge must be translated into action. Scientists must take an active part in political work and plunge themselves into the class struggle.

5. Most of the contemporary scientists are products of bourgeois society. Their bourgeois mentality contains many concepts and attitudes inimical to proletarian science. Among such attitudes are individualism, love of so-called objectivity of study, adherence to bourgeois standards of scholarship, dislike of the class struggle, indifference to politics, lack of experience in labor and production, worship of the United States, and suspicion of the Soviet Union. A thorough thought reform is necessary in order to cleanse the scientists of their bourgeois ideas and attitudes.

6. Thought reform is not an easy matter. It requires a violent struggle with one's inner self, a complete break with the past, and the abandonment of self-pride, self-confidence, even self-respect. Humiliating self-denunciations and confessions are a necessary prelude to firm pledges of "heart surrender" and ideological conversion.

7. There are two roads for science, the bourgeois road and the socialist road. Scientists must turn away from the bourgeois road and take the socialist road. To do this, they must seek the guidance of the Communist Party, and they must carry on their scientific work in such a way that it will directly serve the needs of production and socialist construction.

8. It is not enough to be expert. One must be a "Red expert." "Redness" and "expertness" must go together, but if at any given time it is necessary to sacrifice one for the other, the expertness must take a secondary place with "politics (or Redness) in command."

If the overall viewpoint and general policy are as stated above, what is the status of scientists and scientific research in Communist China today? A few general observations may be made.

1. The Big Leap Forward in ecomomic and material construction has its counterpart in science. The call for a "grand march on science," together with the use of "shock" methods, intensive drives, and emula-

tion campaigns, has produced new interest and new activities in science and technology. Technical institutes on secondary and higher levels are turning out a vast army of engineers, technicians, and technologists.

2. Better facilities for scientific study have been provided since the scientists aired their criticisms in 1956–1957. Living conditions for scientists have been improved; their salaries are in the highest brackets of the national scale. Library and laboratory facilities have been expanded beyond the narrow limits of Soviet publications and equipment of earlier years. Within the area of technical study, the scientists enjoy a broader scope than was permitted before 1956.

3. Preponderant emphasis is put on applied science and technology. Science is closely linked with production and industrialization. Violent opposition to "science for its own sake" tends to push theoretical science into an inconspicuous corner. Some lip service is paid to theoretical science,[77] but scientists are subject to constant pressure to make practical application and avoid being called impractical theorists. The engineer, the inventor of new tools, or the person who introduces innovations in production methods wins immediate acclaim while the theoretical scientist may be criticized for his bourgeois scholasticism.

4. Scientists are so hemmed in by political restrictions and so burdened with political requirements that there is little room for initiative and creativity in their work. The Communists urge that scientists should learn the "Communist way of bold thinking and bold action." What they mean by boldness refers to bold departure from old methods in order to devise new methods, new tools, new machines, etc.—in other words, bold technological innovations and applications. Any departure from the prescribed plans, however, might be attacked as "disobedience to leadership."

5. The technical revolution in China today is not empty talk. There is a mushroom growth of new ideas in technical innovations and improvements. Numerous new tools and machines have been invented. In the machine building industry alone, 180,000 manual operations are said to have been replaced by mechanical and semimechanical devices.[78] "A colossal mass movement for technical innovations and technical revolution is sweeping China with hurricane force," declared one writer.[79] No such claim, however, could be made for the study of basic science or theoretical research.

6. The Communist method of intensive drives and "shock" campaigns does produce spectacular results, but these are often accompanied by letdowns and dislocations in other areas. The steel drive of 1958, for example, did raise the production of steel to amazingly high figures, but concentration on steel caused neglect of agriculture and the breakdown of the transportation system, and even the steel that was

produced was found to be unusable in many instances. Likewise, the Big Leap Forward in science is producing spectacular results in technical innovations, but from the standpoint of a long-range and balanced program of scientific development, much is lacking.

7. The Communists have critized scholars who spend too much time in research and writing "in order to gain fame or to bring personal profit." Scholars have been censured for spending time in writing at the expense of political activities.[80] The primary duty of scientists is considered to be the training of a host of cadres to take up the technical tasks of the industrial revolution and research on projects of definite benefit to socialist construction.

8. If the scientists accept without question the Communist ideology and the centralized planning of the state, if they are content to work in prescribed areas, if they submit to thought reform and pledge allegiance in approved manner, they can manage to get along quite well. They can enjoy material living conditions considerably better than those available to the rest of the population. If their work is judged to be of value to the state, they may be provided with fairly liberal library and laboratory facilities and research funds.

9. Science in Communist China is a handmaid of politics. As long as it serves politics, it is well supported and scientists may be strongly motivated by the clear and specific goals presented to them by powerful propaganda. But the scope of science is definitely restricted. Within the prescribed limits the scientists may enjoy some degree of freedom in the pursuit of their activities, but few dare venture beyond the limits. To what extent this strict political control inhibits the creativity of scientists and shackles the development of science, the future will tell.

10. Having no alternative but to comply, China's scientists have made abject public confessions pleading guilty of individualism, selfishness, disregard of public welfare, neglect of politics and the class struggle, pro-Americanism, and many other "bourgeois sins." They have vowed to make a clean break with the past and to start anew under the close guidance and direction of the Communist Party. They pledge to study Marxism-Leninism, to work for the proletarian-socialist revolution, and to "surrender their heart" to the Communist Party. Have they really surrendered? The incisive criticisms they made of the Communist regime during the brief "hundred flowers" season of 1957 and the complaints they voiced against political control of science at that time give us reason to believe that underneath the outward conformity there still exists in the hearts and minds of the scientists a good deal of resistance to Communist pressure. The events of that short season provided ample evidence that China's intellectuals still had plenty of fighting spirit and unconquered integrity despite the years of political

indoctrination and "thought reform." China's intellectuals have bowed low but their backs are not broken, and the day may yet come when they will stand erect and make their contributions as free men in an atmosphere of intellectual freedom.

NOTES

1. New China News Agency release, December 29, 1951. See *Current Background* (U.S. Consulate General, Honk Kong) No. 153, pp. 37-38, January 20, 1952. Attention is called to the Communist use of the term "working class" as practically synonymous with the Communist Party. The Communists call their Party the vanguard—hence the symbol —of the working class, and vice versa.

2. The confessions of these four scientists appeared in *Chiao Shih Ssu Hsiang Kai Tsao Wen Hsuan* (Selected Documents on the Thought Reform of Teachers), Vol. 1. Chung Nan Jen Min Ch'u Pan She, Kankow, 1953.

3. See Liang Ssu-cheng's article in *Jen Min Jih Pao* (People's Daily in Peking), December 27, 1951.

4. *Chiao Shih Ssu Hsiang Kai Tsao Wen Hsuan,* Vol. I, p. 49.

5. Ch'en's article in *Ssu Hsiang Kai Tsao Wen Hsuan,* Vol. IV; the others in *Chiao Shih Ssu Hsiang Kai Tsao Wen Hsuan,* Vol. II.

6. *Chiao Shih Ssu Hsiang Kai Tsao Wen Hsuan,* Vol. I, p. 96.

7. *Chiao Shih Ssu Hsiang Kai Tsao Wen Hsuan,* (Selected Documents on the Thought Reform of Teachers), Vol. I, p. 32. Chung Nan Jen Min Ch'u Pan She, Kankow, 1953.

8. *Ssu Hsiang Kai Tsao Wen Hsuan,* Vol. II, p. 27.

9. *Jen Min Jih Pao,* March 12, 1952.

10. Liang Ssu-cheng in *Jen Min Jih Pao,* December 27, 1951.

11. *Ssu Hsiang Kai Tsao Wen Hsuan,* Vol. V, p. 15, 42.

12. See statements in a forum of the Academy of Sciences in Peking, as reported in *Kuang Ming Jih Pao,* March 21, 1958.

13. The complete slogan was: "Let a hundred flowers blossom together, let a hundred schools contend."

14. *Survey of China Mainland Press* (American Consulate General, Hong Kong) No. 1541, p. 9, May 31, 1957.

15. See his article in *Jen Min Jih Pao,* January 22, 1956.

16. *Current Background,* No. 257, p. 6, August 25, 1953.

17. Report in *Tzu Yu Chen Hsien* (Freedom Front Weekly, published in Hong Kong) Vol. XIII, No. 5, pp. 16-17, February 27, 1953.

18. Statement made in forum on scientific publications in *Kuang Ming Jih Pao,* June 10, 1957.

19. *Survey of China Mainland Press,* No. 1541, p. 19, May 31, 1957.

20. Interview with Prof. Li Pei-lin, published in *Kuang Ming Jih Pao,* May 18, 1957.

21. *Survey of China Mainland Press,* No. 1541, p. 9.

22. Interview reported in *Jen Min Jih Pao,* December 18, 1956.

23. Statements of Lin K'ung-hsiang in *Kuang Ming Jih Pao,* June 12, 1956, and June 1, 1957.

24. Report in *Kuang Ming Jih Pao,* August 21, 1958.

25. The "two roads" refer to the road of socialism and the road of capitalism.

26. *Kuang Ming Jih Pao,* August 30, 1958. A similar "uncompromising struggle" against bourgeois ideology in mathematics in Szechwan University was reported by the same newspaper on October 17, 1958.

27. Minister Yang Hsiu-feng's speech in National People's Congress, reported in *Kuang Ming Jih Pao,* April 9, 1960.

28. Another speech by the Minister, reported in *Jen Min Jih Pao,* June 15, 1960.

29. *Fu ki chu i* means the "ism" of restoring the ancient, or the past.

30. *Jen Min Jih Pao,* February 4, 1960.

31. Reports of the attack on psychologists in the Peking Normal University appear in *Kuang Ming Jih Pao,* August 15, and 25, 1958.

32. *Survey of China Mainland Press,* No. 2196, p. 9, February 15, 1960.

33. See article on psychologists in rural study, in *Kuang Ming Jih Pao,* June 12, 1959.

34. Text of proposal in *Kuang Ming Jih Pao,* June 9, 1957.

35. Text of Kuo's address in *Jen Min Jih Pao,* July 6, 1957.

36. *Kuang Ming Jih Pao,* July 14, 15, 25, 1957.

37. *Kuang Ming Jih Pao,* July 27, 1957.

38. *Kuang Ming Jih Pao,* July 31, 1957.

39. *Kuang Ming Jih Pao,* January 22, 1958.

40. Two of his students, for example, wrote an article eulogizing his scientific achievements. It was published in *Kuang Ming Jih Pao,* December 18, 1956.

41. In any attack against an errant, it is customary Communist practice to launch a campaign in which the friends and associates of the errant are urged to indicate their own stand by helping to expose the objectional words and deeds of their condemned colleague. Those who are slow in participating are subject to possible identification as sympathizers of the reactionary element.

42. Reports of Tsinghua mass meetings against Ch'ien appear in *Kuang Ming Jih Pao,* June 6, 1957, and *Jen Min Jih Pao* of the same date.

43. Charges of his anti-Soviet statement are found in *Kuang Ming Jih Pao,* July 21, 26, 1957, and June 6, 1957.

44. *Survey of China Mainland Press,* No. 1581, p. 7, July, 31, 1957.

45. *Kuang Ming Jih Pao,* July 14, 1957.

46. *Kuang Ming Jih Pao,* July 21, 1957.

47. According to a recent dispatch, the Central Committee of the Communist Party decided to remove the "Rightist" label from several hundred persons, and Ch'ien was one of them. UP dispatch from Tokyo, *Chinese World* (San Francisco) November 25, 1960.

48. Report of meeting of physicians in *Chinese World,* April 7, 1958.

49. *Survey of China Mainland Press,* No. 1721, p. 1, February 28, 1958.

50. *Survey of China Mainland Press,* June 21, 1957.

51. *Survey of China Mainland Press,* July 13, 1957.

52. *Survey of China Mainland Press,* June 26, 1957.

53. Hua's speech in the National People's Congress, *Survey of China Mainland Press,* July 14, 1957.

54. This is a continuing campaign which has not yet ended.

55. *Survey of China Mainland Press,* November 24, 1957.

56. Report on forum of Academy of Sciences. *Survey of China Mainland Press,* July 6, 1958.

57. *Survey of China Mainland Press,* September 7, 1957.

58. Kuo's speech reported in *K'o Hsüeh T'ung Pao* (Science Journal), No. 9, 1959; also *Hsin Hua Pan Yüeh K'an,* No. 157, p. 155, June 10, 1959.

59. *Kuang Ming Jih Pao,* March 13, and 15, 1958.

60. This is in line with the Communist policy of wiping out the barriers between the intellectuals and the masses and of destroying the sense of superiority on the part of the intellectuals.

61. *T'u Chuan-chia.*

62. *Kuang Ming Jih Pao,* May 11, 1958.

63. See, for example, article by Lin K'ung-hsiang in *Kuang Ming Jih Pao,* November 15, 1956.

64. Report in *Kuang Ming Jih Pao,* November 14, 1958.

65. See article on "The Mass Movement for Agricultural Research" in *Extracts from China Mainland Magazine* (American Consulate General, Hong Kong) No. 211, pp. 34, 37, May 16, 1960.

66. *Jen Min Jih Pao,* May 22, 1958.

67. Contest reported in *Kuang Ming Jih Pao,* October 30, 1958.

68. See report on scientific research in higher institutions in *Kuang Ming Jih Pao,* August 11, 1958.

69. Speech reported in *Kuang Ming Jih Pao,* March 15, 1958.

70. *Kuang Ming Jih Pao,* August 30, 1958.

71. Article by Ch'en Shun-yao, in *Jen Min Jih Pao,* January 27, 1960.

72. *Kuang Ming Jih Pao,* January 5, 1958.

73. *Kuang Ming Jih Pao,* January 7, 1958.

74. *Kuang Ming Jih Pao,* January 30, 1958.

75. *Kuang Ming Jih Pao,* April 9, 1958.

76. *Kuang Ming Jih Pao,* January 31, 1959.

77. See Liu Hsien-chou's remarks on theoretical research that is of benefit to socialist construction, *Jen Min Jih Pao,* December 19, 1957. Also *Jen Min Jih Pao* editorial, March 11, 1959, condemning the "bourgeois view" that stresses abstract theories divorced from practice.

78. *Peking Review,* p. 4, April 26, 1960.

79. *Peking Review,* pp. 19-21, March 29, 1960.

80. See criticism of faculty of Putan University in report on heart surrender of university faculties, in *Survey of China Mainland Press,* No. 1821, p. 9, July 29, 1958.

SAL P. RESTIVO

The Ideology of Basic Science*

Basic, or pure science as an activity is the pursuit of science (knowledge, explanation, understanding) for its own sake.[1] Applied science, in contrast, involves applying basic scientific principles to practical (e.g., technological or clinical) problems. In terms of the relationship between scientific and other activities in society, the commitment to basic science represents a claim by scientists for (1) autonomy in determining scientific procedures, (2) the right to select and formulate problems themselves, and (3) the right, among themselves, to decide whether theories are acceptable or not.[2] The distinction between basic and applied science is based on the judgment that scientific activity can be, and *basic* science *is*, free of instrumental or utilitarian values and norms; this

*This previously unpublished paper is based on my "Visiting Foreign Scientists at American Universities: A Study in the Third-Culture of Science" (Ph.D. diss., Michigan State University, 1971). Research for the dissertation was supported by a National Science Foundation Dissertation Grant. I gratefully acknowledge the contributions of my coworker on this project, Professor C. K. Vanderpool, and the critical assistance and encouragement of Professors John and Ruth Hill Useem through all phases of my dissertation research.

334

judgment is expressed in the idea of a value-free or value-neutral form of inquiry.[3]

Increasingly, students of science and society have come to question the utility and the reality of the distinction between basic and applied science. In some cases, critics have affirmed the possibility of basic science, but asked that given scientific activities be tested for "purity." Physicist Charles Schwartz, for example, has reminded his colleagues that "in order to decide whether some given organization is in fact free, pure, and disconnected from the troubles of the world is a matter for objective evaluation, not wish fulfillment."[4] Some scientists, in their efforts to promote social responsibility in science, have drawn attention to facts such as the scarce fifty years that separate Becquerel's discovery of radioactivity (1896) and Hiroshima.[5] Barry Commoner, noting how narrow the gap between basic discovery and scientific application has become, argues that scientists "can no longer evade the social, political, economic, and moral consequences of what they do in the laboratory."[6] The implication here is that even if the distinction between basic and applied science was in some earlier period valid, it no longer is.

Social philosophers and social scientists have constructed a more radical criticism of basic, pure, disinterested inquiry. Marx, for example, who established some of the basic premises of what has come to be called the sociology of science, wrote:

> Even when I carry out *scientific work,* etc., an activity which I can seldom conduct in direct association with other men—I perform a *social,* because *human,* act. It is not only the material of my activity—like the language itself which the thinker uses—which is given me as a social product. My *own* existence *is* a social activity.[7]

More recently, Marcuse has argued that the scientist is socially responsible *as a scientist* "because the social development and application of science determine, to a considerable extent, the further conceptual development of science. The theoretical development of science is thus in a specific political direction, and the notion of theoretical purity is thereby invalidated."[8]

The sociological perspective on purity in science is sketched with admirable clarity in Merton's critique of G. N. Clark's defense of disinterestedness. Merton notes that a range of motives, from "the desire for personal aggrandisement to a wholly 'disinterested desire to know' " is compatible with "the demonstrable fact that the thematics of science in seventeenth century England were in large part determined by the social structure of the time."[9] Merton notes further that even Clark's thesis concerning disinterestedness as a motive "is debatable in view of the explicit awareness of many scientists in seventeenth century England concerning the practical implications of their research in pure science."[10]

There is a clear imperative in the preceding arguments for a problematic view of basic science. My intent in this paper is to explore the thesis that the commitment to basic science *can* be ideological. The rationale for this thesis is rooted in (1) principles of cognition and motivation, and (2) the concept of science as a social activity and social process. My study of visiting foreign scientists at several midwestern universities yielded data which support and illustrate the idea of basic science as ideology.[11] My discussion of these data is followed by some general observations on science, social organization, and ideology.

SCIENCE: COGNITION, MOTIVES, AND SOCIAL CONTEXT

Abraham Maslow conceives science as a cognitive activity which has as its goal knowledge, understanding, and explanation.[12] In general, cognitive activities can be instigated by anxiety, or by anxiety-free curiosity.[13] Thus "pure" or "basic" science *can* function as a calming mechanism, a means available to the human organism for lowering "the level of tension, vigilance, and apprehension." What appears to be a pursuit of science for its own sake *may* be an attempt by the scientist to "detoxify" his environment, to make it something that he need not fear. Prediction, control, proof, and other scientific objectives can become pathologized in the scientist's efforts to avoid and control anxiety. Scientific objectives "may be mechanisms for detoxifying a chaotic and frightening world *as well as* ways of loving and understanding a fascinating and beautiful world."[14]

R. W. Friedrichs argues that, from a motivational viewpoint, there can be no distinction between basic and applied science: both satisfy needs. In applied science, the needs are those of non-scientists and are rooted in utilitarian market demands; in basic science, they are the curiosity needs of scientists. The motivational perspective confronts us with the issue of self-interest. According to Friedrichs, in basic *and* applied science, "one declares for *self* and in doing so inevitably denigrates the interests of the other."[15] The preference for basic science is simply "the preference for satisfying one's own needs rather than the needs of others."[16] On the basis of this idea, Friedrichs concludes that the abdication of responsibility for, and indifference to, scientific application cannot be justified in terms of a value-free, value-neutral commitment.

Friedrichs carries his analysis further, and places the commitment to purity in science in a societal context:

Value-neutrality is a concept devised by man to assist him in achieving certain ends. It should never be taken simply at face value. Rather, one must assess the larger context that includes motivation and the use of scarce resources—the society's investment in the nurture of the child-become-scientist as well.[17]

He suggests that the term "pure" in "pure science" give way to the phrase, "purely to the self-interest of the scientific community." This situation "cannot be equated with the implication that the scientist is somehow outside of the undulating stream of value-commitment that courses through the public domain. Indifference to all but one's own interests is not neutrality."[18]

The sociological view of science has inevitably led to analyses of the impact of certain social proceses on scientific activity and ideas about science. Friedson, for example, has pointed out some of the dysfunctional effects of professionalization in medicine. He argues that while professional autonomy may have facilitated the development of knowledge about disease and treatment, it "seems to have impeded the improvement of the social modes of applying that knowledge."[19] Horowitz is more specific in establishing the connection between professionalization and the issue of social responsibility in science:

> ... professionalization, by virtue of its grim fight for status, ironically permits a kind of irresponsibility with respect to the future of the social world. The professional can, by virtue of his professionalism, exempt himself as a scientist from responsibility for the ends to which his scientific findings are put.[20]

Maslow's psychology of science, and the concept of science as a social activity and social process, provide a rationale for considering the relevance of ideology in explaining and understanding certain aspects of scientific activity, in particular the commitment to basic science. Friedson, for example, notes that occupational training "can constitute part of an ideology, a deliberate rhetoric in a political process of lobbying, public relations, and other forms of persuasion to attain a desirable end —full control over its work."[21] In applying ideological analysis to scientific activity, Haberer has distinguished between the theology of science, which stresses the intrinsic values of science, and the actual history of science, a movement that has "from the beginning leaned very heavily in the direction of an instrumentalism. . . ."[22] The application of ideological analysis to scientific activity is not, however, a simple matter. Sociologists have not been consistent in their usage of ideology. Shils, for example, concludes that science *cannot* be an ideology,[23] while Apter argues that there *is* an ideology of science.[24]

SCIENCE AND IDEOLOGY

Shils argues that the sciences cannot be considered ideological for the reason that they are "genuinely intellectual pursuits, which have their own rules of observation and judgment and are open to criticism and revision."[25] The relationship between science and other social activities is, in Shils' view, likely to intensify, with scientific knowledge becoming more predominant and exerting even greater influence than previously on societal patterns of belief. Ideologies, outlooks and creeds, systems of movement and thought, and programs are among the comprehensive societal patterns of beliefs about man, society, and the universe that Shils has identified. He conceives ideology to be the antithesis of science. The disciplined pursuit of truth, the conception of autonomy in the spheres and traditions of disciplined intellectual activities, and the independence of man's cognitive powers and strivings are all antithetical to the totalistic demands of ideologies.[26] Shils contends that "fragments" of ideologies and quasi-ideologies have occasionally influenced scientific knowledge. But the basis for his rejection of assertions to the effect that science is an ideology is his view that "science is not and never has been an integral part of an ideological culture."[27]

Apter, by contrast, identifies science as one form of ideological thought; the second form is dogma. Apter conceives the ideology of science in terms of "high information and practical realism;" it is "not merely a style of thinking about problems, nor is it solely a derivation from the functional significance of science in an industrialized world, although this is clearly the origin of its power:" the ideology of science is "the application of rational methods and experimentation to social affairs."[28]

Apter summarizes the characteristics of the ideology of science as follows: (1) norms of empiricism, predictability, and rationality, (2) the acceptance of social science as scientific, and increasing acceptance of scientific norms in social conduct, (3) "a universal trend toward planning, calculations, and rationalistic goals concerned with the future," (4) in developing areas, adoption of science by vulgar ideologies, "through some form of socialism in association with the national independence movement," and (5) in the developed areas, the emergence of a meritocracy.[29] Why these characteristics are considered aspects of an ideology when they appear to be nothing more than a reiteration of the normative-evaluative system in science posited by Merton and others (with the confusing addition of certain references to societal development) is clear given Apter's conception of ideology. According to Apter, ideology "links particular actions and mundane practices with a wider set of meanings, giving social conduct a more honorable dignified complexion.

. . . From another viewpoint ideology is a cloak for shabby motives and appearances. . . ."[30] But in distinguishing science from dogma, Apter, like Shils, neglects the fact that science is a social activity and social process; in deemphasizing ideology as "a cloak for shabby motives and appearances" Apter, like Shils, ignores the distinction between the normative view of science, and a realistic, empirically-grounded view of science. The pervasive notion that of all human activities, only science is cumulative and progressive obscures the simple observation that science is subject to changes in social structure and idea systems; and such changes may be functional *or* dysfunctional for scientific growth.[31] Professionalization and bureaucratization, for example, can introduce a rigidity into science which is accompanied by a dogmatization, or ideologicalization of ideas, theories, and methods. Apter's view, incidentally, appears to provide for the transformation of science into dogma. The adoption of science by "vulgar ideologies" seems to entail a potential for something akin to the scientistic metamorphosis of science in twentieth-century China.[32]

Shils' contention that the sciences cannot be ideological can only be valid for an activity detached from the spheres of cognition, motives, and social organization. It is possible to maximize the extent to which scientific activity proceeds free of internal and external obstacles in the search for understanding, explanation, and knowledge. There is no reason to quarrel with Shils' argument that the "detachment" necessary to transcend "the deformation of the ideological orientation" can follow from appropriate training and discipline.[33] But we cannot assume that such training and discipline are possible under any and all psychological, institutional, socio-cultural, and historical conditions.

Hagstrom's discussion of science and ideology represents a typical approach to this problem in the sociology of science, and more generally in the sociology of occupations and professions. Hagstrom assumes that every established scientific discipline possesses an ideology, "a more or less explicit justification of the privileges and the claims it makes upon the scientific world and the larger society."[34] But his focus is on the articulation of established disciplines "with groups and organizations in their environments;" their ideologies are therefore "restricted in scope and oriented to specific audiences, primarily within science and scientific organizations."[35] But this approach does not permit attending to ideology in the context of relationships between science and other social activities.

Dibble's thesis is suggestive: he hypothesizes that (1) "higher ranking occupations are more likely to have highly developed ideologies," and (2) the ideologies of higher ranking occupations are likely to be less parochial than the ideologies of lower ranking occupations. Such ideologies function as systems of justification and explanation for the individual in terms of self (I'm doing the right thing for myself) and in terms of society (I'm doing the right thing for society). The ideology of science is not, in Dibble's terms, parochial—it does "include ideas which are relevant to the concerns of laymen entirely apart from their dealing with the occupation in question."[36] It is not clear, however, that Dibble recognizes those aspects of ideology in science which offer scientists a rationale for not acting with reference to "laymen."

I conceive ideology, following Marx and others, as (1) the system of beliefs produced by false consciousness; (2) a deliberately misleading system of ideas; and (3) a system of ideas generated by both false consciousness and a more or less conscious, more or less deliberate attempt to mislead. Ideologies are systems of explanation and justification[37] which identify and establish claims to control over domains of interest such as occupational or professional autonomy.

To conceive an "ideology of science" means that there exists a dogmatization in support of science as a style of life, and a collective cultivation of a false consciousness which conceals from scientists the psycho- and socio-cultural foundations and consequences of their activities; following Mannheim, it should be noted that false consciousness can take "the form of an incorrect interpretation of one's own self and one's role."[38] Little empirical material has been assembled to support and illustrate ideological aspects of scientific activity.[39] In the following sections, I discuss data from a survey of visiting foreign scientists which support the idea of basic science as ideology.

CHARACTERISTICS OF THE SURVEY

A colleague[40] and I conducted in-depth interviews with a non-random sample of eighty-two visiting foreign scientists at three midwestern universities.[41] The universities selected as research sites are major teaching-research centers with established graduate and professional schools. They are characterized by (1) some disparity in "quality rating" based on an index derived from the Carter report ratings,[42] (2) some diversity in community settings, and (3) accessibility in terms of time and travel funds available for the study. Following completion of the interviews, a mailed-questionnaire was designed to broaden the data base on selected topics from the interview schedule. Lists of visitors in residence were obtained from the chairmen of science departments at four additional midwestern universities.[43] Questionnaires were mailed to all the individuals named on the lists returned by the department chairmen. Time and financial limitations prevented following up on the original mailing. The return rate anticipated under these conditions was 50 percent; the actual rate was 53.0 percent, or 140 returned, usable questionnaires. In terms of the limited information available on the lists submitted by the department chairmen—fields of study, departmental affiliation, and universities—no systematic bias distinguished respondents from non-respondents. The distribution of interviewees and questionnaire respondents by level of development of home countries[44] fairly represents the distribution for the national population of visiting foreign scientists. In addition, it should be noted that approximately three-quarters of the visiting foreign *scholars* in the years between 1965 and 1972 have been in the physical and life sciences, medical sciences, social sciences, and engineering; approximately half of these scientists are in the physical and life sciences.[45] The sample included 92 physical scientists, 113 biological scientists, and 12 social scientists.[46]

"Visiting foreign scientists" were defined as follows: all foreign citizens not considered students (e.g., visiting professors, lecturers, instructors, advanced research and teaching fellows and associates, visiting scholars, academic guests, specialists, and all such foreign senior participants in educational programs) who are physical, biological, or social scientists; who were in residence at a university we selected as a research site; who were on campus for one month or longer during the academic year 1969-1970, and the 1970 summer session; and who are permanent residents of a foreign country.[47]

Data were available in some cases only for interviewees, or only for questionnaire respondents due to differences in the interview schedule and mailed-questionnaire. Interpretations are thus based on a variable data base.[48]

Not unexpectedly, men outnumbered women in the sample by about ten to one. The mean age of all respondents was 32.0 years; 76.0 percent of the respondents were 35 years old or younger. More than half of the visitors had advanced degrees (Ph.D. or M.D.) awarded in 1966 or later.[49]

THE COMMITMENT TO BASIC SCIENCE

The research described here was not designed as a study in the ideology of science. It was designed to explore the conditions under which visiting foreign scientists work at American universities. The concept of the commitment to basic science as an ideological commitment emerged from an analysis of several items related to respondents' definitions of their work, and their comments on the question of social responsibility in science.

More than 90 percent of the interviewees in the sample (N=61)[50] indicated that their present research has no relationship to applied so-

cial, economic, political, or physical problems in their home countries
or elsewhere. Among *all* respondents, more than half described the
work they were involved in before coming to the United States as "basic
research." Thirty-two of the eighty-two interviewees were unable to
respond to the question, "How do your colleagues define your professional role?" Of the fifty scientists who did respond, nearly two-thirds
said their colleagues considered them "basic researchers."

Only about 20 percent of the respondents (N=190) indicated that
their work had clinical aspects, and about 30 percent (N=200) felt that
their work had technological aspects. By contrast, the proportion of
respondents who indicated that their research is "primarily" or "to some
extent" concerned with basic theory was, respectively, 31.9 percent and
44.7 percent (N=201). Only about 23 percent indicated that basic theory
was "not at all" part of their research concerns (N=201), whereas approximately 80 percent (N=190) and 70 percent (N=200) noted, respectively, that clinical and technological concerns were "not at all" part of
their scientific work.

Basic science was "very important" or "somewhat important" for,
respectively, 77.5 percent and 18.1 percent of questionnaire respondents
(N=138) in determining the problems they selected for research. While
nearly 80 percent rated basic science "very important" in determining
problem selection, 29.3 percent rated problems facing their home countries as "very important" determinants (N=133), and 30.3 percent rated
problems facing mankind "very important" determinants.

Science was a primary reference group and scientists were "very
significant others" for the respondents. Among interviewees, more than
half (N=74) said they would like to be remembered primarily for a basic
contribution to science; and nearly two-thirds (N=53) said they would
like to be remembered by scientists, as opposed to family, friends,
relatives, or "mankind." Nearly 70 percent of questionnaire respondents
rated the opinions of their wives and children "very" or "somewhat"
important determinants in their choices of job locations. Salary and
country were also considered important. But the basic science commitment manifested itself in the fact that nearly 90 percent of the questionnaire respondents rated "quality of scientific facilities" and "quality of
scientists" very or somewhat important determinants of work location
choices.[51]

Finally, approximately two-thirds of all respondents indicated that
their primary purpose in coming to the United States was to learn
and/or improve basic research techniques and skills.

The preceding summary illustrates the commitment among these respondents to basic science. In the next section, the ideological function
of this commitment is illustrated and discussed.

BASIC SCIENCE AS IDEOLOGY

"Science for its own sake" is the core of the norms of science identified by Merton and other sociologists of science, and articulated by scientific spokesmen. From the normative perspective, respondents in this survey appeared to have deep roots in the social system of science. The figures cited in the preceding section, however, do indicate considerable deviation from the basic science norm; respondents gave evidence of being oriented to and directed by norms other than science for its own sake. The deviation from the general normative framework of basic science was most unambiguous in the responses of the visitors to the question, "Do you feel a sense of responsibility for the possible social consequences of your research?" Nearly 60 percent expressed a "definite" sense of social responsibility; another 30 percent expressed "some" sense of social responsibility (table 1).

TABLE 1   Expressed Sense of Social Responsibility—All Respondents

Response	Number	Percentage of Total
Definite	125	58.4
Some	65	30.4
Not at all	24	11.2
Total	214[*]	100.0

[*]N = 222, number of "no response" = 8.

All of the questions posed for respondents that I have discussed to this point, including the question on social responsibility, are very general, and not specifically tied to the present research activities of the visitors. Where present research activities were concerned, the basic science commitment was much more strongly and unambiguously manifested. This is the key to the ideological function of the commitment to basic science. Many interviewees, along with a high proportion of questionnaire respondents, expressed a sense of social responsibility when asked the question on general sense of social responsibility discussed above (table 1). But when the interviewees were asked how this sense of responsibility was reflected in their actual research activities, they said they were engaged in "basic research," and therefore social responsibility was either (1) inherent in what they were doing, or (2) irrelevant because there was no way to predict the consequences of their work. One visitor, for example, said simply that he "believed in the efficacy of well-done research"; he believed, he said, in the "goodness of basic research."

While nearly 60 percent of the questionnaire respondents felt a "definite" sense of social responsibility, and about 30 percent felt "somewhat" responsible, only 6 percent felt their research might have an adverse effect on mankind (table 2). Judging from discussions with interviewees, responses to this question reflected adherence to the idea that the basic scientist is by definition socially responsible.

TABLE 2   Anticipated Social Consequences of Present Research—Questionnaire Respondents

Response	Number	Percentage of Total
Will be of great benefit	64	48.9
Will have no foreseeable effect	59	45.0
Will have an adverse effect	8	6.1
Total	131*	100.0

*N = 140, number of "no response" = 9.

This idea is integrated into a web of faith and optimism that has its roots in the ill-fated "idea of progress." More than three-quarters of the interviewees believed that scientists are important for achieving the ideal future they foresaw for their home countries (table 3), and nearly three-quarters felt that the ideal future they foresaw for their home countries would be achieved (table 4). Interviewees expressed a comparable optimism about developments in the world as a whole over the next decade (table 5).

TABLE 3   Rated Importance of Scientists in Future Development of Home Countries—Interviewees

Rated Importance of Scientists	Number	Percentage of Total
Important	65	82.5
Not important	9	11.4
Uncertain	5	6.2
Total	79*	100.0

*N = 82, number of "no response" = 3.

It is not surprising, given the strong commitment to basic science and the faith and optimism about science and society expressed by all the respondents, to find that they are only peripherally involved in extra-scientific activities. Many questionnaire respondents agreed in principle with the idea that every scientist and scholar should be directly in-

TABLE 4 Views of Development of Home Countries during the Next Decade—Interviewees

Response	Number	Percentage of Total
Optimistic	49	70.0
Pessimistic	12	17.2
Uncertain	9	12.8
Total	70*	100.0

*N = 82, number of "no response" = 12.

TABLE 5 Future of World during Next Decade—Interviewees

Response	Number	Percentage of Total
Optimistic	53	68.0
Pessimistic	12	15.4
Uncertain	13	.16.6
Total	78*	100.0

*N = 82, number of "no response" = 4.

volved in national decision-making processes (table 6). But the respondents are, in general, distinguished by their lack of involvement in extra-scientific organizations and activities at *all* socio-political levels—neighborhood, community, and national (tables 7 and 8). Still another indication of their "lack of involvement" is the fact that only about

TABLE 6 Direct Participation of Scientists and Scholars in National Decision-Making Processes—Questionnaire Respondents

Statement: Every scientist and scholar should be directly involved in national decision-making processes in his home country.	Number	Percentage of Total
Strongly agree	21	15.6
Agree	54	40.3
Neither agree nor disagree	25	18.7
Disagree	26	19.4
Strongly disagree	8	6.0
Total	134*	100.0

*N = 140, number of "no response" = 6.

8 percent of the interviewees are members of professional associations organized around the goal of promoting social responsibility in science (table 9).

TABLE 7   Involvement in Extra-Scientific Organizations—
All Respondents

Involvement in extra-scientific organizations or activities (last five years)	Number	Percentage of Total
Involved	74	34.4
Not involved	141	65.6
Total	215*	100.0

*N = 222, number of "no response" = 7.

TABLE 8   Involvement in Social and/or Political Change-Related Activities—Interviewees

Involvement in change-related activities	Number	Percentage of Total
Involved	8	9.8
Not involved	74	90.2
Total	82	100.0

TABLE 9   Membership in Professional Associations Concerned with Social Responsibility in Science—Interviewees

Membership status	Number	Percentage of Total
Member	6	7.4
Not a member	75	92.6
Total	81*	100.0

*N = 82, number of "no response" = 1.

The pursuit of science for its own sake appears, in most cases, to require a commitment to work and profession (no doubt stimulated and reinforced by the less enlightened aspects of professionalism, e.g. the "publish or perish" imperative, and the competitive struggle for upward career mobility) that makes it difficult or impossible to find time for "outside" activities. Indeed, it would not be hazardous to speculate that only the intellectually and energetically exceptional scientist (and the independently wealthy scientist, who may or may not be "exceptional") can "make it" in the profession, and at the same time be active in a variety of other, extra-scientific activities. As an ideology, basic science can provide support for a scientist's narrow life-and-work-decisions, decisions he may have to make, consciously or subconsciously, to survive as a professional scientist.

Ideology is not an individual phenomenon. It develops in relation to the social organization of human activities. In concluding, I would like to link basic science as ideology to certain changes in the social organization of scientific activities.

SCIENCE, SOCIAL ORGANIZATION, AND IDEOLOGY

The emergence of science as an autonomous, functionally differentiated social activity (beginning in sixteenth-century Europe) and the emergence of the scientific role were preconditions for the continuous, cumulative, evolution of modern science.[52] The institutionalization of scientific activity, in its early stages, may have made it possible to approximate the "pure," self-correcting, rational, and tentative model of science to some degree. But, under conditions of material and non-material, real and culturally-conditioned scarcity, the functional differentiation of scientific activity eventually placed science in a competitive relationship with other more or less differentiated and autonomous social activities for access to and control over scarce resources. Under these conditions, institutionalization has proceeded in a manner which has increasingly rigidified the scientific enterprise. In conjunction with the professionalization and bureaucratization of science, competition for resources has stimulated the emergence of a set of ideas which explain and justify the scientist's demands on society's reservoir of resources.

It *would not* be surprising to find the search for knowledge, explanation, and understanding proceeding without interference and with maximal detachment from dysfunctional cognitions, motives, and organizations in Utopia. Neither *should it* be surprising that life in an era which must be characterized as more dystopia than utopia tends of necessity to distort, undermine, and even relegate to the realm of the impossible the higher levels of cognition, the finest motives, and humane, cooperative, progressive organizations, i.e., conditions which are linked to evolutionary potential.[53] The convergence of the dysfunctions of professionalization and bureaucratization tends to increase specialization and over-specialization in a conflictful division of labor, thereby decreasing individual and organizational adaptive potential. Occupational and organizational closure is generated; the internalization of "shabby motives" becomes more likely; humane creative, critical intelligence is eroded. Ultimately, the ability to perceive "truth" is threatened.[54] The final price of uncontrolled professionalism in conjunction with the bureaucratization of personality and social organization must be the loss of the capacity to be human, the routinization of rationality, and the deterioration of the capacity to learn and to want to learn in an anxiety-free, meta-motivated, cooperative way. To the extent that this is not, or cannot be, achieved, purity in science—at best misleading regardless of how carefully it is applied given the psychological and social relations of science—must inevitably root itself in the struggle for status, institutional survival, and anxiety reduction. Ideology can provide its only sustenance.

The visiting foreign scientists in my sample have encountered the ideology of basic science in their pre-United States experiences. In the United States, they encounter that ideology in great intensity. The United States is a major, if not the major, center of world scientific activity, and a primary source of the ideology of basic science. Whatever the claims to the contrary, scientific activity is and must be "closely tied to the vocabulary and needs of the body politic."[55] Elsewhere, I have noted that the experience of the visiting foreign scientist is not conducive to his education as a human being capable, in terms of the image of man we find in Marx and Maslow (among others), of infinite growth and development; nor does the experience appear to contribute significantly to the visitor's development as a creative, critical scientist. His experience does not actively contribute, except in the most accidental way, to the creation of closer, cooperative links among societies. He is generally isolated from extra-scientific activities, and his role repertoire becomes intensified, i.e., the scientific role is stimulated in a way which accords it increasing predominance over other actual and potential roles.[56]

I am impressed by the extent to which the scientists in my study manifest Gouldner's warning that the liberating "good news" of the scientific revolution has given way to "the hostile information" of science—professionalized, bureaucratized, ideological—as an instrument of the state in a death-culture. It is easy to succumb in the face of this reality to the seductive arguments of technological fatalism, crackpot technological realism (to paraphrase Mills), or counter-culture revolt.[57] The "good news," I believe, survives—however precariously—in critical science, radical science, and the idea of a technologically inefficient society.[58] There is still a glimmer of hope for generating what Francis Bacon referred to as a charitable inquiry—science for life.[59]

NOTES

1. B. Barber, *Science and the Social Order*, rev. ed. (New York: The Free Press, 1952), pp. 134-142; R. K. Merton, *Social Theory and Social Structure*, enlarged ed. (New York: The Free Press, 1968), p. 597; G. A. Theodorson, and A. G. Theodorson, *Modern Dictionary of Sociology* (New York: Thomas Y. Crowell Company, 1969), pp. 369-370; S. Z. Nagi and R. G. Corwin, "The Research Enterprise: An Overview," in *The Social Contexts of Research*, ed. S. Z. Nagi and R. G. Corwin (New York: Wiley-Interscience, 1972), p. 6; H. M. Vollmer, "Basic and Applied Research," in *The Social Contexts of Research*, ed. Nagi and Corwin, p. 67.

2. W. Hagstrom, *The Scientific Community* (New York: Basic Books, 1965), p. 108.

3. Merton, *Social Theory and Social Structure*, p. 597; J. Haberer, *Politics and the Community of Science* (New York: Van Nostrand Reinhold, 1969), p. 321; R. Friedrichs, *A Sociology of Sociology* (New York: The Free Press, 1970), p. 170.

4. C. Schwartz, "Professional Organization," in *The Social Responsibility of the Scientist*, ed. M. Brown (New York: The Free Press, 1971), pp. 25-26.

5. J. Shapiro, L. Eron, and J. Beckwith, in a letter to *Nature*, December 27, 1969, quoted in *The Social Responsibility of the Scientist*, ed. Brown, pp. vii-viii.

6. B. Commoner, "The Ecological Crisis," in *The Social Responsibility of the Scientist*, ed. Brown, p. 174.

7. K. Marx, *Selected Writings in Sociology and Social Philosophy*, trans. T. B. Bottomore, ed. T. B. Bottomore and M. Rubel (New York: McGraw-Hill, 1956), p. 77. The quotation is from Marx's *Economic and Philosophic Manuscripts*, originally published in 1844.

8. H. Marcuse, "The Responsibility of Science," in *The Responsibility of Power*, ed. L. Krieger and F. Stern (New York: Doubleday-Anchor, 1969), p. 477.

9. Merton, *Social Theory and Social Structure*, p. 662.

10. Ibid., p. 662.

11. S. P. Restivo, "Visiting Foreign Scientists at American Universities: A Study in the Third-Culture of Science" (Ph.D. diss., Michigan State University, 1971), pp. 111-156.

12. A. Maslow, *The Psychology of Science* (Chicago: Henry Regnery Company, 1966), p. 20.

13. Ibid.

14. Ibid., p. 30.

15. Friedrichs, *A Sociology of Sociology*, p. 163.

16. Ibid.

17. Ibid., p. 164.

18. Ibid., p. 165; see also Barber, *Science and the Social Order*, p. 55.

19. E. Friedson, *Profession of Medicine* (New York: Dodd, Mead and Company, 1970), p. 371.

20. I. L. Horowitz, "Mainliners and Marginals: The Human Shape of Sociological Theory," in *The Sociology of Sociology*, ed. L. T. Reynolds and J. M. Reynolds (New York: David McKay, 1970), p. 345; originally published in *Sociological Theory: Inquiries and Paradigms*, ed. L. Gross (New York: Harper and Row, 1967), pp. 358-83.

21. Friedson, *Profession of Medicine*, p. 80.

22. Haberer, *Politics and the Community of Science*, p. 321.

23. E. Shils, "The Concept and Function of Ideology," in *International Encyclopedia of the Social Sciences*, ed. D. L. Sills (New York: Macmillan and the Free Press, 1968), pp. 73-74.

24. D. Apter, *The Politics of Modernization* (Chicago: University of Chicago Press, 1965), p. 343.

25. Shils, "The Concept and Function of Ideology," p. 73.

26. Ibid.

27. Ibid., p. 74.

28. Apter, *The Politics of Modernization*, p. 343.

29. Ibid.

30. Ibid., p. 314.

31. J. R. Ravetz' *Scientific Knowledge and its Social Problems* (Oxford: Clarendon Press, 1971) is a brilliant dissertation on the problematics of scientific knowledge and social organization; on the possibility that science may, in the long-run, be either self-reinforcing or self-terminating, see R. Richter, *Science as a Cultural Process* (Cambridge, Mass.: Schenkman Publishing Company, 1972), p. 97.

32. D. W. Y. Kwok, *Scientism in Chinese Thought, 1900–1950* (New Haven: Yale University Press, 1965).

33. Shils, "The Concept and Function of Ideology," p. 74.

34. Hagstrom, *The Scientific Community*, pp. 211-12.

35. Ibid., p. 212.

36. V. K. Dibble, "Occupations and Ideologies," *American Journal of Sociology* 68 (September, 1962): 230.

37. H. Blumer, "Collective Behavior," in *Principles of Sociology*, ed. A. M. Lee (New York: Barnes and Noble, 1955), p. 210; R. Bendix, *Work and Authority in Industry* (New York: John Wiley and Sons, 1956), p. 2, note.

38. K. Mannheim, *Ideology and Utopia* (New York: Harvest Books, 1936), p. 96.

39. See S. S. West, "The Ideology of Academic Scientists," *IRE Transactions of Engineering Management* EM-7 (1960), pp. 54-62. The more recent literature exhibits an increasing concern with problems directly or indirectly related to ideology: e.g., Haberer, *Politics and the Community of Science*, Friedrichs, *A Sociology of Sociology*, A. W. Gouldner, *The Coming Crisis of Western Sociology* (New York: Basic Books, 1970), pp. 497-98; H. Rose, and S. Rose, *Science and Society* (Baltimore: Penguin Books, 1970), pp. 159, 179ff., 210ff.

40. C. K. Vanderpool, "Visiting Foreign Scientists in the United States: The Impact of Systemic and Role Circumscription and Dissociative Experiences on the Homogeneity of the International Scientific Community" (Ph.D. diss., Michigan State University, 1971).

41. A stratified random sample was initially drawn from a sampling list made available to us by a national organization. The list was inaccurate, something my colleague and I were not able to discover until we began our field interviews. We decided to simply attempt to reach all those visitors who *were* available and who met our criteria for inclusion in the sample.

42. A. M. Carter, *An Assessment of Quality in Graduate Education* (Washington, D.C.: American Council on Education, 1966).

43. Responses were obtained from all but a few department chairmen; the nonrespondents were associated with a special government-related center, which may have had an impact on their consideration of our request.

44. The classification of respondents' home countries as "developed" or "developing" follows, with some modifications, the schema of F. Harbison and C. A. Myers, *Education, Manpower, and Economic Growth* (New York: McGraw-Hill, 1964), p. 33.

45. *Open Doors* (New York: Institute for International Education, 1954–   ). This is an annual report on educational exchange.

46. The classification of scientific fields according to the categories "physical," "biological," and "social" follows the system used by the National Science Foundation in its classification of scientific occupations.

47. This definition follows the definition of "visiting foreign scholar" used by the Institute for International Education (see footnote 45).

48. Unless otherwise indicated, percentages are based on N = 82 (interviewees), N = 140 (questionnaire respondents), and N = 222 (all respondents). In only three cases are discrepancies the result of missing data; the differences between N's and base N's are generally accounted for by the "no response" category.

49. Thirty-seven respondents failed to indicate their Ph.D. or M.D. status; 5 respondents did not possess the Ph.D., M.D., or an equivalent terminal degree.

50. The large discrepancy between N and base N is due primarily to dropping one item which could not be coded "basic" or "applied" unambiguously.

51. The rating choices on this question were "very important," "somewhat important," "hardly important," and "not important." The base N for "family opinion" is 121; for "quality of scientific facilities" it is 136; and for "quality of scientists" it is 132.

52. H. Karp and S. Restivo, "Ecological Factors in the Emergence of Modern Science," p. 123 in this book.

53. In addition to the arguments on loss of diversity and specialization as factors which decrease evolutionary potential (arguments which appear to be valid for personality and social systems as well as for biological, and ecological ones), I would add an argument for *cooperation* as an evolutionary survival mechanism. Cooperation is always important, but it becomes *critically* important as a socio-cultural system begins to press the limits of available resources in its environment; and it appears to be a critical component of any societal model which is designed to maximize human potential. On conditions for evolution and devolution see, for example, M. D. Sahlins, and E. R. Service, *Evolution and Culture* (Ann Arbor: University of Michigan Press, 1960); T. Dobzhansky, *Mankind Evolving* (New Haven: Yale University Press, 1962); E. Mayr, *Animal Species and Evolution* (Cambridge, Mass.: Harvard University Press, 1963); R. C. Stebbins, "The Loss of Biological Diversity," in *The Social Responsibility of the Scientist*, ed. Brown, pp. 165-71; B. Commoner, *The Closing Circle* (New York: Alfred A. Knopf, 1971); G. Bateson, *Steps to an Ecology of Mind* (New York: Ballantine Books, 1972); A. W. Crosby, Jr., *The Columbian Exchange: Biological and Cultural Consequences of 1492* (Westport, Conn.: Greenwood Publishing Company, 1972).

54. Horowitz, "Mainliners and Marginals," p. 347.

55. Friedrichs, *A Sociology of Sociology*, p. 300.

56. S. P. Restivo, "Visiting Foreign Scientists at American Universities: A Study in the Third-Culture of Science," *Studies of Third-Cultures*, No. 8 (E. Lansing, Mich.: Michigan State University, Institute for International Studies in Education, 1973).

57. On technological fatalism, see J. Ellul, *The Technological Society*, trans. J. Wilkinson (New York: Alfred A. Knopf, 1964); for an excellent review of crackpot technological realism, see E. S. Schwartz, *Overskill* (New York: Ballantine Books, 1972); on the counter-culture revolt, see T. Roszak, *The Making of a Counter-Culture* (New York: Doubleday-Anchor, 1969).

58. On "critical science," see Barry Commoner's works, especially *Science and Survival* (New York: The Viking Press, 1966). Ravetz views critical science as a potentially viable response to the contemporary perils of science: *Scientific Knowledge and its Social Problems*, pp. 424-431. For continuing reports on activities in the radical science movement, see bimonthly issues of *Science for the People*, a publication of Scientists and Engineers for Social and Political Action, Jamaica Plains, Massachusetts. Schwartz, *Overskill*, argues for the development of an "inefficient society," a society based on the ideas discussed in the literature on evolutionary potential referred to in footnote 53.

59. F. Bacon, *Works of Francis Bacon*, ed. J. Spedding, R. L. Ellis, and D. D. Heath, 14 vols. (London: Longmans, Green, & Company, 1857–1874), Vol. IV., pp. 20-21. Quoted in Ravetz, *Scientific Knowledge and its Social Problems*, p. 436.

PART IV

Science and Development

Part four focuses on "Science and Development." Students of what we refer to as the third-culture of science have been consistent in stressing both the actual and normative relationships between science, science-related activities, and development (i.e., industrialization and modernization) and the ecumenization of the world's societies. Educational exchange, institution building, technical and economic assistance, and the development of national scientific communities are viewed as systems or sub-systems "designed to facilitate the process of modernization" (Useem and Useem 1968, p.43). This view is reflected in the rationales for exchange programs and visiting scholar activities. In a recent study of postdoctoral education in the United States, many university administrators expressed the feeling that international education is "a responsibility of the world's richest country" (NAS 1969, p. 209). At a national conference on higher education and development in 1967, a United States State Department official noted that one problem America faces is "how education in America, for the foreign and American student alike, can help bring together the advanced and developing world" (Canter 1967, p. 37).

For many scholars, science is the critical activity in national and world development. Bernal's *The Social Function of Science* established a Marxist-socialist perspective on the relationship between science and society: "It is to Marxism that we owe the consciousness of the hitherto unanalysed driving force of scientific advance, and it will be through the practical achievements of Marxism that this consciousness can become embodied in the organization of science for the benefit of humanity" (1939, p. 415). In a Festschrift for Bernal published a quarter of a century later (Goldsmith and Mackay 1966, p. 55), physicist P.M.S. Blackett, a Nobel recipient in 1948, argued that "the West should make the great experiment of sacrificing some of its immediate prosperity to give massive aid to the have-not countries. . . . Scientists and technologists have a special responsibility in this matter, since it is their genius and their skill which alone can bring the material basis of happiness within the reach of all." Other students of science and society who have not explicitly committed themselves to a Marxist-socialist framework have reached similar conclusions (e.g., Meier 1966; Silvert 1963, pp. 435-36; Apter 1965). In addition, Gilpin has related scientific acitivity to the economic and technological status of developed nations (1968, p. 25).

The theme which unites these and related conceptions of science and society is an unchallengeable awareness of the critical function of organized human inquiry in problem-solving at the national and world level.

Many observers believe that (1) special skills, knowledge, and techniques are appropriate to the solution of developmental problems at subsocietal, societal, and suprasocietal levels, and (2) broad creative and critical scientific imagination is necessary for short—as well as for long—term developmental planning. The function of scientific activity as a link between developed and developing countries is, under these conditions, defined in much more complex terms than simple transfers of technologies and research "know-how" either through transfer of scientific and technological goods and services or training foreign personnel. Calder, for example, writes that "there is every reason why poor countries should attempt to leapfrog over the obsolescent technologies of the rich . . . imitation of the present rich countries may be quite inappropriate for countries with different climates, cultures and interests;" he suggests that the developing countries "formulate their own visions of the future and experiment with novel technologies. . . ." (1970, p. 268). Gilpin argues similarly for the less developed European countries (1968, p. 459).

There has been widespread affirmation of these ideas (e.g., Apter 1966, p. 222; Brown 1954; Brown and Harbison 1957, pp. 78ff; Gruber 1961; Halpern 1967, p. 183; Harbison and Myers 1964, p. 69; Lewis 1962; Meier 1966; Myrdal 1968, pp. 55ff; Perkins 1966, p. 617; Revelle 1963,

p. 138; Shah 1967; Shils 1961, p. 219; Shils 1966, p. 212; Shils 1967, pp. 482f; Stone 1969, p. 1118; UNESCO 1963 and 1970). Some of the questions these ideas raise have been examined in our study of visiting foreign scientists at several Midwestern universities (Restivo 1971; Vanderpool 1970). In general, our conclusion is that the experience of the visiting foreign scientist is *not* designed to facilitate national and/or world development by linking developed and developing countries. Their involvement in basic research, their status as visitors, and a rigorous work schedule are among the factors which tend to isolate visiting foreign scientists from non-work milieux. Their work experience promotes an orientation to science as an autonomous profession. University science departments stress research efficiency and training in skills and techniques rather than creative research and technological innovation. What little research has been carried out on the experiences of visiting foreign scientists in America supports these general conclusions (e.g., NAS 1969), though more in-depth studies are needed to establish firm conclusions.

The selections in this section outline some problems and perspectives on science and development. In part one, we considered why the scientific revolution occurred in Western Europe. In the first article for part four, Basalla takes us from that question to an analysis of the diffusion of modern science from the West. He proposes a three-stage model which describes the introduction of modern science into any non-European nation. In Phase I, the nonscientific society or nation (e.g., China or India) is a source for European science; Phase II is a period of "colonial science;" the third phase is a struggle to achieve "an independent scientific tradition (or culture)." In his conclusion, Basalla reiterates a basic rationale for this text when he notes that "the lack of comparative studies of the development of science in different national, cultural, and social settings can be attributed to the widespread belief that science is strictly an international endeavor." Basalla does not agree with "Nazi theorists" that science is a manifestation of race or nationality; neither does he agree with Chekkov that "there is no national science just as there is no national multiplication table. . . ." He draws attention to the fact that "science exists in a local social setting," but argues weakly that their effects may be "more profound" than simply determining "the number and types of individuals who are free to participate in the internal development of science." We are still left with the general impression that science emerged and developed in the West, and that its diffusion has proceeded without significant changes in organizational characteristics which might affect the nature and products of scientific activity. Basalla's article, finally, raises questions he himself does not address concerning the diffusion of science and technology from the West as an aspect of expansionist policies and the extension of spheres of influence.

Basalla, in discussing his three-stage model, notes that "the struggle to establish an independent scientific tradition, which takes place during the third phase, is the least understood, appreciated, or studied aspect of the process of transference of modern science to the wider world." The paper by Friedmann and the following selections by Apter, Dedijer, and Vanderpool help to fill in some of the details of this "third-phase." Friedmann's discussion of the conflicts between "modernists" and "traditionalists" draws on materials from his studies in Brazil. He focuses on the modernist intellectual as an agent of the diffusion of Western values and institutions; a critical part of this general process is the diffusion of Western science and technology. Friedmann sets the process of modernization in a global context by noting that "political and economic interdependency on a global scale is increasingly becoming part of the human condition; it is *the* historical reality of the contemporary age."

For Apter, "modernizing elites" are agents "in the universalization of a new type of scientific establishment, which, having made its appearance in the most highly developed industrial system, has now become critical in modernizing systems." His discussion in the excerpt from *The Politics of Modernization* (1965) reprinted here is relatively abstract and independent of a substantial empirical foundation. Apter's conception of science as an international community, his indiscriminate use of the concept "ideology" in reference to the values and norms of science, and his generally optimistic forecasts for the future of "representative government" are open to criticism, but do not detract from what is generally a provocative analysis of science and modernization.

Steven Dedijer, an internationally known student of science policy and development, is the author of the next selection. In a paper addressed to "the presidents and prime ministers of those countries where science does not yet exist on any significant scale," Dedijer discusses aspects of science policy which have been insufficiently stressed or otherwise "passed over in silence at international conferences." He argues (from admittedly crude data) that, at least up to the early 1960s, leaders in developing countries exhibited a lower level of awareness of the significance of science policy and the development of research work than their counterparts in the advanced countries. This reflects Dedijer's pessimism about what Basalla refers to as the third-phase in the diffusion of Western science: "underdeveloped decision-makers on science, underdeveloped administrators of science and underdeveloped scientists." His proposals, based on the concept of a "planned national policy," are cogent. His conception of the "scientific community," on the other hand, is idealistic. He recognizes that "the cultivation of science as a collective understanding and success in it depends on an appropriate social structure;" but he fails to consider heterogeneity, and actual and potential changes in that social structure.

It is possible to identify centers of scientific activity which, because of their exemplary research efforts, technical manpower, esteemed scientists, advanced technologies, and prestigious journals and associations, exert a "pull" on scientists outside the centers, i.e., in the periphery. As part of our broader study of visiting foreign scientists, Vanderpool attempted to determine to what extent scientists are conscious of "centers" and "peripheries" in their own fields. Vanderpool notes that his findings and interpretations must be evaluated cautiously given the small number of scientists involved in the study. But his paper suggests that the level of development of scientists' home countries is one source of heterogeneity in the system of scientific activities.

We have not included in this or other sections any articles from the "radical science" movement. While such essays generally manifest a realistic view of the relations between science and society, they tend to be brief essays and to obscure sociological content amidst political rhetoric. Readers interested in the issues raised in this book, and in this section on Science and Development especially, are encouraged to acquaint themselves with this literature. Specifically, we recommend *Science for the People,* the bi-monthly publication of Scientists and Engineers for Social and Political Action (e.g., Bazin 1972).

REFERENCES

Apter, D. E. 1965. *The Politics of Modernization.* Chicago: University of Chicago Press.

———. 1966. Forming a Scientific Culture. In *Education and the Development of Nations,* ed. John W. Hanson and Cole S. Brembeck, pp. 217-24. New York: Holt, Rinehart and Winston.

Bazin, Maurice. 1972. Science, Scientists, and the Third World. *Science for the People* 4:3-7.

Bernal, J. D. 1939. *The Social Function of Science.* Cambridge, Mass.: Massachusetts Institute of Technology Press, 1967.

Brown, Harrison. 1954. *The Challenge of Man's Future.* New York: Viking.

Brown, J. D. and Frederick Harbison. 1957. *High-Talent Manpower for Science and Technology.* Princeton, New Jersey: Princeton University Press.

Calder, Nigel. 1970. *Technopolis: Social Control of the Uses of Science.* New York: Simon and Schuster.

Canter, Jacob. 1967. American Higher Education for Students of the Developing World. In *Higher Education and the International Flow of Manpower: Implications for the Developing World,* pp. 29-37. Proceedings of the National Conference. University of Minnesota, Minneapolis. April 13-14.

Gilpin, Robert. 1968. *France in the Age of the Scientific Estate.* Princeton, New Jersey: Princeton University Press.

Goldsmith, M., and Allan MacKay, eds. 1965. *Society and Science.* New York: Simon and Schuster.

Gruber, Ruth, ed. 1961. *Science and the New Nations.* New York: Basic Books.

Halpern, Manfred. 1967. The Rate and Costs of Political Development. In *Readings on Social Change,* ed. W. E. Moore and R. M. Cook, pp. 182-84. Englewood Cliffs, New Jersey: Prentice-Hall.

Harbison, F., and Charles A. Myers. 1964. *Education, Manpower, and Economic Growth.* New York: McGraw-Hill.

Lewis, W. A. 1962. Education for Scientific Professions in the Poor Countries. *Daedalus* 91 (Spring): 310-18.

Meier, R. L. 1966. *Science and Economic Development.* 2d ed. Cambridge, Mass.: Massachusetts Institute of Technology Press.

Myrdal, Gunnar. 1968. *Asian Drama.* 3 Vols. New York: Pantheon.

National Academy of Sciences (NAS). 1969. *The Invisible University.* Washington, D.C.

Perkins, James A. 1966. Foreign Aid and the Brain Drain. *Foreign Affairs* 44 (July): 608-619.

Restivo, S. P. 1971. Visiting Foreign Scientists at American Universities: A Study in the Third-Culture of Science. Ph.D. diss. E. Lansing, Mich.: Michigan State University.

Revelle, Roger. 1963. International Cooperation and the Two Faces of Science. In *Cultural Affairs and Foreign Relations,* ed. Robert Blum, pp. 128-38. Englewood Cliffs, New Jersey: Prentice-Hall.

Shah, A. B., ed. 1967. *Education, Scientific Policy and Developing Societies.* Bombay: Massakalas.

Shils, Edward. 1961. Scientific Development in the New States. In *Science and the New Nations,* ed. Ruth Gruber, pp. 217-26. New York: Basic Books.

———. 1966. Toward a National Science Policy. In *Education and the Development of Nations,* ed. John W. Hanson and Cole S. Brembeck, pp. 209-17. New York: Holt, Rinehart and Winston.

———. 1967. On the Improvement of Indian Higher Education. In *Education, Scientific Policy and Developing Societies,* ed. A. B. Shah, pp. 475-99. Bombay: Massakalas.

Silvert, K. H., ed. 1963. *Expectant Peoples; Nationalism and Development.* New York: Random House.

Stone, B. C. 1969. Gaps in the Graduate Training of Students from Abroad. (letter to) *Science* 165 (June 6): 1118.

UNESCO. 1963. *Science and Technology for Development, Report on the United Nations Conference on the Application of Science and Technology for the Benefit of the Less Developed Areas,* in several volumes. Complete subject index and list of papers and reports in *Science and Technology for Development,* Vol. 8, "Plenary Proceedings, List of Papers and Index." New York, UNESCO.

———. 1970. *Science and Technology in Asian Development.* Paris. UNESCO.

Useem, John, and Ruth Hill Useem. 1968. American Educated Indians and Americans in India: A Comparison of Two Modernizing Roles. *The Journal of Social Issues* 24 (October): 143-158.

Vanderpool, C. K. 1971. Visiting Foreign Scientists in the United States: The Impact of Systemic and Role Circumscription and Dissociative Experiences on the Homogeneity of the International Scientific Community. Ph.D. diss. E. Lansing: Michigan State University.

GEORGE BASALLA

The Spread of Western Science

A small circle of Western European nations provided the original home for modern science during the 16th and 17th centuries: Italy, France, England, the Netherlands, Germany, Austria, and the Scandinavian countries. The relatively small geographical area covered by these nations was the scene of the Scientific Revolution which firmly established the philosophical viewpoint, experimental activity, and social institutions we now identify as modern science. Historians of science have often attempted to explain why modern science first emerged within the narrow boundaries of Western Europe, but few if any of them have considered the question which is central to this article: How did modern science diffuse from Western Europe and find its place in the rest of the world?

The obvious answer is that, until fairly recent times, any region outside of Western Europe received modern science through direct con-

George Basalla, "The Spread of Western Science," *Science* Vol. 156 (May 5, 1967); pp. 611-621. Copyright 1967 by the American Association for the Advancement of Science. Reprinted by permission of the author and publisher.

tact with a Western European country.[1] Through military conquest, colonization, imperial influence, commercial and political relations, and missionary activity the nations of Western Europe were in a position to pass on their scientific heritage to a wider world. This simple explanation is essentially correct, but it is entirely lacking in details. Who were the carriers of Western science? What fields of science did they bring with them? What changes took place within Western science while it was being transplanted? By what means is a flourishing scientific tradition fully recreated within societies outside of Western Europe? In this article I undertake to incorporate all these questions into a meaningful framework through the means of a model designed to aid our understanding of the diffusion of Western science.

THE MODEL

While making a preliminary survey of the literature concerning the diffusion of Western European science and civilization, I discovered a repeated pattern of events that I generalized in a model which describes how Western science was introduced into, and established in, Eastern Europe, North and South America, India, Australia, China, Japan, and Africa. The model, like the survey that produced it, is preliminary; it is a heuristic device useful in facilitating a discussion of a neglected topic in the history of science.

Three overlapping phases or stages constitute my proposed model. During "phase 1" the nonscientific society or nation provides a source for European science. The word *nonscientific* refers to the absence of modern Western science and not to a lack of ancient, indigenous scientific thought of the sort to be found in China or India; *European,* as used hereafter in this article, means "Western European." "Phase 2" is marked by a period of colonial science, and "phase 3" completes the process of transplantation with a struggle to achieve an independent scientific tradition (or culture).

These phases are conveniently represented by the three curves of Fig. 1. The shapes of the curves were not determined in any strict quantitative way, for qualitative as well as quantitative factors are to be included in the definition of scientific activity. In determining the height of a curve I am willing to consider quantifiable elements—number of scientific papers produced, manpower utilized, honors accorded—as well as the judgments of historians who evaluate, on a more subjective basis, the contributions of individual scientists. Furthermore, the curves describe a generalized process that must be modified to meet specific situations. Japan, for example, had an unusually long, and initially slow-growing, second phase because of the policy of political, commercial, and cultural isolation practiced by her rulers. This long interval quickly reached a peak after the Meiji Restoration (1868), when Japan was fully opened to Western influence.

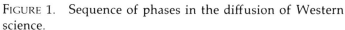

FIGURE 1. Sequence of phases in the diffusion of Western science.

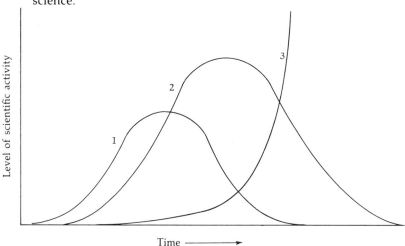

Thus it should be clear that when I refer to the graph of Fig. 1, I will (i) be mainly concerned with the gross features of the curves and (ii) be using the curves to illustrate my discussion and not to bolster it with independent support from empirical sources.[2]

The first phase of the transmission process is characterized by the European who visits the new land, surveys and collects its flora and fauna, studies its physical features, and then takes the results of his work back to Europe. Botany, zoology, and geology predominate during this phase, but astronomy, geophysics, and a cluster of geographical sciences—topography, cartography, hydrography, meteorology—sometimes rival them in importance. Anthropology, ethnology, and archeology, when they are present, clearly rank in a secondary position. These various scientific studies may be undertaken by the trained scientist or by the amateur who, in the role of explorer, traveler, missionary, diplomat, physician, merchant, military or naval man, artist, or adventurer, makes an early contact with the newly opened territory. Training and expertise in a science will increase the European observer's awareness of the value and novelty of his discoveries, but they are not the crucial factors. What is important is the fact that the observer is a product of a scientific culture that values the systematic exploration of nature.

Science during the initial phase is an extension of geographical exploration, and it includes the appraisal of natural resources. Whether the "New World" to be studied is North or South America, Africa, Antarctica, the moon, or a neighboring planet, it is first necessary to survey, classify, and appraise the organic and inorganic environment.[3] If the territory under surveillance is to serve eventually as a settlement for European colonists, the observer will probably follow the advice Sir Francis Bacon offered 17th-century planters of colonies.[4] First, he counseled, "look about [for] what kind of victual the country yields of itself " and then "consider . . . what commodities the soil . . . doth naturally yield, that they may some way help to defray the charge of the plantation." Botany, zoology, and geology have a direct relevance to this search for foodstuffs and exportable natural products.

Phase-1 science is not limited to the uncivilized country where European settlement is the object. It is also to be found in regions already occupied by ancient civilizations, some with indigenous scientific traditions. India and China, two nations in this category, fell under the scrutiny of European scientists when they came into continuous contact with the West. Although the possibilities for trade in exotic items partly explain European interest in the natural history of these countries, commerce did not supply the major impulse. Trade and the prospect of settlement both influence the European observer's investigation of a new land but ultimately his work is to be related to the scientific culture he represents. He is the heir to the Scientific Revolution, that unique series of events that taught Western man the physical universe was to be understood and subdued not through unbridled speculation or mystical contemplation but through a direct, active confrontation of natural phenomena. The plants, animals, and landscape of Europe had revealed their secrets when subjected to this method of inquiry: why should not the flora, fauna, and geology of an exotic land reveal as much or more?

The historical record is filled with examples of European naturalists collecting and classifying the plant and animal life they find in remote jungles, deserts, mountains, and plains and then publishing the results for the illumination of the European scientific community. In the Americas we begin with Gonzalo Fernández de Oviedo, called the first naturalist of the New World, and his book delineating the natural history of the West Indies (1535). From Oviedo in the 16th century, through the 17th and 18th centuries, there is a constant stream of Spanish, French, German, Dutch, Swedish, and English naturalists traveling on scientific expeditions to South America. In the early decades of the 19th century this movement culminates in the work of Alexander von Humboldt and Charles Darwin.[5]

Thomas Harriot, a 16th-century traveler and writer on the natural products and natives of Virginia, is the progenitor of a North American group of collectors, geologists, and surveyors. During the 18th century American colonial naturalists joined their European-based colleagues and continued the task of expanding European knowledge of the natural history of the northeastern and southeastern United States. Mark Catesby, John and William Bartram, Alexander Garden, Peter Kalm, and John Clayton are familiar names in this scientific enterprise. By 1800 the region east of the Mississippi River had been explored: now there was a shift of interest to the Western lands.

The wave of modern science had traveled from Europe across the Atlantic to the eastern and middle-western United States. During the 19th century science maintained its westward thrust as it was carried beyond the Mississippi by a series of government-supported and privately supported exploratory expeditions. From Lewis and Clark to the Colorado River venture of John Wesley Powell (1804–1870), the American West was the scene of phase-1 science. The sponsors of this science, however, did not reside in the older scientific capitals of Western Europe —London, Paris, Berlin—but lived in the eastern United States—in Boston, Philadelphia, and Washington, the emerging counterparts of the older capitals. This region, moving through the second into the third phase of the transmission process, was now in a position to act as a center for the diffusion of modern science. The time lag between the phases in the various geographical sections of the United States has had its effect on the current American scientific scene. The unequal distribution of scientific centers of excellence throughout the nation is due, *in part,* to the fact that some sections began the process of transplanting and nurturing science at a later date than others.[6]

The Pacific Ocean was opened to European scientists by the three exploratory voyages undertaken by Captain James Cook between 1768 and 1780. Cook carried Sir Joseph Banks with him on his initial voyage, and it was the latter who uncovered the botanical, zoological, and ethnological treasures of the Australian continent. Botanist Robert Brown, spurred on by Bank's success, gathered some 3900 species of Australian plants and produced *Prodromus florae novae Hollandiae et insulae Van-Diemen* (1810), a classic in botanical literature. Later in the century Sir Joseph Dalton Hooker and Alfred Russell Wallace were to make significant contributions to science, based on their collecting ventures in, respectively, Antarctica (1839–43) and the Malay Archipelago (1856–62).[7]

China, India, and Japan posed new problems for the spread of Western science. Ancient and civilized peoples inhabited these nations, not the primitives encountered elsewhere. Nevertheless, the first Europeans who visited them began the surveying and collecting of plant and animal life that has consistently marked their early contact with new territory. Natural history was studied in Japan prior to the arrival of the Christian missionaries in the late 16th century, but this native endeavor was soon to be dominated by Europeans, with their superior classificatory systems. Two Germans, Andreas Cleyer and Engelbert Kaempfer, are noted for their botanical work in 17th-century Japan. In the succeeding century Carl Peter Thunberg and Philipp Franz von Siebold, two medical officers in the employ of the Dutch East India Company, made lasting contributions to the study of Japanese natural history.[8]

China, once it was opened to Western ideas by the Jesuits in 1583, provided vast new opportunities for European scientific exploration. One customarily reads of the Jesuit missionaries as carriers of the new astronomy of Copernicus and Galileo to the learned men of China, but their correspondence and memoirs attest to their interest in the biological and geological sciences. The natural history studies of the first missionaries were soon to be expanded as hundreds of European scientists journeyed to China in the 17th, 18th, and 19th centuries. Botany alone caught the attention of so many of these Europeans that over 1000 pages in a two-volume history of European botanical discoveries in China[9] are devoted merely to a listing of their names and accomplishments!

The Portuguese, in pursuit of the spice trade, opened a sea route to India, bringing with them the first European science-collectors to that continent. When, in the 17th century, England replaced Portugal as the major influence in Indian affairs, English missionaries and physicians assumed the task of investigating Indian natural history. In the 18th century the English became masters of Indian trade, and the men attached to the East India Company turned naturalists. They acquired extensive collections of flora and fauna and hired native artists to sketch the specimens in their proper colors and ecological setting. The Company formally acknowledged the economic importance of its servants' botanical labors by establishing the Botanic Gardens at Calcutta in 1787. The travels and writings of Sir Joseph Dalton Hooker, whose botanical expedition to the Himalayas was the basis for his *Flora Indica* (1855) (see *10*), are a reminder that professional scientists were also actively engaged in study of the natural history of India.

The western coastline of Africa was explored by 15th-century Portuguese navigators, but the easy availability of gold and slaves on the coast, and the natural barriers to the exploration of the interior, kept substantial European contact limited to the periphery of the continent until the late 19th century. The Cape area, however, serving as a way station for India-bound vessels, had a European settlement in 1652.

South Africa, offering communities of European settlers and the advantages of a southerly location for telescopic observation, early attracted naturalists and astronomers, who came to observe its plants and animals, its geography and geology, and its heavens.[11]

In the second half of the 18th century a small number of observers from France, England, Sweden, and Denmark began a more intensive investigation of the natural history of the continent of Africa. The most ambitious 18th-century scientific expedition was mounted in 1798 by Napoleon Bonaparte as part of his military campaign in Egypt (1798–1801). Naturalist Geoffroy St. Hilaire, attached to the Napoleonic venture, collected Egyptian flora and fauna, paying special attention to the native fishes. His colleagues, including some eminent French scientists of the day, analyzed the soil and water of Egypt, made astronomical observations, and sketched and gathered Egyptian antiquities, thereby laying the foundations for modern Egyptology.[12, 13]

One should not conclude from this swift survey of world history that phase-1 science is confined to the period beginning in the 16th and terminating in the mid-19th century. Late in the 19th century, when Germany became an imperial power by acquiring territory in Africa and the Pacific, she assessed her colonial wealth in *Das Deutsche Kolonialreich* (1909–10), a work that includes studies by zoologists, botanists, geologists, and geographers. In the first half of the 20th century the polar regions were the goals of the scientific explorer. The current need for natural history studies in underdeveloped regions[14] and the prospects of lunar and planetary exploration promise new tasks for the phase-1 scientist.

All of the plant, animal, and mineral specimens collected in the foreign lands, as well as the information amassed there, were returned to Europe (or, at a later date, to the United States) for the benefit of its scientists. Phase-1 science may be scattered around the globe, but only nations with a modern scientific culture can fully appreciate, evaluate, and utilize it.

As early as the 17th century it was realized that contact with new lands is certain to affect the development of science at home. Bishop Sprat, in his history of the Royal Society of London (1667),[15] wrote that maritime nations were "most properly seated, to bring home matter for new Sciences, and to make the same proportion of Discoveries . . . in the Intellectual Globe, as they have done in the Material." . . . The "matter" sent back by the collectors filled the zoological and botanical gardens, herbariums, and museums of Europe; made obsolete the classificatory systems devised for European flora and fauna; gave rise to the new studies of plant and animal geography; and decisively influenced the Darwinian theory of organic evolution.[16]

The scientist who went out on an exploratory expedition often found that the experience gained from studying natural history in a foreign land modified his own scientific views. Michel Adanson, recalling his stay in Senegal (1749–54), commented:[17] "Really, botany seems to change face entirely as soon as one leaves our temperate countries." And halfway across the earth in Australia, Sir James E. Smith (1793) concurred:[18] "When a botanist first enters . . . so remote a country as New Holland, he finds himself . . . in a new world. He can scarcely meet with any fixed points from whence to draw his analogies." Thus European science, its practitioners forced to come to terms with exotic material at home and abroad, underwent a significant transformation while it was in the process of being diffused to a wider world.

COLONIAL SCIENCE

Colonial science (phase 2) begins later than phase-1 science but eventually reaches a higher level of scientific activity (Fig. 1) because a larger number of scientists are involved in the enterprise. Let me explain this use of the adjective *colonial.* First, as I use the term, colonial science is dependent science. At the phase-2 stage the scientific activity in the new land is based primarily upon institutions and traditions of a nation with an established scientific culture. Second, *colonial science* is not a pejorative term. It does not imply the existence of some sort of scientific imperialism whereby science in the non-European nation is suppressed or maintained in a servile state by an imperial power. Third, phase 2 can occur in situations where there is no actual colonial relationship. The dependent country may or may not be a colony of a European nation. This usage permits discussion of "colonial science" in Russia or Japan as well as in the United States or India.[19]

Natural history and the sciences closely related to the exploration of new lands dominate phase 1. During the early years of phase 2, natural history is still the major scientific interest, with the first colonial scientists joining in the survey of the organic and inorganic environment conducted by the European observers. As colonial scientific activity increases, the range of the sciences studied is expanded and finally coincides with the spectrum of scientific endeavor in the nation, or nations, supporting the activity. There is a possibility that the colonial scientist will extend this spectrum, that he will open up wholly new fields of science, but this is unlikely, not because the colonial scientist is necessarily inferior to his European colleagues but because he is dependent upon an external scientific culture and yet not a fully participating member of that culture.

Who is this colonial scientist? He may be a native or a transplanted European colonist or settler, but in any case the sources of his education, and his institutional attachments, are beyond the boundaries of the land in which he carries out his scientific work. This pattern is found in 18th- and 19th-century North and South America, Russia, and Japan; in 19th-century Australia and India; and in 20th-century China and Africa. If formally trained, the colonial scientist will have received some or all of his scientific education in a European institution; if informally trained, he will have studied the works of European scientists and will have purchased his books, laboratory equipment, and scientific instruments from European suppliers. This training will direct the colonial scientist's interest to the scientific fields and problems delineated by European scientists. Colonial scientific education is inadequate or nonexistent; the same can be said for colonial scientific organizations and journals. Therefore, the colonial scientist seeks the membership and honors of European scientific societies[20] and publishes his researches in European scientific journals.

Does the dependency of colonial science mean that it must be inferior to European science? Any answer to this question must consider the vigor of the scientific culture upon which the colonial science is dependent. Colonial science in Latin America, for example, advanced slowly as compared with developments in Western Europe. Several possible explanations of this lag may be proposed, but included among them must be the realization that modern science had not been extensively cultivated by Spain and Portugal, the colonizers of South America. A case in point is Brazil. Brazilian science received its greatest impetus during the hiatus in Portuguese rule when the Dutch (1624–54) broke the old ties and brought the colony under the full influence of Western European culture.[21]

Having mentioned the special case of Spain and Portugal, let me return to the general question of the inferiority of colonial science. As already noted, the colonial scientist works under handicaps at home and relies upon a scientific tradition located abroad. Although the group of men involved in the enterprise of colonial science is larger than that involved in phase-1 collecting, the number has not yet reached the critical size necessary for reciprocal intellectual stimulation and self-sustaining growth. The weakness, or lack, of colonial scientific institutions tends to cancel the advantages otherwise gained as the group approaches its critical size.

There is one final difficulty. Colonial scientists are oriented toward an established scientific culture but they cannot share in the informal scientific organizations of that culture. They cannot become part of the "Invisible Colleges" in which the latest ideas and news of the advancing frontiers of science are exchanged, nor can they benefit from the "continuing mutual education" provided by these informal groups of scientists.[22] These are some of the disadvantages colonial scientists face even when they are in touch with the superior and vigorous scientific traditions of a France, Germany, Great Britain, or United States.

Colonial science has its drawbacks, but it is in the fortunate position of being able to utilize the resources of existing scientific traditions while it slowly develops a scientific tradition of its own. Although colonial science will rarely create great centers or schools of scientific research, open new fields of science, or completely dominate older areas of scientific inquiry, it does provide the proper milieu, through its contacts with the established scientific cultures, for a small number of gifted individuals whose scientific researches may challenge or surpass the work of European savants. These few men become the heroes of colonial science, and the debt they owe the older scientific traditions is often obscured, as is the fact that they are not representative of the state of colonial science. Benjamin Franklin is such a hero. He was a creative experimentalist and theorizer whose researches on electricity overshadowed the contributions of many of his European contemporaries. However, in praising Franklin we should remember that his intellectual and institutional home was London and Paris, not Philadelphia, and that his model was Sir Isaac Newton.[23] The 18th-century chemist Mikhail V. Lomonosov holds a similar position in Russian colonial science, and, similarly, his intellectual base was outside of Russia, in Germany.[24] Colonial science need not be inferior to European science, and in the hands of a scientific genius it might be superior, but its ultimate strength lies in the growing number of practicing scientists whose education and work are supported by an external scientific tradition.

The United States and Japan provide interesting illustrations of the course and nature of colonial science. The American colonial period of science extended beyond the nation's colonial political status. In 1847 Swiss-born Louis Agassiz criticized American deference to England in scientific matters,[25] and as late as 1922 American physicists preferred to publish in the prestigious English journal *Philosophical Magazine* rather than in the American *Physical Review*.[26] By the second half of the 19th century, Germany and France, not England, had come to hold the greatest attraction for American scientists. The young Josiah Willard Gibbs received his doctorate in science from Yale (1863) and immediately left for Europe to complete his scientific education in Paris and Berlin. Gibbs was not alone. Hundreds of American chemists, physicists, and biologists in the late 19th and early 20th century pursued graduate studies, or gained Ph.D.'s, at Berlin, Leipzig, Göttingen, Heidelberg, Munich, or Paris.[27] In 1904 the president of the American Mathematical Society estimated that 10 percent of its members held Ph.D.'s from German universities, and at least 20 percent had studied mathematics there.[28] American scientific institutions could not provide the training or experience these men needed to bring them to the forefront of scientific knowledge.

"Of all the wonders of the world the progress of Japan . . . seems to me about the most wonderful." So wrote Charles Darwin in 1879 to Edward S. Morse, the American zoologist who had introduced the theory of organic evolution into Japan.[29] The rapid progress of Japanese science that impressed Darwin was of relatively recent origin. Prior to the Meiji restoration (1868) Japanese colonial science grew at a slow pace, thwarted by governmental prohibition, linguistic barriers, and cultural resistance. European science, carried to Japan by 16th-century Jesuit missionaries, was banned in 1636 when the government moved to halt the infiltration of Western religion and thought. This seclusionist policy was not relaxed until the first half of the 18th century, when Western ideas were permitted to enter in the form of Sino-Jesuit scientific treatises and Dutch books. The Dutch, having maintained limited commercial contacts during the period of isolation, provided the only direct channel of communication between Japan and Europe. Once the ban was removed, Japanese scholars took advantage of this channel by translating Dutch books summarizing Western science and learning. These books made available European knowledge of human anatomy and medicine, heliocentric astronomy, and developments in chemistry and physics. The Japanese translators went beyond their linguistic tasks and often repeated the experiments they learned about in their reading. Thus the physician Hashimoto Sokichi, while translating some Dutch books on electricity, decided to confirm their accuracy by repeating the electrical experiments of Benjamin Franklin. . . . [30]

In the Tokugawa era (1600–1868) (see *31*), especially in the later years, a growing number of Japanese savants were attempting to assimilate European science and technology. Nevertheless, Meiji science far surpassed the modest accomplishments of the previous period. After 1868 the Japanese government undertook a deliberate program of modernization, a program that paid special attention to the science of the West. The Japanese imported American, German, English, and Dutch scientists, engineers, and physicians to serve in native universities as teachers of aspiring scientists. Between 1868 and 1912 over 600 students were sent abroad for special training in the scientific and technological centers of America and Europe. Linguistic barriers were overcome by the translation of Western scientific textbooks and by the compilation of a dictionary of technical words (Japanese, English, French, and German). Insofar as her science was concerned, Japan was as dependent upon the Western scientific culture as any of those countries that are conventionally classified as political colonies of the Western nations.[32]

Colonial science begins when a small number of native workers or European settlers in the land recently opened to European science first participate in phase-1 exploration and then gradually shift their interest to a wider spectrum of scientific activity. All this takes place while the colonial scientist relies upon an external scientific tradition. The transition from phase 2 to phase 3 is more complex. Scientists in the third phase are struggling to create an independent scientific tradition; they are attempting to become self-reliant in scientific matters.

What spurs the colonial scientist to move from dependency to independency? Nationalism, both political and cultural, can sometimes be identified as the moving force. After the American Revolution there was nationalistic sentiment in the new nation which encouraged the building of an American science upon a native foundation.[33] Similar sentiment appeared in the South American colonies after their break with Spain.[34] In 1848 Andrés Bello, a Venezuelan thinker and educator, called for a South American science, bearing the stamp of its national origin, that would not "be condemned to repeat servilely the lessons of European science." "European science seeks data from us," he said, and then asked rhetorically, "shall we have not even enough zeal and application to gather it for them?" The answer was that data-gathering was not to be the only job of the Latin-American scientist, for the American republics had a "greater role to play in the progress of the sciences."

Nationalistic feelings may be significant in the transition from phase 2 to phase 3, but there are more fundamental forces working to bring about this change. Colonial science contains, in an embryonic form, some of the essential features of the next stage. Although the colonial scientist looks for external support, he does begin to create institutions and traditions which will eventually provide the basis for an independent scientific culture. A modest amount of scientific education will be undertaken by the colonial scientist, he will agitate for the creation of native scientific organizations, he may work for the establishment of a home-based scientific journal, and he begins to think of his work, and of the researches of his immediate colleagues, as being the product of his own nation. Colonial science has passed its peak when its practitioners begin a deliberate campaign to strengthen institutions at home and end their reliance upon the external scientific culture.

INDEPENDENT SCIENTIFIC TRADITION

The struggle to establish an independent scientific tradition, which takes place during the third phase, is the least understood, appreciated, or studied aspect of the process of transference of modern science to the wider world. Historians and sociologists of science have failed to realize the difficulty of fully integrating science into a society that previously had little contact with Western science. The easy success of colonial science does not adequately prepare a country for the arduous task of creating and supporting native scientific institutions and fostering attitudes conducive to the rapid growth of science. Scientists working in phase 3, and historians who later attempt to plot the development of science during phases 2 and 3, often misunderstand the era of colonial science. In both cases they tend to praise the high level of scientific activity reached in the colonial era and forget that that level of attainment was made possible through a reliance upon an older, established scientific tradition.

The colonial scientist, who was a member of a relatively small group of men oriented toward an external scientific culture, is to be replaced during the course of phase 3 by a scientist whose major ties are within the boundaries of the country in which he works. Ideally, he will (i) receive most of his training at home; (ii) gain some respect for his calling, or perhaps earn his living as a scientist, in his own country; (iii) find intellectual stimulation within his own expanding scientific community; (iv) be able to communicate easily his ideas to his fellow scientists at home and abroad; (v) have a better opportunity to open new fields of scientific endeavor; and (vi) look forward to the reward of national honors—bestowed by native scientific organizations or the government —when he has done superior work. These six elements are more in the nature of goals to be attained than common characteristics of phase-3 science. Since phase 3 is marked by a conscious struggle to reach an independent status, most scientists will not personally achieve all of these goals, but there will be general agreement that an overt effort should be made to realize them.

If a colonial, dependent scientific culture is to be exchanged for an independent one, many tasks must be completed. Some of the more important ones are as follows.

1) Resistance to science on the basis of philosophical and religious beliefs must be overcome and replaced by positive encouragement of scientific research. Such resistance might be ignored or circumvented by the colonial scientist, but it must be eradicated when science seeks a broad base of support at home.

The slow development of science in China can be explained, in large measure, by the inability of modern science to displace Confucianism as the prevailing philosophy. Confucian thought stressed the importance of moral principles and human relationships and discouraged systematic study of the natural world. The Confucian rejection of scientific knowledge is epitomized in a poem written in the early 19th century by a Chinese dignitary:[35]

> With a microscope you see the surface of things.
> It magnifies them but does not show you reality.
> It makes things seem higher and wider,
> But do not suppose you are seeing the things in themselves.

Attitudes of this sort persisted in China until the end of the 19th century, at which time the Confucian ideals were decisively challenged and gradually replaced by value systems closer to the spirit of Western science.[36]

2) The social role and place of the scientist need to be determined in order to insure society's approval of his labors. If science in general, or some aspect of the scientist's work, is considered suitable only for the socially inferior, the growth of science may be inhibited. When Louis Agassiz visited Brazil in 1865 he was surprised to find that the higher social classes held a strong prejudice against manual labor. This prejudice had its effect upon the development of science in Brazil. Agassiz noted[37] that as long as Brazilian "students of nature think it unbecoming a gentleman to handle his own specimens, to carry his own geological hammer, to make his own scientific preparation, he will remain a mere dilettante in investigation." ... The Brazilian naturalists were thoroughly acquainted with "the bibliography of foreign science," but their social mores cut them off from "the wonderful fauna and flora with which they [were] surrounded." Prejudices so deeply rooted in the social structure are not likely to be removed easily, and science is retarded.

3) The relationship between science and government should be clarified so that, at most, science receives state financial aid and encouragement and, at least, government maintains a neutral position in scientific matters. The history of Japanese science affords examples of the several possibilities in a government's response to science. Western science was suppressed by the Japanese government in the 17th century, partially accepted in the 18th century, and then enthusiastically supported after 1868. At no time was the Japanese government reacting to the general will of its people.

In those nations where public opinion is more instrumental in the shaping of government policy, state aid to science will depend upon the citizen's evaluation of the significance of science. This was the case in Australia in the 1830's, when there was some hope for the establishment of a national geological survey. A Sydney newspaper, however, expressed the prevailing sentiment when it declared editorially:[38]

"Zoology, Mineralogy, and Astronomy, and Botany are all very good things, but we have no great opinion of an infantile people being taxed to support them. An infant colony cannot afford to become scientific for the benefit of mankind." Scientists seeking the help and recognition of the state have, until recently, found it difficult to justify the expenditure of public funds to promote scientific research.

4) The teaching of science should be introduced into all levels of the educational system, provided, of course, an adequate educational system already exists. This will entail the building, staffing, and equipping of schoolrooms and teaching laboratories; the training of science teachers and of instructors in supporting disciplines; the production of science textbooks in an appropriate language; and the founding of libraries of science. Education in the sciences is not enough; a parallel program must be instituted to train the "foot soldiers of the scientific army"—the technicians, instrument makers, and their like. These changes can only be made if they are judged worth while by society, and if they are not too strongly opposed by conservative educators who are committed to other educational patterns. In view of the many possible sources of resistance to innovation in education, the creation of adequate programs in the teaching of science must necessarily be a long-term process. It is for this reason that American scientists were still seeking European scientific training after the United States had acquired many of the other features of an independent scientific tradition.

5) Native scientific organizations should be founded which are specifically dedicated to the promotion of science. These would include general professional associations, working for the advancement of the whole scientific profession; specialist societies, serving the particular needs of men engaged in research within a given field of science; and elite, honorific organizations, providing rewards for those who make the greatest contribution to the advancement of science. Scientific societies have always been closely associated with Western science; the foundation, in the 17th century, of the Accademia del Cimento, the Royal Society of London, and the Académie des Sciences is usually cited as one proof of the emergence of modern science in that era.

Napoleon Bonaparte acknowledged the importance of scientific societies when he founded the Institut d'Egypte, patterned after the major contemporary French scientific society, the Institut de France. Determined to bring the science of Western Europe to the ancient Near East, he attempted to recreate the Institut de France in Egypt in the hope that the new organization would play a part in the growth of Egyptian science as important as the part its progenitor had played in France. Military defeat (1801) ended Napoleon's plans, and it is doubtful whether the Institut alone could have carried the burden of introducing

Western science into Egypt.[13, 39] Nevertheless, Napoleon was correct in believing that scientific organizations were crucial to the establishment of modern science in a land hitherto untouched by Western influence.

6) Channels must be opened to facilitate formal national and international scientific communication. This can be accomplished by founding appropriate scientific journals and then gaining their widespread recognition. Many problems are likely to be encountered here.[40] A scientific journal cannot flourish unless there are enough scientists and subscribers to fill its pages and pay its costs. Even if the requisite number of potential contributors exists, there remains the question of its prestige. The colonial scientist, who is accustomed to writing for established European scientific journals, may not wish to jeopardize his international reputation by reporting his work in an unknown native periodical. Will the 18th-century American scientist, or his counterparts in 19th-century India or Australia, whose contributions appear in the *Philosophical Transactions of the Royal Society of London,* be satisfied to write for a natively produced periodical with few readers and little influence?

Finally, there are the difficulties presented by language. Should national pride dictate that the contributions to the new journal be printed in the mother tongue when that language is not familiar to Western Europeans, or should some concession be made in order to gain European readers? This was the question faced by the founders of scientific periodicals in Japan, in China, and (in the case of Rumania) in Central Europe.[41] Despite these problems, it is important that a country struggling to create an independent scientific tradition should publish journals of science filled with the researches of its own scientists.

7) A proper technological base should be made available for the growth of science. Western Europe had reached an advanced state of technical progress when modern science first made its appearance, and since that time it has been assumed that the two are fundamentally related. The exact nature of that relationship has not as yet been revealed by historians of science and technology.

Even without clear guidelines it is possible to indicate some of the links between science and technology that are significant for this discussion. A nation hoping to be self-sufficient in the realm of science certainly must maintain a level of technology that will produce the scientific instruments and apparatus needed for research and teaching. That this level is not to be reached without some difficulty is proved by the American example. European technology was transmitted to North America by the early settlers, but the colonies were slow to develop a craft tradition that specialized in the construction of scientific instruments for purposes other than navigation and surveying. Fine scientific instruments, to be used by American scientists in research, teaching, and exploration, were customarily purchased in England and France until the second half of the 19th century.[42] If America found it necessary to rely on Europe, one can imagine that an African or Asian culture, existing beyond the influence of Western technology, would find it much more difficult to reach the desired technological level and make its own instruments.

Economic determinists, along with some historians of technology, argue that technology has more to offer science than a mere collection of scientific instruments. They say that technology poses the very problems that dominate a scientific field in a given era. For the most part, historians of science reject this external interpretation and concentrate on the internal, conceptual development of science. If technology does direct scientific inquiry, as the first group contends, then it will be the overriding factor in the establishment of an independent scientific culture; if it does not, then it should be reduced to its role of provider of gadgetry for the scientist.

These are the extreme positions, but there is the possibility of a compromise that calls for a recognition of the complexity of the relationship between science and technology and demands a more subtle analysis. A proponent of this compromise will ask that the following investigations be made. First, we should determine to what extent a lifelong familiarity with a variety of machines prepares and predisposes an individual or culture to accept and extend the predominantly mechanical view of the physical universe bequeathed to us by the founders of modern science. Second, we should study the products of technology not merely as mechanical contrivances designed to fulfill specific, limited purposes but as cultural complexes that carry with them the attitudes, skills, and ideas of the culture that produced them. The latter topic has been explored in a recent book on the introduction of steamboats on the river Ganges in the 1830's.[43] These vessels provided far more than a rapid and effective means of transportation. They were vectors of Western civilization carrying Western science, medicine, and technical skills into the interior of India. In exploring these two topics we are likely to uncover the nature of the links between science and technology and learn more about the technological underpinnings of an independent scientific tradition.

Any one of the seven tasks listed above would present major problems for those who wished to gain an independent stronghold for modern science. Collectively, they present so severe a challenge that even a concerted effort on the part of the scientists will not soon bring noticeable results. Because of the difficulties involved in the completion of the tasks, I speak of a "struggle to establish an independent scientific tradition," and I illustrate it by the slowly rising curve of phase 3 (Fig. 1). Note, however, that, if the outcome of the struggle is successful, the curve rises abruptly, signifying the emergence of the nation among the leaders of world science.[44]

If my analysis of phase 3 is correct, then we should find that the non-European nations, after a long period of preparation, have only recently approached the supremacy of Westen Europe in science. The leadership achieved by Western Europe at the time of the Scientific Revolution was not challenged until the United States and Russia emerged as leading scientific nations in the period between world wars I and II. America first gained scientific eminence in the fields of genetics and big-telescope astronomy. In 1921 the English geneticist William Bateson, commenting upon recent American contributions to genetics, was moved to say in his address to the American Association for the Advancement of Science:[45] "I come at this Christmas Season to lay my respectful homage before the stars that have arisen in the West." The stars were the American biologists who had finally attained European recognition with their work in a new field of science. Physics in America came of age within two decades of Bateson's speech. Recalling the state of American physics in 1929, and testifying to the beneficent influence of J. R. Oppenheimer upon its maturation, I. I. Rabi remarked:[46] "When we first met in 1929, American physics was not really very much, certainly not consonant with the great size and wealth of the country. We were very much concerned with raising the level of American physics. We were sick and tired of going to Europe as learners. We wanted to be independent. I must say I think that our generation . . . did that job, and that ten years later we were at the top of the heap."

Contrary evaluations of Soviet science offered by friends and foes of the Communist ideology[47] have made it difficult to determine objectively the state of science in the U.S.S.R. In the 1940's, proponents of planned economy and planned science hailed Soviet scientific achievements, while opponents, pointing to the damaging influence of ideology on the biological sciences, held little hope for science in a totalitarian regime. Russia's advancement in weaponry and space technology during the next decade left no doubt that a strong program in basic science supported these technical feats. Critics might still complain of a bias toward applications in Soviet science, and point to the relatively small number of Russian scientists who have won the Nobel prize, but there is general agreement that the U.S.S.R. has taken her place as one of the leaders of world science.[48]

After several centuries of contact with European science the United States and the U.S.S.R. finally reached, and in some cases surpassed, the science of the Western European nations. This cannot be said of any other land outside of Western Europe. Japan, Australia, and Canada have shown signs of vigorous scientific growth, but they definitely rank below these two nations. China, India, and perhaps some South American and African countries may be placed in a third grouping of nations with great potential for future scientific growth and with major obstacles to be overcome before they establish their independent scientific cultures.

CONCLUSION

There is no need to summarize the features of this simplified model, which describes the manner in which modern science was transmitted to the lands beyond Western Europe. The graph of Fig. 1 and the examples drawn from science in various lands should have made them clear. It may be in order, however, to reiterate that there is nothing about the phases of my model that is cosmically or metaphysically necessary. I am satisfied if my attempt will interest others to go beyond my crude analysis and make a systematic investigation of the diffusion of Western science throughout the world.

Such an investigation would include a comparative appraisal of the development of science in different national, cultural, and social settings and would mark the beginnings of truly comparative studies in the history and sociology of science. The present lack of comparative studies in these disciplines can be attributed to the widespread belief that science is strictly an international endeavor. In one sense this is true. As Sir Isaac Newton remarked in his *Principia,*[49] "the descent of stones in Europe and in America" must both be explained by one set of physical laws. Yet, we cannot ignore the peculiar environment in which members of a national group of scientists are trained and carry on their research.

While I do not hold with the Nazi theorists that science is a direct reflection of the racial or national spirit,[50] neither do I accept Chekhov's dictum[51] that "there is no national science just as there is no national multiplication table. ...." In emphasizing the international nature of scientific inquiry we have forgotten that science exists in a local social setting. If that setting does not decisively mold the conceptual growth of science, it can at least affect the number and types of individuals who are free to participate in the internal development of science. Perhaps the effect is more profound; only future scholarship can determine the depth of its influence.

NOTES

1. Since World War II the United States and Russia have been in a strong position to act as agents for the introduction of modern science to underdeveloped regions, but their

examples do not break the line of transmission I have indicated. It is an easy matter to trace American and Russian scientific traditions back to the nations originally participating in the Scientific Revolution.

2. I. Bernard Cohen's "The New World as a source of science for Europe," *Actes du Neuvième Congrès International d'Histoire des Sciences* (Barcelona-Madrid, 1959), provided the inspiration for the model. Other works which influenced it included "Basic Research in the Navy," *Naval Res. Advisory Comm. Publ.* (1959); G. Holton, *Daedalus* **91**, No. 2, 362 (1962); T. S. Kuhn, *The Structure of Scientific Revolutions* (Univ. of Chicago Press, Chicago, 1962); D. J. de Solla Price, *Little Science, Big Science* (Columbia Univ. Press, New York, 1963); and W. W. Rostow, *The Stages of Economic Growth* (Cambridge Univ. Press, Cambridge, 1960). Donald Fleming's "Science in Australia, Canada, and the United States," *Actes du Dixième Congrès International d'Histoire des Sciences* (Hermann, Paris, 1964), vol. 1, sharpened my analysis, but Fleming and I disagree on fundamental points. I thank Dr. Everett Mendelsohn, Harvard University, and Dr. David D. Van Tassel, University of Texas, for advice on matters relating to phase 3 of my model.

3. Donald Fleming [*Actes du Dixième Congrès International d'Histoire des Sciences* (Hermann, Paris, 1964)] speaks of the "reconnaissance of natural history" as a part of the Age of Discoveries. See also Robert Boyle's "General heads for a natural history of a countrey, great or small." *Phil. Trans. Roy. Soc. London* **1**, 186 (1665); *ibid.* **2**, 315 (1666); *ibid.*, p. 330. Boyle has written a guide for the European observer in a foreign land.

4. Sir Francis Bacon, *Collected Works*, B. Montague, Ed. (Hart, Carey & Hart, Philadelphia, 1852), vol. 1, p. 41.

5. G. Fernández de Oviedo, in *Natural History of the West Indies*, S. A. Stoudemire, Ed. (Univ. of North Carolina, Chapel Hill, 1959); A. R. Steele, *Flowers for the King: The Expedition of Ruiz and Pavon and the Flora of Peru* (Duke Univ., Durham, N.C., 1964); V. Wolfgang von Hagen, *South America Called Them* (Knopf, New York, 1945). I thank Susan Lane Shattuck, School of Library Science, University of Texas, for information she provided on botanical expeditions to colonial South America.

6. For the pre-1800 period, see T. Harriot, *A briefe and true report of the new found land of Virginia* (London, 1588) [facsimile edition with introduction by R. G. Adams (Edwards, Ann Arbor, 1931)]; W. Martin Smallwood, *Natural History and the American Mind* (Columbia Univ. Press, New York, 1941); W. Bartram, *Travels Through North & South Carolina, Georgia, East & West Florida*, Mark van Doren, Ed. (Dover, New York, 1928); T. Jefferson, *Notes on the State of Virginia*, T. P. Abernethy, Ed. (Harper, New York, new ed., 1964); E. Berkeley and D. S. Berkeley, *John Clayton: Pioneer of American Botany* (Univ. of North Carolina, Chapel Hill, 1963). For Western exploration in the 19th century, see H. N. Smith, *Southwest Rev.* **21**, 97 (1935); W. H. Goetzmann, *Exploration and Empire* (Knopf, New York, 1966); S. W. Geiser, *Naturalists of the Frontier* (Southern Methodist Univ., Dallas, 1948).

7. B. Smith, *European Vision and the South Pacific: 1768–1850* (Oxford Univ. Press, London, 1960); F. W. Oliver, Ed., *Makers of British Botany* (Cambridge Univ. Press, Cambridge, 1913), pp. 108-125; L. Huxley, *Life and Letters of Sir Joseph Dalton Hooker* (Murray, London, 1918), vol. 1, pp. 54-167; A. R. Wallace, *My Life* (Dodd, Mead, New York, 1906), vol. 1, pp. 337-384. For early scientific collecting in other areas of the Pacific, see J. A. James, *The First Scientific Exploration of Russian America* (Northwestern Univ. Press, Evanston, 1942); S. P. Krasheninnikov, *The History of Kamtschatka* (Quadrangle, Chicago, 1962), pp. 57-167; P. Honig and F. Verdoorn, Eds., *Science and Scientists in the Netherland Indies* (Board for the Netherland Indies, New York, 1945), pp. 295-308; *Encyclopedia of the Philippines* (Philippine Education Company, Manila, 1936), vol. 7, pp. 52-107, 476-503.

8. Hideomi Tuge, *Historical Development of Science and Technology in Japan* (Kokusai Bunka Shinkokai, Tokyo, 1961), pp. 35-37, 76-77, 82-84; C. R. Boxer, *Jan Campagnie in Japan, 1600–1850* (Nijhoff, the Hague, 1950), pp. 50-52.

9. A. H. Rowbotham, *Missionary and Mandarin* (Univ. of California, Berkeley, 1942), pp. 271-274; P. M. D'Elia, *Galileo in China*, R. Suter and M. Sciascia, trans. (Harvard Univ.

Press, Cambridge, Mass., 1960); E. Bretschneider, *History of European Botanical Discoveries in China* (Russian Academy of Sciences, St. Petersburg, 1898).

10. E. Hawks, *Pioneers of Plant Study* (Sheldon, London, 1928), pp. 151-156; H. J. C. Larwood, *J. Roy. Asiatic Soc.* **62** (1962); M. Archer, *Natural History Drawings in the India Office Library* (Her Majesty's Stationery Office, London, 1962); L. Huxley, *Life and Letters of Sir Joseph Dalton Hooker* (Murray, London, 1918), vol. 1, pp. 223-365.

11. E. A. Walker, Ed., *Cambridge History of the British Empire:* vol. 8, *South Africa, Rhodesia & the High Commission Territories* (Cambridge Univ. Press, Cambridge, 1963), pp. 905-914.

12. P. D. Curtin, *The Image of Africa* (Univ. of Wisconsin Press, Madison, 1964), pp. 3-27; A. A. Boahen, *Britain, the Sahara, and the Western Sudan: 1788–1861* (Oxford Univ. Press, London, 1964), pp. 1-28.

13. E. L. Schwartz, "The French Expedition to Egypt, 1798-1801: Science of a Military Expedition," senior thesis, Harvard University, 1962. I thank Dr. Thomas deGregori, University of Houston, for his advice on science in Africa.

14. For science in the new German colonies, see *Sci. Amer.* **111**, 450 (1914). Proof that natural history studies are important to underdeveloped regions today can be found in J. P. M. Brenan, *Impact Sci. Soc.* **13**, 121 (1963); C. F. Powell, in *The Science of Science*, M. Goldsmith and A. Mackay, Eds. (Souvenir, London, 1964), p. 82.

15. T. Sprat, *History of the Royal Society* (London, 1667), p. 86.

16. For the impact of specimens collected abroad upon European science, see W. Coleman, *Georges Cuvier, Zoologist* (Harvard Univ. Press, Cambridge, Mass., 1964), pp. 18-25; B. Smith, *European Vision and the South Pacific: 1768–1850* (Oxford Univ. Press, London, 1960), pp. 5-7, 121-125; W. George, *Biologist Philosopher: A Study of the Life and Writings of Alfred Russell Wallace* (Abelard, New York, 1964), pp. 122-155; L. Eiseley, *Darwin's Century* (Doubleday, New York, 1958), pp. 1-26.

17. See F. A. Stafleu, in *Adanson* (Hunt Botanical Library, Pittsburgh, 1963), vol. 1, p. 179.

18. See B. Smith, *European Vision and the South Pacific: 1768–1850* (Oxford Univ. Press, London, 1960), p. 6.

19. Compare this description of colonial science with the one found in D. Fleming, *Actes du Dixième Congrès International d'Histoire des Sciences* (Hermann, Paris, 1964), vol. 1, pp. 179-196, and in *Cahiers Hist. Mondiale* **8**, 666 (1965).

20. As an example, consider the American colonial members of the Royal Society [R. P. Stearns, *Osiris* **8**, 73 (1948)].

21. J. H. Elliott, *Imperial Spain: 1469–1716* (Arnold, London, 1963), pp. 291, 338, 362-364, 381; C. R. Boxer, *The Dutch in Brazil* (Oxford Univ. Press, Oxford, 1957), pp. 112-113, 150-155.

22. The importance of informal communication is discussed by D. J. de Solla Price, *Little Science, Big Science* (Columbia Univ. Press, New York, 1963), pp. 62-91, and L. K. Nash, *The Nature of the Natural Sciences* (Little, Brown, Boston, 1963), pp. 299-311.

23. See *Quart. Rev.* **10**, 524 (Jan. 1814); I. B. Cohen, *Franklin and Newton* (American Philosophical Society, Philadelphia, 1956).

24. See B. N. Menshutkin, *Russia's Lomonosov* (Princeton Univ. Press, Princeton, 1952), pp. 23-37; A. Vucinich, *Science in Russian Culture* (Stanford Univ. Press, Stanford, 1963), pp. 105-116.

25. E. C. Agassiz, *Louis Agassiz: His Life and Correspondence* (Houghton Mifflin, Boston, 1885), pp. 435-436.

26. J. H. Van Vleck, *Phys. Today* **17**, 22 (1964).

27. L. P. Wheeler, *Josiah Willard Gibbs* (Yale Univ. Press, New Haven, Conn., 1952), pp. 32-45; J. M. Cattell, *Science* **24**, 732 (1906).

28. D. E. Smith and J. Ginsburg, *A History of Mathematics in America Before 1900* (Open Court, Chicago, 1934), pp. 111-114.

29. F. Darwin, *More Letters of Charles Darwin* (Appleton, New York, 1903), vol. 1, pp. 383-384.

30. See Hideomi Tuge, *Historical Development of Science and Technology in Japan* (Kokusai Bunka Shinkokai, Tokyo, 1961), pp. 80-81.

31. See Yabuuti Koyosi, *Cahiers Hist. Mondiale* 9, 208 (1965); Shuntaro Ito, *Actes du Dixième Congrès International d'Histoire des Sciences* (Hermann, Paris, 1964), vol. 1, pp. 291-294; Nakayama Shigeru, *Sci. Papers Coll. Gen. Educ. Univ. Tokyo* 11, 163 (1961).

32. Watanabe Minoru, *Cahiers Hist. Mondiale* 9, 254 (1965); Nakamura Takeshi, *ibid.*, p. 294; Masao Watanabe, *Actes du Dixième Congrès International D'Histoire des Sciences* (Hermann, Paris, 1964), vol. 1, pp. 197-208. I call special attention to the work of Eri Yagi (Shizume) on the introduction of modern science into Japan: *Sci. Papers Coll. Gen. Educ. Univ. Tokyo* 9, 163 (1959); *Actes du Dixième Congrès International d'Histoire des Sciences* (Hermann, Paris, 1964), vol. 1, pp. 208-210. She specifically speaks of a colonial period in Japanese science, but limits it to the years 1870–1895, and draws a graph showing the production and education of native physicists in Japan during the period 1860–1960. I thank Dr. Nakayama Shigeru, University of Tokyo, and Dr. William R. Braisted, University of Texas, for their help on developments in Japanese science.

33. See B. Hindle, *The Pursuit of Science in Revolutionary America* (Univ. of North Carolina Press, Chapel Hill, 1956), pp. 380-385.

34. See L. Zea, *The Latin American Mind*, J. H. Abbott and L. Dunham, trans. (Univ. of Oklahoma, Norman, 1963), pp. 100-103.

35. Quoted from H. Bernard, *Yenching J. Social Studies* 3, 220 (1949).

36. For opposition to Western science, see Y. C. Wang, *Chinese Intellectuals and the West: 1872–1949* (Univ. of North Carolina Press, Chapel Hill, 1966), pp. 3-37; G. H. C. Wong, *Isis* 54, 29 (1963). Religious beliefs may act to spur on the acceptance and growth of science. Max Weber and others have attempted to link the initial appearance of modern science with the prevailing Protestant ethic. See R. K. Merton, *Osiris* 4, 359 (1938).

37. L. Agassiz, *A Journey in Brazil* (Houghton Mifflin, Boston, 1884), p. 499. Agassiz's remarks and the illustration of Fig. 4 were first brought together by T. B. Jones, *South America Rediscovered* (Univ. of Minnesota, Minneapolis, 1949), pp. 193-195.

38. A. Mozley, *J. and Proc. Roy. Soc. N. S. Wales* 98, 91 (1965). I thank Mrs. Mozley for information regarding the growth of science in Australia.

39. Y. Salah El-Din Kotb, *Science and Science Education in Egypt* (Columbia Univ. Press, New York, 1951), pp. 80-83.

40. B. Silliman's struggle to establish the *American Journal of Science* serves as a case study of the difficulties involved in the foundation of a new scientific journal [see J. F. Fulton and E. H. Thomson, *Benjamin Silliman, Pathfinder in American Science* (Schuman, New York, 1947), pp. 117-129].

41. For language problems and science in China and Japan, see, respectively, J. Needham, *Science and Civilization in China* (Cambridge Univ. Press, Cambridge, 1954), vol. 1, pp. 4-5, and Eri Yagi (Shizume), *Sci. Papers Coll. Gen. Educ. Univ. Tokyo* 9, 163 (1959). A French edition of the official organ of the Rumanian Academy of Science was published up until the year 1948 [see R. Florescu, *Ann. Sci.* 16, 46 (1960)].

42. D. J. de Solla Price, *Science Since Babylon* (Yale Univ. Press, New Haven, 1961), pp. 45-67; S. A. Bedini, *Early American Scientific Instruments* (Smithsonian Institution, Washington, 1964), pp. 3-13. The Wilkes Expedition (1838–42), America's first venture in government-sponsored maritime exploration, was outfitted with European scientific instruments [see D. E. Borthwick, *Proc. Amer. Phil. Soc.* 109, 159 (1965)].

43. H. T. Bernstein. *Steamboats on the Ganges* (Orient Longmans, Calcutta, 1960).

44. Certain factors affecting the acceptance or rejection of modern science have not been listed here because they are so fundamentally related to the culture within which the struggle for an independent scientific tradition takes place that the participants in the struggle are not in a position to alter them, even if they are aware of their existence. Furthermore, we know very little about these factors. The following are some examples.

(i) *Language*. B. L. Whorf [*Language, Thought, and Reality* (M.I.T. Press, Cambridge, 1966), pp. 57-64, 207-219] has argued that a people's view of physical reality is conditioned by the structure of their language. Scientific concepts developed in one culture might be rejected, or misunderstood, in another. According to Whorf, the Hopi language embodies a metaphysics of space and time that is opposed to the classical Newtonian world view. (ii) *Childrearing patterns*. A Japanese sociologist has suggested that "the introduction of modern science over the last century [has] been especially accelerated because Japanese culture values childhood curiosity and, unlike some other societies, does not attempt to repress it" [*Science* **143**, 776 (1964)]. (iii) *Political nature of a society*. A. de Tocqueville [*Democracy in America*, P. Bradley, Ed. (Knopf, New York, new ed., 1948), vol. 2, pp. 35-47] believed that a democratic nation, fostering the individual's pursuit of profit and power, would necessarily excel in technology and applied science, while aristocratic societies would be predisposed to cultivate the theoretical aspects of science.

45. See *Science* **55**, 55 (1922).

46. See *In the Matter of J. Robert Oppenheimer: Transcript of Hearing Before Personnel Security Board* (Government Printing Office, Washington, 1954), pp. 464-465. Remarks made by two other scientists on the late emergence of American science may be found in J. H. Van Vleck, *Phys. Today* **17**, 21 (1964), and F. Seitz, *Science* **151**, 1039 (1966). Many historians claim that American science achieved maturity at an earlier date, sometime in the 19th century. D. Fleming does not think so [see *Cahiers Hist. Mondiale* **8**, 666 (1965)].

47. For the controversy over "free" and "planned" science in Great Britain, see N. Wood, *Communism and British Intellectuals* (Columbia Univ. Press, New York, 1959), pp. 121-151, 190-193.

48. For recent evaluations of Soviet science, see *Survey: A Journal of Soviet and East European Studies* **1964**, No. 52 (July 1964); J. Turkevich, *Foreign Affairs* **44**, 489 (1966).

49. I. Newton, *The Mathematical Principles of Natural Philosophy*, F. Cajori, Ed. (Univ. of California Press, Berkeley, 1934), p. 398.

50. See R. K. Merton in *The Sociology of Science*, B. Barber and W. Hirsch, Eds. (Free Press, Glencoe, Ill., 1962), pp. 19-22.

51. See *The Personal Papers of Anton Chekhov* (Lear, New York, 1948), p. 29.

JOHN FRIEDMANN

Intellectuals in Developing Societies

The intellectual's contribution, and particularly that of the "modern" intellectual, is made principally in three ways, each of which is essential to the process of cultural transformation: he mediates new values, he formulates an effective ideology, and he creates an adequate collective self-image. . . .

MEDIATING NEW VALUES

By virtue of his education and training, the intellectual is the most exposed among the population to the cross-currents of ideas that move the world. He is an urban resident; he travels; and he has easy access to communication media. Above all, his social status as an intellectual

John Friedmann, "Intellectuals in Developing Societies," *Kyklos* 13 (1960): 524-40 (abridged). Reprinted by permission of the publisher.

obliges him to be at least superficially concerned with values and ideas and forces him to react to, and to take a stand either for or against, the important new ideas that come to his attention. What he cannot do, in contrast to other segments of the population, is to remain indifferent.

The manner in which he reacts to these ideas will very much depend on his initial orientation, on whether he is "modern" or "traditional." It will also depend on his self-image. But where the traditionalist will have only a vague conception of himself, absorbed as he is in the collective life of his community, the modernist knows himself to be the carrier of new ideas, and there is a sense of great purpose about him.

The "modern" intellectual's self-image will be found to vary from country to country. A quick sketch of the Brazilian variant, however, may help to bring the portrait into focus.[1]

The modernist (in Brazil) conceives of himself as the herald of a new future. Self-consciously, he will profess the life of reason and of impersonal order. He will also avow a belief in experimental science, efficiency, and performance. Politics he will judge to be rotten and corrupt: salvation is rather to be found in management and administration—preeminently rational activities. The New Man is proud to be a specialist, a true professional, who perhaps knows only a little, but this little very well. He deprecates dilletantism, encyclopedism, and rhetoric. His ruling passion is professional achievement, and because of this he will often express a desire to leave his country and to work, perhaps, for an international organization or, at any rate, to live in a social environment that recognizes more fully the merits of his peculiar talent. For the Brazilian stage has grown too confined and too confining, his audience too small. Scientific inquiry as a distinctly superior value is not yet widely appreciated in Brazilian society: essays, history, *belles lettres*—not econometrics or a treatise on the chromosome—are still the high roads to popular acclaim and honor.

One might conclude, because of the moral isolation of the modernist, that his cocky self-assurance would soon be eroded by the more powerful currents of traditionalist thought. This is not so. What at first may seem to be a disadvantage, is in fact a gain. His self-image is reinforced in many ways, but principally through the solidarity characteristic of his small, exclusive group. Modern intellectual life is concentrated in Rio de Janeiro and Sao Paulo, and this facilitates personal contact and encourages a lively *esprit de corps.*

The intellectuals who thus make up the core of modernism in Brazil have many essential attributes in common: youth, intellectual background, and a passionate belief in their own cause. Where the traditionalists look inward upon their own community, the modernists look outward to foreign lands for inspiration. If in nothing else, they are united in this basic orientation.

What are the values for which the "modern" intellectual stands and which give resolution to his spirit? As suggested in the foregoing description of the Brazilian modernist, they are the secular values which an aggressive, Faustian West has acclaimed for the past 150 to 200 years and to which it is still committed.[2]

1. That constant striving is the *summum bonum* of existence and that the result of such striving is progress infinitely extended into the future;

2. That environment can be successfully mastered and turned to man's advantage through the practice of science and scientific technology;

3. That significant truth is established only empirically, and that achievement in performance is a major standard of practical as well as moral excellence;

4. That social and political equality is either a natural right or an historical inevitability;

5. That suffrage should be extended to all adult persons possessing minimum qualifications, and that participation in at least some of the processes of governing a nation is a right no less than an obligation of each citizen.

These ideas, so fundamental to the social order in the West with which we are familiar, are turned into powerful weapons of attack in the hands of a few "modern" intellectuals in societies whose central value system is constructed on quite different assumptions. By the new standards, traditions suddenly become devalued; the possibility of an harmonious adjustment between man and nature is destroyed; philosophical speculations and literary-aesthetic pursuits fall before the sword of empirical research; and feudal, autocratic, and colonial patterns are attacked in the name of democracy and the right of self-determination. Political ideology may endow these values with a special meaning, and certainly their institutional embodiment may vary from culture to culture, though perhaps not as much as one would be inclined to believe. Yet through all the diversity of forms, some common, basic elements persist and distinguish these values unmistakably as "modern" in the sense in which the term is used in these pages. Progress conjures up the ethos of the times; science and technology are the means by which progress is achieved; the theorems of social equality give rise to movements of reform; and democracy—whether of the people's variety or liberal—establishes the necessary political setting. These values are interrelated and mutually supporting. They go hand in hand with the profound economic changes from an agrarian to an urban way of life, and the institutions of urban-industrial society are unthinkable apart from them.[3]

There appears to be a sequence of actions that is normally followed in the attempt to substitute, for the inherited ideas of right and wrong, new values. First, the assault will be upon the existing native traditions only. It will then be extended into a global criticism of the social order from the vantage point of the new values, culminating in an insistence upon the necessity for fundamental reform or revolution. In the following sketch of this action-sequence, no pretense is made to be historically specific. Although unique deviations from the general pattern can be expected for every country, the schema which follows may be nonetheless useful as a heuristic device in the study of the process that leads to social transformation.

The modernist begins by turning against his fellow-intellectuals, the traditionalists who, under the persistent fire of verbal attack, emerge as the principal defenders of the customs and values handed down through the generations. This initial skirmish may end in the complete capitulation of the traditionalists, especially in countries where indigenous intellectual traditions are not highly developed. But it may also lead to remarkable reinterpretations of some of the leading traditions which will render them more suitable to the present and blunt the edge of the modernist's tongue. The varieties of traditionalist response are outlined, for China, by Levenson[4] and for Japan, by Sansom[5]; they do not need to be repeated here. On the other hand, in most countries, the traditionalist faction of the intellectual elite will have been so thoroughly discredited by its recognized failure in the leadership of the country, that it can at best fight a delaying action.[6] Popular feeling will be easily aroused against it and nothing in this regard is likely to be left undone. In any event, the result of this initial proselytizing stage is the widening influence in intellectual circles of a western mode of thought and valuing, as well as the growing admission of the western superiority in practical matters. The gate is thus thrown open for the "demonstration effect" to do its work.

The second stage is reached when the modernist's polemic is turned forcefully, directly against the entrenched powers. To this, the government's reaction may be so determined, that some of the more articulate intellectuals may actually be forced to "go underground" or into involuntary exile. But lacking in popular support, the government's response may also be quite feeble, giving way to the new challenge without a serious struggle. In either event, the ground will be slowly prepared for a basic change in the institutional structure of the country. In periodicals, newspapers, and pamphlets, the existing regime and its supporters will be subject to a baiting kind of criticism, month after month, year after year. Scandals will be brought to light; shameful social conditions will be uncovered; economic mismanagement will be highlighted. Each and any action of the regime will be exposed to cutting ridicule. In the meantime, more or less openly or in conspiratorial silence, the plotting will go on, with perhaps an occasional assassination and other acts of terror, until the proper moment—some bumbling failure, some final stupidity of the regime—arrives.

The third and concluding stage—that of the actual "takeover"—is reached when the "modern" intellectual leads a revolutionary or, more rarely, a reform movement to unseat the regime and commence the great task of national reintegration around new value principles. The revolutionary action may be directed against an existing colonial regime, as in Indonesia, India, or French Indo-China, or a feudalistic indigenous government, as in Turkey, China, or Egypt. But in the end, the result will be the same: a new elite of intellectuals will have risen to power under the guiding hand of charismatic leaders: Soekarno, Nehru, Ho-Chi-Min, Ataturk, Sun yat-sen, Nasser, Lenin, Mao-tse-tung, Vargas, Nkrumah . . . great men of our time, regardless of where they stand (or stood) politically. They are acclaimed the fathers of their nations.

Only now can a beginning be made in the slow and tedious struggle for reform and national reintegration around the focus of new values.[7] Province by province, profession by profession, activity by activity, the same process must be repeated: traditionalists must be won over; the existing order must be challenged; reform must be initiated. It is the process by which urban institutions and urban values spread out to organize society in a new way.[8]

Precisely because it is slow and tedious, this process may lead to feelings of frustration and bitterness among the "modern" intellectuals. Again, the Brazilian situation can be enlightening, showing the possible range of reactions to the new situation. The period to be discussed is post-revolutionary and post-Vargas. Brazil is making rapid progress in industrialization, but the country is so huge, communications are so backward, and modern industrial centers so few, that large regions of the country have been left behind in the race towards a brighter future. The following observations are taken from notes made during my stay in Brazil between 1955 and 1958.

> The modernist levels a constant fire of criticism, much of it purely destructive, against the existing order. To the outsider, it is often frightening with what sardonic delight he is told tales of official ignorance and cupidity, of public immorality and professional incapacity. Instance is piled upon instance until a truly monstrous picture emerges. The picture is obviously distorted, but it is the reality by which a large number of intellectuals lives.

It is also a fashionable reality. Like Cato crying the repeated refrain concerning Carthage, the modernist ends his litanies with the remark that the mentality of the Brazilian people must be changed. Now, the analysis of national character has perforce something glamorous about it, and the less flattering the portrait, the more credit goes to the painter for his presumably astute and penetrating perceptions. It is therefore not always easy to know whether the bitterness so freely worn on the sleeve is posed or genuine. At any rate, it is interesting to observe that this profound dissatisfaction with the national way of life is already of long standing, and that its expression continues to be something of a mark of intellectual integrity.[9] Already Capistrano de Abreu born in Ceara in 1853 wrote in this modish vein:

"The jaburú is the bird that for me symbolizes our native land. He is big and imposing, with coarse legs and strong wings. And he passes his days, one leg crossed over the other, sad, sad, with that austere and vile sadness."

In view of this continuing attitude, it is small wonder that the solutions which are frequently proposed are radical: self exile, revolution.

The new men are impatient. For the most part, they are blind to the constructive forces of gradual change. The dissatisfactions are great, and their frustrations still greater. Exaggerations become accepted in this context. And the tragic sentiment of life from which even the modernist has not entirely extricated himself, adds the last touch. The argument is pressed to its ultimate tragic conclusion: The vicious circle is brought into play and flight and violence appear as the only reasonable ways to break through the magic of the spell.

FORMULATING AN IDEOLOGY

One of the signs of the disintegration of a social order is the widely felt loss of moral direction. As the past, with its store of accrued values, customs, and traditions, is losing its intrinsic significance for the individual and leaves him wondering about the present and the future, the pursuit of narrow self-interest may be the only alternative left to him and, indeed, may appear to be a necessary measure of protection against the predatory actions of others. The result will be a disjunction of the social will; a certain aimlessness; a lack of stability in social institutions; and a widespread shunning of responsibilities beyond those to oneself and one's immediate family. If this situation is inimical to the social good, it is likewise prejudicial to economic progress. For instead of reintegration of the social system, there is only division upon division in the social, economic, and political life of the community.

It is the intellectuals who are chosen for creating, out of this incipient chaos, the foundations for a new moral order. Somehow, the cataclysmic changes of the present which, taken by themselves, appear to be so utterly senseless, must be rendered meaningful: they must be interpreted within a larger and inclusive framework of objectives that paints in imaginative strokes the vision of a better world. Unless an adequate ideology is forthcoming, sufficiently convincing to appeal to large numbers of the population, the ascendency of strong men relying principally on force and on the fear it inspires, is a near certainty. Every large-scale and lasting social transformation must be grounded in an ideology.

In developing societies, this ideology is economic development. The formulation, elaboration, and promulgation of this ideology is one of the principal tasks of "modern" intellectuals in developing countries. Politically, it is an altogether neutral ideology which may be held with equal fervor by Communists and Liberals alike.

The Communists, in particular, have been acutely aware of the essentiality of an ideology of economic development and have directed their appeal in the underdeveloped countries primarily at the intellectual who is both more susceptible to this approach than others and more likely to use it as a tool for social change on a grand scale. According to Morris Watnick,

> [the Communist parties in colonial areas] have acquiesced in the primacy of the intellectuals in the movement because the acceptance of any alternative leadership coming from the ranks of the peasantry or the industrial workers (assuming the possibility of such leadership) would entail sacrifice of prime objectives of the party—viz., the seizure of power and the launching of a long-range plan for internal planning and reconstruction. Gradual and piecemeal reforms and certainly basic reforms designed to bring immediate economic relief to the masses (for instance, in the credit structure of the area) undertaken by non-Communist regimes would be welcomed by the masses of the peasantry because they are in accord with their immediate and most pressing interests. A program of seizing political power followed by prolonged industrialization, economic planning, recasting the social structure, realignment of the country's international position in favor of the U. S. S. R.—these are considerations of the type which can attract intellectuals only.[10]

In other words: reform is empirical, pragmatic, and piecemeal; it does not require ideological support. But wholesale social transformation, especially where it is to be achieved with great rapidity, cannot occur without an explanation and a broad-scale justification of what is happening, expressed in symbols that claim to be universally valid. It is the intellectuals who alone are capable of such an explanation, of what Plato, in quite another context, referred to as the Big Lie. For admittedly, an ideology can only be "true" in the ideal realm, but never in the arena of everyday affairs, where it must forever appear as a subtle mockery of the facts of existence.

Essentially, what the ideology of economic development preaches is the doctrine of the unlimited expansion of a nation's capacities for production and consumption. A whole range of values, images, and ideas is implied in this conception. It is not simply a useful device for ordering the data of experience so that social theory may be advanced. Everyone who thinks in terms of economic development, automatically divides the world into two unequal parts, the one developed and the other under-developed, and no terminological contortions can obscure the implied value that somehow the under-developed areas should come up to the standards of the developed world. The ideal image of a "developed" society which is progressing along a path of unlimited expansion is the standard by which a country is classified as either under-developed, less developed, backward, or developing, all adjectives, however, meaning the same thing. As Gunnar Myrdal has observed: "The use of the concept 'the under-developed countries' implies a value judgement that it is an accepted goal of public policy that the countries so designated should experience economic development. It is with this implication that the people in the poorer countries use the term and press its usage upon people in the richer countries."[11]

But "poor" and "rich" are hardly satisfactory criteria in this context. Poverty and wealth refer to given *conditions;* they are essentially static conceptions. The ideology of economic development, on the other hand, is a dynamic conception: the developed countries have achieved the break-through to cumulative growth; the under-developed countries still remain in the nether regions of economic and social stagnation. The "truth" or validity of this picture is irrelevant; what is important is the belief in such an imagery. Accordingly, economic development is equivalent to the path which a society describes as it changes from one historical configuration of social relationships to another, as it "takes off"—the metaphor is characteristic—into the rarified attitudes of cumulative progress.

Underlying these changes are the Faustian values of the West, the values of perpetual striving, science, egality, and popular democracy. These may be viewed as simply a continuation of the 19th century philosophy of progress in the 20th—an instance of a curious cultural lag. All doubts regarding the fundamental reality of progress are cast to the wind, and for a country to remain outside the pale of progress is considered plainly irresponsible. For the morality of our time preaches the great munificence of the West which alone is held capable of bringing happiness among the people. That happiness is universally desired is, of course, a questionable proposition. In certain languages, the word does not even exist (for instance, in Japanese). But the problem reaches still more deeply. For the West the value of progress is essentially indeterminate and open-ended: the good life is not this or that, but any life which is constantly endeavoring to change itself. According to this doctrine, happiness is found in *doing* only; it is the ethic of work that is praised as the fundamental meaning of existence. And thus examined closely, the ideology of unlimited expansion turns out to be simply a logical extension of the Calvinist doctrine of salvation through works. It is essentially a theological conception that is being foisted upon the rest of the world.

That the ideology of economic development has its origin in the historical experience of western nations gives a certain flavor of reality to the idea and presses it into a particular mould of institutional and behavioral patterns from which but few deviations are possible. As Rostow[12] has pointed out, the issue at stake is "the long and fluctuating story of sustained economic progress in the West." And the United States, and lately Russia, is taken as the model by which the efforts in the rest of the world are judged.

A recent publication of an American research institution formulates the ideology of unlimited expansion in this way: ". . . in North America, and some other places, the material betterment of human living has been proceeding for at least 150 years . . . Economic growth is a well-established process . . . The revelations . . . afforded by statistical knowledge disclose unmistakably a long trend of growth. Since the main causes identified with growth continue to operate, that trend can reasonably be expected to continue. A momentous conclusion follows: *that if we conduct our economic affairs with intelligence, we can reach a general level of material well-being higher than was ever before conceived of as possible.*"[13] Progress, in this view, becomes merely a matter of intelligent behavior. Few are the "modern" intellectuals anywhere today who would not support this affirmation as describing the essence of their faith.

The near-universality of the ideology of economic development is itself evidence of the creative role of ideas in historical evolution and evidence also that the community of "modern" intellectuals is becoming international. The language of economists in both East and West abounds with such suggestive terms as break-through, take-off, and cumulative growth. In this view, problems are automatically defined as merely "temporary obstacles" to unlimited expansion that can be overcome by appropriate action. One may therefore be permitted to wonder whether the Development Plans that are becoming standard instruments of government policy in certain countries are not basically statements of ideological purpose rather than aids to economic decision-making. Are not such plans primarily expressions of intent, setting forth the immediate objectives and the means for attaining them, one Plan succeeding another, with the objectives being pushed further and further into the future, as it were, without limit? Do they not attempt to justify present sacrifices in the name of a future prosperity and national strength? And do they not promise the eventual attainment of that ultimate bliss, the propulsion into the hyperbola of economic growth? Economists may be debating the relative merits of industry vs. agriculture, but this is a dispute over means, not ends. The end is tacitly agreed to from the very outset. It is the image of a materialistic, urban-industrial civilization reaching for ever higher levels of prosperity and a constantly increasing measure of rational control over environment. Political differences may obscure this agreement, but among the "modern" intellectuals it is pervasive and compelling.

The "modern" intellectual in developing societies has taken over from the West his values and the ideology of economic development. But he serves as more than merely a transmitter of ideas which he has learned and made his own. He also applies them to the specific instance of his country; he interprets them; he formulates policies in terms of them. As an ideology, economic development becomes a rallying cry for the masses, a political issue of compelling urgency that no government, in the long run, can ignore. It also represents a vantage point from which social criticism may be levelled and serves as a schema for explaining the often contradictory experiences of a society in transition. But above all, it indicates the direction in which such societies must move if they are to be admitted into the charmed circle of the "progressive" (developed?) nations and be respected by the countries they would imitate.

CREATING A NATIONAL SELF-IMAGE

Among the intellectuals in developing societies, a singular, intensive preoccupation may be discerned: the shaping of a national self-image. Again and again, as if ridden by a great anxiety, they will repeat their questionings: Who are we? What is unique about us? What is our destiny as Japanese, Koreans, Indians? As the former certainties fall by the wayside, and Western values gradually encroach upon many of the old traditions, there is great danger that the sense of national identity will be lost, if indeed it ever existed. Some countries may be altogether bewildered by the consequences overtaking them, by the clash of cultures and the paradoxes this engenders, an intellectual bewilderment initially, but one which has far-reaching consequences for all parts of society. More than do others, intellectuals need some sense of moral direction, some self-confidence that is more than merely a reliance on one's faculties of reason, but must be grounded in a profound sense of national identity. French, German, or American intellectuals may occasionally aspire to be universal spirits, untrammelled by a provincial outlook; but they are also deeply intricated in a national way of life. While they may not always take pride in it, they cannot help but think, speak, and *be* either French, German, or American. But what, by contrast, does it mean to be an Indian or a citizen of Ghana? In countries which, like India, are composed of a patchwork of local cultures; or like the new nations of Africa, are lacking in an eloquent and universally admired high cultural tradition; or like Indonesia, are missing a common national past reaching back over the centuries; or like Puerto Rico, are being overwhelmed by cultural influences from a more powerful neighbor, the quandary of the intellectual, and especially the "modern" intellectual, is very great, indeed.

In such countries, the urge to discover a truly national consciousness, a holy mission capable of inspiring both the elite and the masses with a sense of potential greatness and of unifying the nation's culture, is especially great. A simple form of nationalism is hardly, in itself, an adequate means for this purpose. The standard brand of nationalism is rather noted for its defensive qualities: it is not pro-anything, but virulently anti-foreign. Only when a satisfactory self-image has at last been found, will nationalism in its more extreme varieties shrink into insignificance. As a defensive ideology, auxiliary to the primary ideology of economic development, nationalism may be considered as a temporary expedient, as a way to gain time during which a positive self-image, with all the powers of a collective myth, can be shaped. Just as Americans will frequently search out the place of their European origin and trace back their ancestry to this or that region, so a young nation will seek to explain itself in terms of a self-image which is capable of forging a link with the past and of revealing the true genius of its people.

For some, a revitalized religion may be a sufficient source of self-identification. U Kyaw Thet, for instance, notes the new use of Buddhism for national identity in Burma:

> Painfully aware that their national pride—even their continued existence—was manifestly debatable, the Burmese had to produce something tangible and traditional to justify their future as a separate entity. They found what was needed in Buddhism. The assorted Europeans might be richer, stronger, better trained, but it was comforting to know that all this was as nothing because they did not possess the jewel of the true faith. Buddhism began at this stage to acquire nationalistic overtones and, at the same time, its individualism became increasingly significant.[14]

Similar efforts to make religion the basis of a revitalized cultural tradition, capable of standing up to the West, appear to be made in certain Moslem countries, for instance Pakistan. But it is certainly open to doubt whether such efforts can succeed in the long run. The spiritual values of religion are everywhere conservative and contradictory to the basic Western values that are preached as the new gospel. Only by a set of fortuitous circumstances did Calvinist doctrine once fit so beautifully the demands of a rising capitalist middle class. The phenomenon is not likely to be repeated elsewhere. Moreover, many of the leading "modern" intellectuals are, if not actually atheistic, at least indifferent towards religious beliefs in the customary sense. A man like Nasser will look to Islam mainly as a source of potential political unity in a pan-Islamic empire, but not as a source of living values and national identity for Egypt.

In most instances, the process of arriving at an adequate collective self-image will not succeed until a considerable amount of social and economic transformation has already occurred. When this has happened, the "modern" intellectual will be able to shift from a predominantly critical tone to a more positive statement of ideas, and his portrayal of national genius will have a basis in recent accomplishment and fact.

The effect of a national self-image, once it enjoys wide enough currency, tends to be cumulative. It reinforces the initial tendency towards national unification and integration which, in turn, helps in the further elucidation and sharpening of the image. In a sense, what is here involved is the gradual discovery of a nation's cultural assets. Seen by itself, the results of such a search might appear to be esoteric. But in the context of a developing society, the search for a national self-image and the discovery of the unique resources of a culture, plays an essential role.

The most successful attempt in this direction has perhaps been undertaken in Brazil, where the search for a national self-image dates back for several decades.[15] As early as 1942, Gilberto Freyre attempted to formulate an image of Brazil as a " luso-tropical civilization."[16] This attempt has been carried further in his recent book, written in English, and significantly entitled, *New World in the Tropics.*[17] Whether it is wholly justified or not, whether his thesis is part myth or not, Freyre's courage to face his country's future must be acknowledged when he writes (p. 146):

> One thing at least is true: the development of a modern civilization in Brazil is becoming more and more the development of a new type of civilization which makes the Brazilians, already considerable pioneers in

their history, pioneers of a new and even more exciting future. More than
any other people, they are developing a modern civilization in the tropics
whose predominant traits are European, but whose perspectives—and I
should insist upon this point—are extra-European.

It would be inappropriate to attempt to summarize his thesis in a few
words. And it would be wrong to suppose that Freyre's particular version has found great favor among Brazilian intellectuals. And yet, it
appears to be the most successful attempt so far to formulate in positive,
convincing language, the essence of Brazilian culture. Although his
concept of a luso-tropical civilization is not very popular at present,
perhaps because many Brazilians do not care to be reminded of their
Portuguese ancestry, though they may be very much attached to their
mother country through sentimental links, many of Freyre's detailed
comments on the Brazilian scene are already widely accepted and constitute a growing body of shared doctrine among Brazilian intellectuals.
As a result, the next few decades promise to witness a heightened degree
of self-consciousness and increasingly independent behavior on the part
of the Brazilian people.

CONCLUSION

To understand more thoroughly the part played by intellectuals in social
transformation, is to gain a deeper insight into the cultural history of
our time. This history is distinguished from all previous history by
virtue of being in process of becoming truly universal. This means:

1. Western values and institutions are replacing native traditions in
nearly all parts of the world, resulting in the emergence of an international urban culture which, though it may have many local variants, will
have a common ideological, institutional, and even political orientation.

2. Decisions in any part of the world are having significant repercussions in all other parts: political and economic interdependency on a
global scale is increasingly becoming part of the human condition: it is
the historical reality of the contemporary age.

3. Historical research has encountered this reality and is beginning to
work towards an interpretation of world events that does justice to the
growing convergence of the separate historical destinies of human societies: The Archimedean point being located *above* the earth, the perspective becomes global.[18]

The foregoing discussion of the intellectual in developing societies
ought to have made clear that his is largely a contribution to the emergence of a genuine ecumenical history. It is the intellectual who, in a

fundamental way, creates the basis in values and ideology for such a history. At the same time, he also works towards national and cultural identity within the context of a common civilization. The search for a national self-image and the efforts to establish a viable ideational link with one's historical past is nothing else but this.

Finally, the decisions of public policy which are so instrumental in effecting basic social changes are, in a large way, subject to influence by intellectuals: it is they who define the problems, formulate the goals, and indicate the alternative means for attaining intermediate objectives. The balance of power between modernists and traditionalists, and their particular reading of the current problems, will be strongly reflected in the course of the country's development within the modern world. The task of the intellectual is thus to clarify the basic issues confronting the society; to define the choices that can be made; and, in the final analysis, to help implement these choices through political action. In their approaches to this task, traditionalists and modernists will differ markedly from one another. . . .

I can think of no better way to summarize and conclude this discussion than by quoting Florian Znaniecki. "New ideals," he writes, "are initiated by thinkers who become aware of present conflicts within or between human collectivities and believe that the realization of their ideals would substitute harmony for conflict. And this is what creative reorganization of cultural systems is intended to achieve."[19]

NOTES

1. The following remarks are taken from an unpublished study by the author of "Intellectuals and Economic Development in Brazil," and are the result of almost three years of field work in that country (1955–1958).

2. Sjahrir . . . has written: "For me, the West signifies forceful, dynamic, and active life. It is a sort of Faust that I admire, and I am convinced that only by utilization of this dynamism of the West can the East be released from its slavery and subjugation.

The West is now teaching the East to regard life as a struggle and striving, as an active movement to which the concept of tranquillity must be subordinated. Goethe teaches us to love striving for the sake of striving, and in such a concept of life there is progress, betterment, and enlightenment. The concept of striving is not, however, necessarily connected with destruction and plunder, as we now find it. On the contrary, even in *Faust,* striving and struggle have the implication of constructive work, of undertaking great projects for the benefit of humanity. In this sense, they signify struggle against nature, and that is the essence of the struggle: man's attempt to subdue nature and to rule it by his will. The forms that the struggle takes indicate the development and refinement of the individuals who are engaged in the effort. . . . The East must become Western in the sense that it must acquire as great a vitality and dynamism as the West. Faust must reveal himself to the Eastern man and mind, and that is already going on at present." [Soetan Sjahrir, *Out of Exile,* New York, 1949, p. 144-147.] Paradoxically, some western intellectuals have recently displayed a yearning for the quietism of the East!

3. Sansom . . . opposes this thesis by suggesting that Japan's recent development was "but a natural process of evolution which produced results similar to those which had arisen in the West out of similar circumstances." [G. B. Sansom, *The Western World of Japan: A Study of the Interaction of European and Asiatic Cultures,* New York, 1959, p. 35.] Yet already the Meiji charter seemed to provide, at least on the surface, for popular participation in government, social egality, and education for science. While the intent may have been less sweeping than a first reading of the text might suggest, the charter was sufficiently vague to be misleading in precisely this sense and to provide an ideological weapon for the liberal opposition to the Government. Japan's development down to the present certainly suggests that basic western values have had an important and increasing role in reshaping traditional society.

4. [J. R. Levenson, " 'History' and 'Value': The Tensions of Intellectual Choice in Modern China," in Arthur W. Wright (ed.), *Studies in Chinese Thought,* Chicago 1953, p. 9.]

5. *op. cit.*

6. Underlying this statement is the writer's belief that the historical process of westernization—in the sense of the adoption of the principal western values listed—is inevitable. In the increasingly interdependent world society of today (and yesterday), the life of pastoral and agricultural peoples under a system that may be loosely described as "feudalistic" is experienced as an anachronism, that sooner or later must be eliminated from the map. No wonder that "modernism" is allied with "nationalism" in such countries. For pride in the nation can be upheld only if it is moving sufficiently in the direction indicated by the western values in question to be recognized as in some manner "equal" to the dominating western European and American powers. Russians are emulating Americans; Chinese are emulating Russians. The western values are coming to represent more and more a world-wide ethos.

7. It must be supposed that the revolutionary movement that brought the modernists to power had a wide base in popular support. This, for instance, was the mistaken belief of Nasser. . . . [Gamul Abdul Nasser, *Egypt's Liberation: The Philosophy of Revolution,* Washington, 1955.] On the contrary, the real revolution does not begin until power has been taken over. Initially restricted to a few urban-bred "modern" intellectuals, the movement must spread out to encompass and sweep along with it the broad masses of the population.

8. See John Friedmann, "The City and Economic Growth," to be published in *Comparative Studies in Society and History,* Autumn 1960.

9. See Gilberto Freyre, *Casa Grande e Senzala,* 4th ed., Rio de Janeiro 1943; Paulo Prado, *Retrato do Brasil,* 5th ed., Sao Paulo 1944; Jorge Amado, *O Pais do Carnaval,* Sao Paulo 1955; Vianna Moog, *Bandeirantes e Pioneiros,* 2nd ed., Rio de Janeiro 1955.

10. Morris Watnick, "The Appeal of Communism to the Underdeveloped Peoples," in . . . [Bert F. Hoselitz (ed.), *The Progress of Underdeveloped Areas,* Chicago 1952, p. 165.]

11. Gunnar Myrdal, *Rich Lands and Poor: The Road to World Prosperity,* New York 1957, p. 7.

12. W. W. Rostow, "The Take-Off into Self-Sustained Growth," *Economic Journal,* March 1956, p. 29-30.

13. Committee for Economic Development, *Economic Growth in the United States: Its Past and Future.* A Statement of National Policy. New York, February 1958, p. 12-13.

14. U Kyaw Thet, "Continuity in Burma: The Survival of Historical Forces," *Atlantic,* February 1958, p. 119.

15. For instance: Euclydes Da Cunha, *Os Sertoes,* Sao Paulo 1954 (original edition 1912); Sergio Buarque De Hollanda, *Raizes do Brasil,* Rio de Janeiro 1936; Alberto Torres, *O Problema Nacional Brasileiro,* 3rd ed., Sao Paulo 1938: Nelson Warneck Sodré, *Orientaçoes do Pensamento Brasileiro,* Rio de Janeiro 1942; Fernando De Azevedo, *A Cultura Brasileira,* Rio de Janeiro 1943.

16. Gilberto Freyre, *Uma Cultura Ameaçada; A Luso-Brasileira,* 2nd ed., Rio de Janeiro 1942. Luso-tropical=Portuguese-tropical.

17. Gilberto Freyre, *New World in the Tropics: The Culture of Modern Brazil,* New York 1959.

18. "Modern natural science owes its great triumphs to having looked upon and treated earth-bound nature from a truly universal viewpoint, that is, from an Archimedean standpoint taken, willfully and explicitly, outside the earth." (Hannah Arendt, *The Human Condition,* Chicago 1958, p. 11.) See, also, Karl Jaspers, *Vom Ursprung und Ziel der Geschichte,* Frankfurt 1955.

19. Florian Znaniecki, *Cultural Sciences: Their Origin and Development,* Urbana (Ill.) 1952.

DAVID APTER

The Role of the New Scientific Elite and
Scientific Ideology in Modernization

Until recently the norms of science have supplied its ideology. Such norms have an honorable tradition, associated as they are with such names as Diderot and d'Alembert, Holbach and Helvetius, Condillac and Voltaire, Rousseau and Hume. What all these thinkers shared was a belief in the natural goodness of men, who, under the proper conditions of knowledge and understanding of the natural causes of their behavior, would be able to change their situation and improve it. Such notions relied on principles of natural reason, which was assumed to form the basis of our understanding of the laws of nature—laws, one might add, which were common to man as well as to things.

The norms of science suffered somewhat during the nineteenth century when they became mixed with metaphysics. Science was suffused with mysterious implications even by those who, like Comte, argued loudest in its defense. When science became romantic, it moved away

David E. Apter, *The Politics of Modernization* (Chicago: The University of Chicago Press, 1965) pp. 434-450 (abridged).

from the practical workman's universe of the earlier period. Thus, science, and also philosophy, lacked a sufficiently clear and unambiguous basis to be useful during the sudden rise to political importance of a scientific establishment. In particular what was lacking was a theory and a doctrine with which to establish an appropriate relationship between the scientist and his society. It is therefore not surprising that the one explicitly antimetaphysical yet powerful nineteenth-century theory, Marxism, which successfully combined a moral aim, a universal explanation of social change, and a hortatory doctrine, attracted the attention of scientists concerned with their new role in society. When applied to science, Marxism illuminated what had become obscured, namely, the common basis of the natural and the social sciences. Moreover, it prophesied a more powerful political and social framework than capitalism, more rational and scientific. This is the reason so many socially aware and responsible scientists have been (and still are) Marxists, particularly in England and France. In the United States, in contrast, but for the same reason, scientists have tended to participate in new trends in behavioral science. These new approaches, however, lack the characteristic that remains fresh in Marxism, a particularized moral center and a universalized goal. These allow the scientist power together with morality.

With the increasing pace of scientific innovation in both industrialized and modernizing countries, it would merely compound our problems to rely on theories formulated in the nineteenth century as the basis of a contemporary relationship between science and society. Required is a more elaborate set of theories, more comprehensive and less innocent. For this purpose, Marxism is too narrow because of its teleological and economic orientation. Today a scientific theory must include an adequate explanation of choice and the conditions of choice, both within political systems and between them. A new synthesis must take into account the more recent developments in the social sciences.

The scientific ethic is based on the need for free exchange of knowledge and information. This is particularly necessary in modernizing societies, where, although their numbers are small, scientists, social scientists, and technicians are the modernizers. They occupy a set of elite career roles. They train manpower, educating it to occupy modern functional roles. They are responsible for new towns, dams, highways, fiscal reform, and the like. Of course, by being dispersed through the system, they may be controlled and manipulated by the political authorities, as occurs, for example, in totalitarian societies. One method of control in totalitarian regimes is to create pockets of freedom in which scientists have access to necessary materials that are sealed off from the rest of the community; but this is costly. The more often scientific bodies are directly linked to the political life of the country through technical agencies in government, the more dependent the latter become on the skills of the scientists. As a result, the scientific career occupies an exceptional place in a modernizing society. Not likely to be taken in by propaganda, the scientist takes for granted the existence of a variety of political forms. Scientists today want to know where their societies are going.[1]

Moreover, as has been suggested before, the community of science is world wide. The universities of the world are linked together by special bonds. People read each other's works and conduct research on each other's problems. The academic world is moving closer together all the time, and exchanges of teaching and research personnel are increasing. In addition to conducting research and providing for communications, universities serve to screen the community and select those best qualified for scientific work and the scientific role. This function is most significant in modernizing countries where universities are almost the only instrument for selecting and recruiting scientists.

Nor is the scientific spirit limited to scientists. By subscribing to norms of modernization, the professionals, lawyers, journalists, and indeed, all those who are educated, accept the scientific ethic, which includes an emphasis on rationality and the importance of empirical research, an awareness of what constitutes valid evidence, and a feeling that what is known must be susceptible of verification. The scientific spirit is the basis of an ideology that provides a measure of identity for those who subscribe to it and a measure of solidarity for the members of a society in the midst of change.

Today the ideology of science includes social science. The application of scientific techniques to social problems first became possible in economics, coinciding with the enormous growth in productive capacity that had been made possible by new technological developments.[2]

The evolving position of social science provides exceptional opportunities for social scientists who occupy roles half-way between research and policy. These roles are increasingly played by those in universities and related research bodies. Indeed, the emphasis on research in the modern university has itself done much to extend the norms of science to the social sciences and has, as well, attracted government support for policy research on a very large scale.

To discover which roles are critical we must examine the career roles of the modernizing elites. (The exceptional characteristic of the modernizing role is that it is not confined to a particular modernizing or industrial society but participates in a common dialogue with its counterparts in other societies regardless of their stage of development.) Indeed, the common aims of industrial and modernizing societies are a reflection of the values and objectives held in common by such elites (particularly the "scientific" and political elites). The world-wide scientific community subscribes to the same general ideology of science. If the elites in the modernizing nations are to maintain their expertise, they need to keep abreast of the new developments occurring in their specialties. Today's careers are part of a world-wide intellectual community, no matter how provincial their immediate locale. In order to resist provincialism, these elites must increase their functional expertise as agronomists, mathematicians, engineers, and teachers but must also become aware of the interrelationships of careers. In addition, they need to raise their sights to the political level in order to generate power and attain their goals. If successful, the modernizing elites could constitute a crucial link between modernizing and industrializing countries.[3]

The result of the efforts of the modernizing elites is the universalization of a new type of scientific establishment, which, having made its appearance in the most highly developed industrial systems, has now become critical in modernizing systems. This new establishment is composed of a set of interlocking career roles that screen relevant information received from the society and at the same time create new information for government. The establishment may be subject to checks in systems with a high degree of accountability, or it may remain relatively autonomous. In either case its significance goes well beyond the technical skills it commands. The real significance of the scientific establishment derives from its ability to remain outside, to some extent, the relationship between information and coercion that we have discussed at length above. It can avoid some of the pitfalls of high coercion systems since it is able to draw on information outside the system. It has a fund of professional knowledge at its disposal that does not vary inversely with the amount of coercion in the society.

This characteristic puts the new establishment, as it emerges in modernizing countries, in a peculiarly sensitive position. Insofar as it is an information-creating elite, it needs freedom; however, the information it obtains as a consequence of its freedom may be used by government for coercive purposes.

LEGITIMACY AND THE SCIENTIFIC IDEOLOGY

The consequence of rationalistic planning, with its stress on the role of the technician, is to create a new basis for political legitimacy. A new *Stand*—that of planners—slowly emerges as a result of modernization.

Not always subordinate to the political leaders, the new technological positions define an entire sector of modern life. Such positions are powerful and attractive, and those who aspire to them must possess exceptional technical understanding and training. Linked to the educational system, they represent a hierarchy of talent based on egalitarian access. In addition, they exert a strong influence on the more parochial political ideologies (Nkrumaism, Manipol, and so on) because of their direct or indirect support of the scientific ideology. Some modern development communities have accorded the social engineer almost a monitor role in political life. This is not because he possesses a kind of Platonic predisposition for the role of leader; quite the contrary, he usually shares with his fellow citizens a feeling of ambiguity about his role. But he often gains a superior insight into the conduct of his fellows that gives him the power to create a new role, an ideology that follows from it, and a hierarchy of power and prestige based on intellectual ability—what, in its extreme form, Michael Young has called a "meritocracy." Indeed, once the social-scientist discovered that there was a discrepancy between observed behavior and felt behavior, between the act and the rationalization, between the conscious and the subconscious, and between virtue and conduct, he fashioned a new role for himself—the theoretically omniscient observer. He came to believe that human mysteries were technical problems.

Such a position is dangerous for many reasons. The most immediate reason, however, is that it causes a running battle between scientists and ideologues, which is the more intense because of the interdependence of politicians and scientists. Political leaders need to tread lightly in the region between the alienated ideologues and those who desire to apply science to human affairs.

If . . . the dialogue in the modernizing societies is between nationalism and socialism, the conflict in societies moving toward industrialization will be between science and the other ideologies. The ideologue will manipulate slogans. The scientist will ignore him. In such a situation political leaders will have to learn how to rely on the latter without unduly arousing the former.

The tendency of the scientific role is to undermine even the most extreme mobilization system—even when the scientists have been persuaded by the ideologues, at least for a time, to accept a political religion along with their scientific ideology. How many scientists among Poles, Czechs, Russians, and Chinese would secretly agree with Milovan Djilas when he writes:

> The internal monolithic cohesion which was created in the struggle with
> the oppositionists and with the half-Communist groups is transformed

into a unity of obedient counselors and robot-bureaucrats inside the movement. During the climb to power, intolerance, servility, incomplete thinking, control of personal life—which once was comradely aid but is now a form of oligarchic management—hierarchical rigidity and introversion, the nominal and neglected role of women, opportunism, self-centeredness, and outrage, repress the once-existent high principles. The wonderful human characteristics of an isolated movement are slowly transformed into the intolerant and Pharisaical morals of a privileged caste. Thus, politicking and servility replace the former straightforwardness of the revolution. Where the former heroes who were ready to sacrifice everything, including life, for others and for an idea, for the good of the people, have not been killed, or pushed aside, they become self-centered cowards without ideas or comrades, willing to renounce everything—honor, name, truth, and morals—in order to keep their place in the ruling class and the hierarchical circle.[4]

But the matter does not end there. Djilas himself is an illustration of that. The long-run need of the scientific elite is to reduce coercion and to increase information in order to take advantage of the modern advances in science and social science. To reduce coercion, there must be decentralization in economic controls, wider opportunities for local entrepreneurship, and greater social mobility. An important characteristic of the scientific role is the occupant's concern with achievement and with the exploration of his identity through the making of choices. When these concerns predominate, choices are no longer between systems but within the system. The legitimacy of the system depends on the way the opportunities for choice are optimized. And this optimization depends in turn on knowledge. That is the reason the dynamic factor in industrial societies is information. But even more important is the relationship between information and performance. Information leads to achievement, and there is a high correlation between achievement and continued economic growth. D. C. McClelland has shown that people work hard when the results of their labors are given public recognition and esteem. Only then will they prefer risk and adventure to safety, traditionality, and bureaucracy.[5]

The scientific elite has revolutionary potentialities during modernization. Its role is critical even in the kind of situation found in present-day China, for example, where modernization and industrialization proceed side by side, and the government exercises complete control over press, education, art, and literature. As one observer commented recently, "It must be realized that while persuasive and coercive communications are powerful tools of political development and social control, they are subject to definite limitation." One of the limitations he suggests is "double-talk." "One therefore wonders," the observer concludes, "whether the tremendous power unleashed by the Communists and the new forces now set in motion by the Peking regime may not eventually prove too great for the manipulators to handle."[6] When China succeeds in industrializing, the conflicts between these forces ought to become much sharper and the role of the scientist more influential. Already the fields of pure science and atomic research appear to be relatively free of coercion.

The need for information and some relaxation of coercion is not merely a need attached to the role of scientist. The political implications go deeper and rest on the unavoidable contradiction between the exercise of coercion and the maintenance of a high degree of economic efficiency, regardless whether the economic system is socialist or capitalist. The more a society modernizes, the greater need it has for information to spot lags and weaknesses in growth. All modernizing elites, not only scientists and technicians, need knowledge. Moreover, what is necessary in modernizing societies is even more necessary in industrial ones. One of the best indicators of the effects of coercion in industrial societies is the condition of agriculture.

In successful industrial countries, agricultural productivity is very high. But in countries where coercion is high, productivity is low. China, the U.S.S.R., and Czechoslovakia, for example, are high coercion systems, and they all have experienced lags in agriculture. Contrast the tremendous agricultural output of the United States in 1950, with only 12 per cent of the population engaged in agriculture, with that of the Soviet Union in the same year, with 50 per cent so engaged.[7] Such inefficiency extends into virtually all activities, with the exception, perhaps, of those in which the scientific elite has freedom, that is, pure science, mathematics, medicine, and other purely technical subjects. (In these areas the efficiency of the U.S.S.R. is formidable indeed.)

In advanced industrial societies the labor force consists mainly of persons employed in trade, finance, education, and the public service. In such societies the labor force thus "has a large number of diverse employing units in contrast to the underdeveloped country in which government dominates the modern sector."[8] The existence of the large number of employers eventually propels the country in the direction of a reconciliation system. As Frederick Harbison and Charles A. Myers have pointed out in an extremely important passage in their book on human resource development, this produces a trend—a trend toward democracy.

> Basically the *trend* in countries at this level is democratic, with a high degree of participation in the political process by the adult population. In part this is a consequence of widespread public education, education for

citizenship in a democracy. But these countries have differed with respect to the nature of their leadership groups, particularly in their earlier political development. For example, some countries have had middle-class industrializing elites, with broad democratic and egalitarian philosophies during much of their economic development. Others have had a more rigid class structure inherited from a pre-industrial period, and some have begun their march toward industrialization under what might be called dynastic elites. Finally the Communists or revolutionary intellectuals have provided an ideology which prescribes strategies and policies different from those of the other industrializing elites.[9]

The point is that if the scientific elite joins forces with other modernizing elites it immediately acquires critical political importance. This has already begun to occur within the industrialized countries of the eastern European bloc. A wide variety of systems has developed that could be placed along a continuum between the reconciliation and mobilization extremes.* Peter Wiles has gone so far as to suggest that Yugoslavia represents a system in which a regulated market determines resource allocation—a regulated market, but a market, nevertheless, and not a command economy.[10] Already, the Yugoslav pattern is being repeated in selected sectors of the command economy in the Soviet Union, particularly in education. Wiles points out that the U.S.S.R. is shaping the educational system according to functional needs, with these needs thus representing an educational "market." He calls this a "Communist secret weapon."[11] Not only is this change an opening wedge for the market principle of allocation, but it also indicates that educated and trained manpower—in other words, a scientific elite—is the principal by-product of industrialization. It is only a matter of time before the Communists come to recognize that what Wiles suggests is true, that their allocation policy, education policy excepted, is "irrational and therefore a brake on growth. . . ."[12]

Therefore, it may be suggested that four more or less inevitable consequences follow from the growing complexity of industrial societies: First, there is a need for more and more information. Second, there is the growth of an elite that uses and manufactures information and that is critically functional to the continual development of the society. Third, there is the formation of pluralist groups almost in classical Durkheimian terms.[13] Fourth, there is the assumption of leadership by

---

*Apter defines a mobilization system as a social system which has a hierarchical structure and is oriented toward sacred consummatory values. Mobilization systems are "sacred-collectivity models" of the process of modernization. The polar opposite of the mobilization system is the reconciliation system, a "secular-libertarian model" of modernization, which has a pyramidal structure and stresses secular instrumental values. See: David Apter, *The Politics of Modernization* (Chicago: University of Chicago Press, 1965), pp. 24-25.—Eds.

the scientific elite over other modernizing groups. These developments, taken together, form the basis of the long-term trend in political development toward the reconciliation system, as Harbison and Myers have suggested. Already the second development, the information-using elite, is rapidly becoming a *Stand* in virtually all industrial countries, including the Soviet Union. As Alex Inkeles has remarked,

> Although the vital decisions remain the monopoly of the party leaders, the growing importance of technical problems in the governing of the state has forced an informal sharing of power with the many important scientists, engineers, managers, and other crucial technically skilled personnel. Certainly the recent reorganization of administration on a more decentralized basis represents a diffusion of decision-making power, however much the central elite may take precautions not to lose the decisive initiative and control. The full extent of such sharing has by no means been reached.[14]

And a Yugoslav economist has suggested that with a

> centralized system of management and administrative intervention of the state apparatus, the problem of initiative in economic enterprises had to become increasingly acute. Lines and tasks laid down by central authority do not leave much room for free choice in making decisions within an enterprise. The maximum energy could be used only for the purpose of fulfilling to the utmost possible degree the tasks of the plan, while leaving no great possibilities for correcting eventual discrepancies, or—which is even more important—for an effective utilization of the possibilities and advantages which a centralized plan could not foresee. In an underdeveloped economy, where the whole development is concentrated on a few basic tasks, this problem does not acquire its acute form at the beginning. However, if the economy develops, and if development proves to be an increasingly complex process, then strictly centralized planning and administrative management become less and less capable of making effective use of the numerous possibilities offered by the complex life of a modern economy. To the same extent the problem of insufficient initiative becomes increasingly painful and acute.[15]

These comments are, from my viewpoint, particularly appropriate for industrializing countries. If my assumptions are correct, we may view mobilization systems as the "best" political forms for converting poor but modernized countries into steadily growing economies, since they will also create their own "contradictions," to use the Marxian term, and as a result tend to move in the direction of a reconciliation system. If socialism and nationalism, or their modern combinations, are understood in this light, the mobilization system becomes a temporary form of politics, initiating activity but subject to change because of its success; it will lose the loyalty of its people by becoming inefficient.

If the evidence supports the hypothesis that increasing complexity results in greater pluralism within a society, we will need to explore in particular the question whether or not the scientific elites are the focal points of the development of such pluralism by virtue of their leadership of groups functional to the industrialization process. The further question whether or not they become functionally more and more significant and thereby enlarge the area of their power is one that may be empirically answered. Does their ideology become more explicit within the context of the scientific ethic? Indeed, does this ethic place a high value on knowledge and the free exchange of ideas? If the answers are yes, the pluralistic interplay of groups, under the general influence of scientists, will create a structural basis for a liberal society; pyramidal rather than hierarchical authority will slowly spread by means of the decentralization of decision-making, reinforced by the need to maximize efficiency. If this theory is correct, the long-run outlook for representative government under such conditions is that it will be a progressive rather than a conservative force, revolutionary in its implications and attractive to youth who are otherwise cynical. Of course, not all reconciliation systems will become democracies. The obstacles are better known, it would seem, than the general tendency.[16]

NOTES

1. See Don K. Price, "The Scientific Establishment," in Robert Gilpin and Christopher Wright (eds.), *Scientists and National Policy Making* (New York: Columbia University Press, 1964), p. 19. See also Jacques Ellul, *The Technological Society* (New York: Alfred A. Knopf, 1964), *passim*.

2. Durkheim remarked about the pioneering role of economics: "For two centuries economic life has taken on an expansion it never knew before. From being a secondary function, despised and left to inferior classes, it passed on to one of first rank. We see the military, governmental and religious functions falling back more and more in face of it. The scientific functions alone are in a position to dispute its ground . . ." (Emile Durkheim, *Professional Ethics and Civic Morals* [Glencoe: Free Press of Glencoe, Ill., 1958], p. 11).

3. Davidson Nicol has put the matter with characteristic clarity. "If the intellectual community is not encouraged or has not got its roots in institutes or universities which have links and connections with the outside world, it is likely to be crushed between politicians and a rising nationalistic middle class. It is also unlikely that without encouragement creative productivity will flourish in those writers, thinkers, and artists who form the majority of the intellectual community" (Nicol, "The Formation of a West African Intellectual Community," in *The West African Intellectual Community*, Proceedings of the Congress for Cultural Freedom [Ibadan: Ibadan University Press, 1962], p. 14).

4. Djilas, *The New Class* (New York: Frederick A. Praeger, 1957), p. 155.

5. McClelland, *The Achieving Society* (Princeton, N.J.: D. Van Nostrand Co., Inc., 1961), *passim*.

6. Frederick T. C. Yu, "Communications and Politics in Communist China," in Lucian Pye (ed.), *Communications and Political Development* (Princeton, N.J.: Princeton University Press, 1963), pp. 296-97. See also Franz Schurmann, "China's 'New Economic Policy' Transition or Beginning," *China Quarterly*, XVI (January—March, 1964).

7. Harbison and Myers, *Education, Manpower, and Economic Growth: Strategies of Human Resource Development* (New York: McGraw-Hill Book Co., Inc., 1964), p. 133.

8. *Ibid.*

9. *Ibid.,* pp. 133-34.

10. Wiles, *The Political Economy of Communism* (Cambridge, Mass.: Harvard University Press, 1962), p. 71.

11. *Ibid.,* p. 329.

12. *Ibid.,* p. 329. See also the discussion of education and professional employment in the U.S.S.R. by Alexander King, "Higher Education, Professional Manpower and the State," *Minerva,* I (Winter, 1962), 182. King points out that "the two aims of Soviet higher education—the inculcation of a political and social ideology and the training of personnel for an advanced society and economy undergoing rapid industrialization in a time of exceptional technological innovation—are not always in harmony. While the former always has some influence on the latter, the strongly functional trends of education demanded by economic realities are often not acknowledged or legitimated by ideological dogma."

13. See Emile Durkheim, *The Division of Labor* (Glencoe: Free Press, of Glencoe, Ill., 1949), *passim.*

14. Inkeles, "Summary and Review: Social Stratification in the Modernization of Russia," in Cyril E. Blank (ed.), *The Transformation of Russian Society* (Cambridge, Mass.: Harvard University Press, 1960), p. 246. See also George Fisher, *Science and Politics: The New Sociology in the Soviet Union* (Ithaca, N.Y.: Cornell University, Center for International Studies, 1964); this is a very important study of the new role of social science in the U.S.S.R.

15. Borivoje Jelic, "Characteristics of the Yugoslav Economic Planning System," *Socialist Thought and Practice* (June, 1961), p. 63. See also Michel Crozier, *The Bureaucratic Phoenomenon* (Chicago: University of Chicago Press, 1964), *passim.*

16. See the excellent discussion of these matters in Carl G. Rosberg, "Democracy of the New African States," in Kenneth Kirkwood (ed.), *African Affairs* ("St. Antony's Papers," Vol. 15, no. 2 [London: Chatto & Windus, Ltd., 1963]).

STEVAN DEDIJER

Underdeveloped Science in Underdeveloped Countries

I

I am writing this paper in the hope that it will come to the attention of a select audience of the presidents and prime ministers of those countries where science does not yet exist on any significant scale.

Roughly five out of six prime ministers in the world belong to this group. Today between 15 and 30 of the 120 countries of the world, with less than one-third of its population, possess practically all of its science. They spend more than 95 per cent of the world's research and development funds in order to produce, first, practically all of the world's research output in the form of research papers, technical reports, discoveries, patents and prototypes of new products and processes, and second, most of the new generation of trained research workers in science and technology. Furthermore, these countries reaped in the past and are

Stevan Dedijer, "Underdeveloped Science in Underdeveloped Countries," *Minerva* 2 (Autumn 1963): 62-81. Reprinted by permission of the author and the publisher.

now reaping most of the direct economic, political, social and general cultural benefits of scientific research. Finally, during the past twenty years it is mainly these countries which have made the almost simultaneous invention of national research policy as a new institutional mechanism for the development and the use of science to achieve their national objectives.

The other countries—approximately 100 in number—with about two-thirds of the world's population, share in various degrees the remaining one-twentieth of the world's science. They are countries which, either in an absolute or in a relative but very significant sense, have no science.

It has become difficult for these countries to ignore the fact that research is no more than a negligible category in their national division of labour. They cannot avoid being aware that they are essentially pre-research cultures. All kinds of forces, domestic and foreign, political and economic, moral and historical, are acting on the governments of these countries with the inexorability of a law of nature to take some sort of action to promote the development of science in their own countries.

During these past few years it has come to be realised that underdeveloped countries are also countries without science. The evidence is presented in Table 1.

The distress over being a scientifically underdeveloped country is beginning to approximate the distress over being an economically underdeveloped one. The growing consensus on the need for simultaneous action on both these problems is reflected in very tangible political actions. The tension between developed and underdeveloped countries is beginning to focus on the scientific gap, in a manner similar to the tension arising from the cleavage between rich and poor countries.

The worldwide discovery that the problem of national development must be coupled to the development of an indigenous science is very recent. Yet, this discovery is now entering into a worldwide consensus.[2] This can be seen, for example, from the activities of many national and international organisations, from the 1,200 papers presented to the UN conference of over 80 countries held on this problem in 1963 in Geneva, from the papers of similar conferences held in Moscow in 1962 and in Rehovoth in 1960, and from numerous papers appearing elsewhere in the world on this subject.

In spite of all the activity and interest in this problem, there is still today a dearth of systematic information and knowledge on some of its basic aspects. Furthermore, since the development of science in countries without it has become a political problem for the advanced countries, political constraints have strongly influenced the mode of its

TABLE 1.[1]

COUNTRY	EXPENDITURES ON RESEARCH AND DEVELOPMENT		CONSUMPTION OF COMMERCIALLY PRODUCED ENERGY PER CAPITA 1960 (TONS EQUIVALENT COAL)
	% of GNP	$ per capita	
U.S.A.	2.8	78.4	8.0
U.S.S.R.	2.3	36.4	2.9
U.K. (1961)	2.7	35.0	4.9
France	2.1	27.0	2.5
Sweden	1.6	27.0	3.5
Canada	1.2	21.9	5.6
W. Germany	1.6	20.0	3.6
Switzerland	1.3	20.0	1.9
Netherlands	1.4	13.5	2.8
Norway	0.7	10.0	2.7
Luxembourg	0.7	9.3	—
New Zealand	0.6	8.9	2.0
Belgium	0.6	7.5	4.1
Japan	1.6	6.2	1.3
Hungary	1.2	—	2.5
Poland	0.9?	5.3?	3.2
Australia	0.6	5.3	2.2
Italy	0.3?	1.8	1.2
Yugoslavia	0.7	1.4	0.9
China	—	0.6	0.6
Ghana	0.2	0.4	0.1
Lebanon	0.1	0.3	0.7
Egypt	—	0.3	0.3
Philippines	0.1	0.3	0.2
India	0.1	0.1	0.1
Pakistan	0.1	0.1	0.1

presentation, especially at international conferences. Calling a spade a spade at these conferences has rarely been considered politically advisable either by the participants coming from the pre-research cultures or those from the developed countries.

II

In this paper I intend to deal with those aspects of research policy for countries without science which, in my opinion, either have been stressed insufficiently in the many otherwise excellent papers or have been passed over in silence at international conferences.

One of these questions is what the principal decision-makers of underdeveloped countries should do about science. Decision-making on science in every social system from the highest to the lowest is difficult and it is most difficult when it involves decisions affecting the scientific life of an entire national society. It is also the least studied and the most neglected aspect of national policy even in developed countries. In countries without science, it is very much more difficult, very much less studied and very much more neglected. The future of science in countries at present without it, and its development and use to achieve the objectives of national leadership depend, much more than in any other field of their national policy, on the interest which they personally take, on their own understanding of it, their own strength of will and their own exertions.

The first effective steps along the road of national development are unthinkable today without using the results of research from the start. It is impossible to estimate your starting degree of development, it is impossible to define your objective, it is impossible to make each step from the first to the second without research in the natural, social and life sciences. An objective estimate of the human and material resources available and necessary for the very first and each subsequent step in development demands the solution by scientists of a series of problems in statistics, demography, sociology, economics, geology, hydrology, geodesy, geography, etc. The efficient development of these resources, whether human, animal or vegetable, demands from the very first a continuous production of scientific knowledge about their specific properties and potentialities. Practically every decision in any field of national endeavour, whether it is the improvement of the trade balance or community development, requires not only know-how but also scientific knowledge produced by research performed in the local environment. Every aspect of national development policy depends on research conducted within the country, although it must, of course, be based on the achievements of, and conform with, the standards of international science. National development requires a large and continuous production of scientific results; the importation of foreign specialists to produce them is politically and economically intolerable as a long-term arrangement. The development of a national research potential, i.e., qualified scientists, scientific institutions and equipment and a scientific culture within those circles must be created in order to carry out other national policies with any degree of effectiveness. The development of this potential must be regarded from the first not as a luxury but as an inseparable part of the general programme of development. Hence, a policy for the development and the use of science must be from the start an integral part of the national development policy. Science policy must be as important a part of the national development policy as economic and educational policy and, perhaps, more important than foreign, military and other policies. To neglect a planned and vigorous development of indigenous research in the physical, life and social sciences endangers the whole process of development.[3]

Such is the task. How seriously do the underdeveloped countries take that task? Tables 2 and 3 aim to show that the development of science, as previously defined was, right up to 1963, both absolutely and relatively, a much smaller public concern in underdeveloped countries than in advanced countries.

In Table 2 the degree of existence of "science" as an object of concern is roughly estimated for the three countries by the number of papers, articles, books, etc., published in those countries during 1960 on the development of national science. Table 3 compares the awareness of science for 80 countries which participated in the Geneva conference. In this table, the participation in terms of number of delegates and papers submitted to the conference has been used to compare the degree of concern with scientific policy and the application of science to social development. Though these measurements are extremely rough and apply only very generally, one can see from them that actual effort on behalf of science (as manifested in expenditure on science) and a general extra-governmental awareness of its importance are closely associated with each other.

In underdeveloped countries there is less awareness in general public opinion of the importance of science and this is intimately and reciprocally connected with the low priority given to science in development policy and to the carelessness about the cultivation of a scientific potential necessary to produce that science. The farmer, the craftsman, the educator, the civil servant and the politicians in these countries do not see the relevance of science to their concerns. And of course, in underdeveloped countries, there is not a scientific public. There are few scientists or persons who have some measure of scientific education and who follow other professions. There is therefore no representation of the interests of science, no "pressure group" for science, no one to remind the influential section of the population of the need to develop and to apply science. This vicious circle which links the unawareness of the importance of science with its feeble institutional existence and can be broken only by decisions and actions emanating from the central institutional system of the society, from the political elite or from those closely bound up with them and capable of influencing them. There will be no science in the underdeveloped countries unless their political elite become aware of the need for it for their national progress and come sufficiently to appreciate the conditions under which it can be successfully implanted. The first task of the government of such a country, and of its prime minister, is to take the decisions necessary for the creation of scientific institutions, adequately staffed and equipped, and to place

TABLE 2

COUNTRY	POPULATION (Yugoslavia = 1.0)	RESEARCH EFFORT % GNP spent on R & D	AWARENESS amount of published discussion of social, cultural and economic importance of science (Pakistan = 1.0)
Pakistan	4.5	0.1	1
Yugoslavia	1.0	0.7	50
United Kingdom	2.5	2.3	300
All data for 1960			

TABLE 3.[4]

NATIONAL INCOME PER CAPITA ($)	% GNP FOR R & D (estimates)	PARTICIPATION IN GENEVA CONFERENCE		
		No. countries	Delegates (per million population)	Papers submitted (per million population)
100	0.1	34	0.14	0.33
200	0.5	18	0.45	0.90
<500	<1.0	17	0.78	1.00
>500	>1.0	15	1.28	1.24

them into the right relationships with the educational system, government departments, economic institutions and the organs of public opinion.

Yet, everything so far points to the fact that the prime ministers of countries without science have not yet as a rule (which as every rule has its exceptions) learnt any of the above lessons; they have not even discovered the problem. As a rule, they consider science policy and the development of research work a much less important national problem than their opposite numbers in the advanced countries. One need only compare the programmatic speeches and declarations of prime ministers of countries without science with those of the leaders of such developed countries as the U.S.A. or U.S.S.R. to find out how little attention they have devoted to science and research policy. A similar comparison made between the major political parties and parliamentary bodies of underdeveloped and advanced countries provides similar evidence that there is far too little awareness of the importance of science among the political leaders of the former and the institutions which they control. The development of science in the underdeveloped countries needs, together with much else, a matrix of affirmative and informed opinion within and around the political élite.

III

How can this opinion be created where it is now lacking? I would like to make the following suggestion: each prime minister should establish in his office a secretary for science. An intelligent young man with a good undergraduate training in science should be sent to spend a year at the science office of the OECD, at one of the national science policy bodies of the advanced countries (such as the National Science Foundation or Office of the Advisor of Science and Technology in the United States, or at the Science policy seminars at Harvard or Chicago, or, were it possible, to their counterparts in the Soviet Union, and other countries). He could then be expected to undertake the following tasks: prepare information on problems of the development of science abroad and on the state of science within the country and its problems of growth. He should cooperate with the prime minister's chief of cabinet to ensure that his summary reports on science policy questions constitute an important part of the reading material for the prime minister, that domestic and foreign scientific personalities take more than a negligible part of his time and that questions dealing with science policy are frequently placed on his cabinet's agenda.

Furthermore, steps should be taken to ensure that all members of the cabinet, all branches of government and all leaders of political parties (if more than one) should have access and be exposed to material on the country's scientific policy problems. Even short courses on the importance and modes of interaction of science and society in general and in their own society in particular, held for cabinet members, key parliamentarians and leaders of the key economic sectors would not be amiss.[5]

Journalists and broadcasters should be urged to use their media of communication to arouse the interest of the broader, more or less educated sections of the public on the value of science. It should be arranged that the embassies, high commissions, etc., obtain all publications, reports, etc., concerning science policy originating in the countries to which they are accredited. Copies of these should be sent to the prime minister's secretary for scientific affairs to foster the increase of the social status of the country's scientists and to establish personal and wide contacts with the nascent scientific community.

Those responsible for the industrial and agricultural development of the country should be helped to become aware of the importance of the results of scientific research for the solution of their current problems.

A central research organisation should be established, which will not only have the responsibility of fostering research, but which will also be placed in a position to increase the awareness of the importance of science among all the important decision-making sectors of society, and to establish and support within the universities the academic study of problems of the growth of science within the country.

The government of each country should ensure that foreign aid programmes include systematic advice and material help in the establishment and cultivation of scientific research within the country.

Of course all these actions will be of consequence only if the most strenuous exertions are made to promote the conduct of creative scientific research. Scientific teaching and research in the university must be accorded the highest priority. Financial support, administrative facilities, freedom from red tape in the importation of equipment must be provided. Special care must be taken within universities and research organisations to see that creative research is not sacrificed to administrative protocol, to see that young scientists fresh from their advanced degree research are not sacrificed to older men whose interests are largely administrative. Great care will have to be taken to see that the small scientific community and its even smaller subdivisions are not allowed to wither away in isolation, to lose contact with each other, with their peers in neighbouring countries and with their colleagues, teachers and former fellow-students in the wider international scientific community.

IV

"The soil of X is deeply inimical to the growth of science," a scientist from a highly developed country recently stated in a private letter after visiting an underdeveloped country known for its prime minister's awareness of the importance of science. His statement expresses what every scientist or scientific administrator I have met has said about the situation in underdeveloped countries. I have repeatedly met scientists from underdeveloped countries, who have made such statements as: "My country wants neither me nor any other scientist," or "The government of my country is inimical to science," etc. The reluctance of the highly trained young scientist from an underdeveloped country to return to his own country upon completion of his training is not simply attributable to deficient patriotism or enslavement to the money bags and flesh-pots of the advanced countries. In many cases it is motivated, at least in part, and in some cases it is entirely motivated by the knowledge that it is difficult to do good research in their own countries. Not only is equipment and financial provision incomparably poorer than it is in the advanced countries, but scientific administration is usually far more bureaucratic and antipathetic to the needs of scientists for freedom from petty controls. Moreover at home, scientists are few and isolated; there are too few for stimulating interaction and there is none of the atmosphere of excitement which arouses curiosity.

The sources—institutional, cultural, psychological, economic and political—which give rise to such widespread beliefs about the uncongeniality of the condition of the underdeveloped countries must be understood and dealt with if science is to grow there. Although widespread public understanding is desirable, it is especially urgent that the head of government of such a country understands the necessities of a scientific policy, for he has at his disposal the only institution and the only resources that can initiate and support the institutions necessary for the growth of science.

The cultural obstacles to the growth of science in the underdeveloped countries come from a plurality of sources, traditional and modern, indigenous and exogenous. To begin with there was no such thing as modern systematic, theoretically oriented science or scientifically based technology in the traditional indigenous cultures, even of those countries inheriting the great world religions. There was some empirical medicine, some astronomy and mathematics, but little else. Certainly what we call the scientific outlook, the belief in the value of systematic and persistent observation as a means of discovering the coherence and determinateness of the natural order of existence, was not widely diffused in any of these traditional cultures. Such indigenous science as had once existed has long since died away and no traces of it are still active in the contemporary form of these traditional cultures. In the modern sector of the cultures science has not played a large part. The educational system, beside the fact that it left most of the population untouched, had very little scientific content. The universities were primarily literary and abstract in their orientation, the civil services were modelled on the metropolitan services which stressed humanistic, legal and administrative studies (occasionally mathematics) as preparation for entry, and politics naturally had no place for science—even radical and socialist politics which spoke of planning and of "scientific socialism." This was approximately the cultural situation of science on the accession of independence and it has not changed greatly since then.

Curiosity, the pleasure of discovery, the readiness to discard previously held views in the face of new observations and new theories have not begun to spring from minds still rooted in the indigenous traditional and colonial modern cultures. The institutions which generate such dispositions and which keep them alive are difficult to create. They involve an intricate structure of relations within departments, of departments within faculties, faculties within universities, universities in relation to public authority, governmental research institutions and many others. These are only the most external aspects of the system. More fundamental are the normative and motivational systems, the models of action in past and present, which underlie and permeate the social relationships. In any case, they scarcely exist at present in underdeveloped countries.

Underdeveloped countries are pre-research cultures lacking the institutional and motivational elements. Hence, they are basically alien or hostile to almost every aspect of research and the utilisation of its results. The embryonic science developing in this environment will show in every one of its cells, that underdeveloped countries have underdeveloped decision-makers on science, underdeveloped research councils and science advisers, underdeveloped administrators of science and underdeveloped scientists. It is not that scientists in underdeveloped countries are technically untrained or technically incompetent; it is rather that, being a part of their national culture, they will themselves lack, or will not be able to impose or recreate in their society and culture, so alien to science, those fundamental orientations (if they have them) which are necessary for really productive research.

It is under such conditions that political leaders will have to promulgate policies and execute decisions on science. They will be without the support of these three major, more or less autonomous sectors of society which, in the advanced countries, contribute so much to the growth of science:

(1) A scientific community with its own institutions of training, research and communication and its own scientific tradition.

(2) A government apparatus—politicians and civil servants—with a tradition in dealing with science, making some provision for it or at least appreciating its intellectual and practical value.

(3) Industrial, agricultural, commercial, educational, medical, military and other institutions, which have learnt the value of the results of research and have learnt to make more or less reasonable demands for them on the scientific community and the government.

There is another constraint further complicating the conditions under which the political leaders of underdeveloped countries must work in general and on science in particular. Foreign pressures and domestic centrifugal forces give rise to grave political, economic and social strains and instabilities in every underdeveloped country. In those where the prime ministers are endeavouring to modernise, these strains and instabilities are incomparably greater. The political elite are especially prone to experience these strains which threaten all their past and future achievements. Their response to these threats is to block free communication, with the result that they fail to learn what science can offer them, and they also render the conditions in their countries even more unattractive to their own and expatriate scientists.

So we see that, in dealing with science, a series of pitfalls and traps are set for the political elites of underdeveloped countries. Their own lack of knowledge, experience, the underdeveloped cultural environment of which they are parts and which influence the formulation and execution of their decisions, and their own sensitivity to real and imagined threats to the stability of their countries, all help to obscure their vision.

Because of these factors the political elites of underdeveloped countries will be even more prone to make mistakes about science than their colleagues in the developed ones, no matter how able they are personally or how much they are aware of the general importance of science. Even when they try to make research activity a part of the national division of labour, they can foster a series of malformations in their scientific institutions, which might take a long time to diagnose and cure.

V

The present supermarket of world science contains so many attractive goods so expensive for the purse of the underdeveloped country, that only through judicious budgeting of their resources can they buy those most suitable for their capacities and useful for their particular practical and scientific purposes. Every decision on science because of the nature of research and our ignorance is liable to be much more complicated, much more uncertain and hazardous than almost any other type of decision. Even in countries with the most developed science, decisions about it are bound to be leaps into the dark by more or less blind men. Then, of course, in addition to the sheer element of unpredictability inherent in scientific research, many other factors enter into decisions in scientific policy. Ideology, concerns of political and economic advantage, even the pressure group-like action of particular sectors of the scientific community can influence decisions in scientific policy. In underdeveloped countries, ignorance, prejudice and the absence of sources of reasonable advice render such decisions much more difficult, their success much more problematic. There, because of the lack of check and balance within and by the scientific community and other institutions, e.g., industrial and military, with some experience in dealing with science and because of the lack of "personal," if not technical knowledge, of the nature of research work, decision-making on science is liable to go even more astray than in advanced countries.

The small number of scientists, the gerontocratic tendencies of the traditional culture, the hierarchical civil service traditions of the modern sector of the society, the concern for national prestige and for "monuments," the preoccupation with metropolitan models, plain fraudulence among many of those who offer themselves as advisers, are only a few of the many impediments to rational science policies. In underdeveloped countries, powerful and misinformed military, economic or political interests, native scientists with real or fictitious scientific achievements and no experience in science management are capable of sending science expenditures down the drain for years on completely unrealistic projects, simply because the decisions were made *ad hoc;* without an open discussion on the basis of widely gathered information and advice from home and abroad.

To reduce the probability of such outcomes every decision on science must be part of a national plan for the development and use of the results of research. Science must be looked upon as part of a planned national policy. The formulation and control of the execution of research policy and its continuous improvement has to be one of the constant major tasks of the highest political leader of an underdeveloped country. This means that he must participate in the basic decisions on all of the key components of a national science policy. This means that he must take personal responsibility for the principal national objectives to be reached by the existing research potential of the country; the planning of the growth of various elements of this research potential, such as scientific and technological manpower, training and research institutions, equipment and buildings, scientific publications, financial support; long-range research programmes; the distribution of scientific potential with respect to the prospective productive forces of the country and with respect to the principal goals to be reached through science. This calls for decisions on priorities in the establishment of institutes for such basic survey services as geodesy, meteorology, demography, geology, pedology, hydrology, etc.; for institutes in basic and applied research and development in natural, social and life sciences; research institutes for work in special fields like nuclear energy, automation, military research, etc.

The political leader of an underdeveloped country must also see to it that distribution of the research effort and the research potential between the government, the economy and the universities is an optimal one which serves the needs of the former and does no harm to the latter. He must also attend to the phasing of the scientific effort and of the utilisation of the scientific potential. In other words, he must give personal attention to the drafting of a plan for the development of science and its continuous revision in the light of past mistakes and emergent tasks.

Finally, the president or prime minister must initiate or support a series of decisions on measures which aim at making research productive of intellectual and technological results. To be able to accomplish this tremendous task, it will be incumbent on the political leader to remember that there is no such thing as spending too much on research and development.

VI

Investments in science, though giving abundant returns of all kinds including money, are not "get-rich-quick" investments. The hope, most often unrealised, for quick and bountiful returns from investments in apparently more urgent undertakings gives rise to a situation in which research expenditures are pushed lower and lower on the list of investment priorities of the underdeveloped countries. As may be seen on Table 1, the more underdeveloped a country, the smaller the proportion of its income invested in scientific research. That is why, as I pointed out several years ago,[6] the rate of development of science in underdeveloped countries probably is lower than the growth of their economy. As a result, the underdeveloped countries condemn themselves to lag further and further behind in the growth of science and the utilisation of its results. This means that they can expect to be beaten on the competitive world market even in the future by countries which invest more in research. A twofold increase in research expenditure every two years for the next decade will barely permit an underdeveloped country to keep up with the growth of investment in research in the developed countries. If he decides, as he should, on such a course, a president or prime minister of an underdeveloped country will have a hard time with his finance and other ministers. He must, however, persist, keeping his eyes on the future and pointing out that many, if not most, of the "get-rich-quick" investments in industry, agriculture, power, mining, etc., have not realised the the hopes which were placed in them.

Just as he must insist with unremitting obstinacy on expanding the allocations for science and technology, so too he must insist that his country cannot have too many scientists. Underdeveloped countries have so few scientists, engineers and doctors per million population, compared with the developed ones, that however great an expansion they undertake, they will still lag far behind the more advanced countries for a long time to come. Furthermore, as it has been shown, the number of science and technology students per million population is at present about a tenth as large in countries with an income of $100 per head as it is in those of $500 or more. Endeavours to correct this balance must constitute one of the central, if not the central, science policy tasks of political leaders.

In the countries with an income of $300 and less per head, for which such data are obtainable, i.e. about 15, one notes that in all of them without exception, what research does exist in the country is isolated from the universities, where students can specialise in science. As a matter of fact, one could almost define an underdeveloped country as a country in which all research is performed, not in industrial laboratories or in the universities, but in large, almost hermetically sealed, government institutes. The arguments for the relatively small number of science, technology and medical students and for this divorce of research from training in scientific research work have been repeated so often that they have become almost dogmas. Political leaders must demand a careful review of this situation. There is a certain blindness to the importance of scientific manpower in underdeveloped countries. It therefore seems acceptable to isolate so much of the training capacity of the country, however meagre that capacity may be, from the prospective trainees. The reasons for this neglect are obscure. Perhaps it is because training has come to be thought of as a postgraduate activity into which the universities of underdeveloped countries have not felt it incumbent on themselves to enter, partly because the political and journalistic elites have seldom been postgraduate students themselves. But whatever the cause, manpower plans, even those which devote much attention to engineering, medicine and teaching, do not pay much attention to the output of scientists and even less to the output of scientists who can train more scientists.

The determination of the proper balance in a national research programme in basic and applied physical, social and life sciences represents one of the most difficult research policy problems for the underdeveloped countries. Yet, there is a simple rule which can guide the decision-maker in facing this task.

The need to survey the human, mineral and vegetable resources of the country should constitute the first claim and should determine what kind of research institutes the country should invest in first of all. A country with abundant agricultural resources of a given kind and with the prospect of developing them, would do best to concentrate intense research efforts in all branches of science, life, natural and social, basic and applied, which would increase the knowledge bearing on agriculture. The development of basic research in these circumstances is indispensable from many standpoints; it is necessary for broadening the range of practical possibilities available in the given complex of resources: it is necessary for the maintenance of research standards and research morale, etc.

The survey and development of resources should definitely include the development of human resources. For this reason, a vigorous development of the social sciences must not be neglected. The social sciences in the developed countries, both East and West, play an ever increasing and indispensable role in the formulation of national policies and in the evaluation of their execution. In underdeveloped countries, the social sciences have the special task of bringing the mirror to the face of the nation to show what it is: an underdeveloped culture, and to illuminate, as much as its present methods permit it, the rough road of the cultural revolution, which is necessary if the country is to develop.[7]

The political leadership of an underdeveloped country, even if it wishes to develop science as vigorously as possible, must confront not only its own ignorance of the problem but also the difficulty of getting informed and disinterested advice. Eagerness to develop science, ignorance, numerous distracting preoccupations all make it easy for cranks and promoters to come forward and to acquire influence over scientific policy.

VII

Certain of the malaises of the scientific life of the underdeveloped countries are well known and we have already referred to them. But there has not yet been sufficient analysis of them. For example it is frequently, and on the whole correctly, asserted that a tradition of scientific research is lacking in underdeveloped countries. We have, however, still to understand what is implied by this observation which is of the greatest importance. It is a matter of attitudes towards discovery, capacities to perceive problems, to sense the relevance of theories to observations, and vice versa. It is in part a frame of mind, which emanates from one person, which is received, perhaps through identification, by another person and which is constantly renewed and revised. The conditions of this transference of a general outlook, the internal arrangements within an institution, the optimal combination of different generations—on such matters we as yet know very little. Yet if scientific policy is to be made effectively it must take into account the necessity of implanting and continuously reproducing the tradition of science, that paradoxical combination of continuity and innovation which is at the centre of scientific growth.

How to create something important which is lacking is certainly difficult. It should not stand in the way of clearing the ground of obvious malformations, the presence of which can only impede the growth of science and which affect almost every detail of the formulation and execution of national science policy. Such malformations, some of which will be referred to later for illustrative purposes, affect the work of the individual research worker, of the directors of laboratories and the highest scientific administrators in the country. They are expressed in the national research programme and in the outlook of research workers and in their relationship with each other. Those which I mention are only a few of the many which exist; all are worthy of detailed analysis because, unless they are well understood, their pernicious character will not be appreciated and their elimination will be delayed.

There often exists a belief that the development of a particular branch of science, or of a particular project, will enable the country "to jump generations," to turn it overnight from an underdeveloped to a developed culture, or to make it militarily vastly superior to its enemies. Although I have encountered only two leaders of underdeveloped countries making such statements openly and repeatedly, I conclude from many interviews with scientific policy officials and from the policies themselves that the belief is fairly widespread.

Another malady of science policy is to concentrate most of the national research effort on one major project and to plan its development far beyond the actual capacities or the need of the country for this particular branch of research or its results.

The tendency to devote resources to research programmes which confer prestige internationally is another vice of contemporary science policy in the countries we are considering, much more so than in the developed countries.

Even countries with the minimal provision for science are bound to work through a number of institutions, agencies, departments, etc., which compete for the meagre funds available from the science budget. Very often these agencies are concerned with particular projects arising out of the real (or imagined) needs of the country. They are often ruled by powerful personalities, who are imbued with a missionary spirit on behalf of their particular branch of science, and who act as "scientific empire builders." Each of them rules his "empire" with an iron bureaucratic hand, keeping most of his staff in misery and eager to leave the institution and even the country at the first opportunity. Sometimes the aspiring "scientific imperialists" make pacts with other scientific empire builders not to recruit staff from each other, thus reducing mobility and demoralising their "captive scientists." This feature is accentuated by the civil service-like bureaucratisation of scientific institutions in the underdeveloped countries, which, among other rigidities, prevent scientists from attempting to transfer to another research institution unless they have the permission of their present superiors.

More generally, civil service procedures result in almost unbelievably frequent and changing demands on scientific institutions and individual scientists for "proper procedure"; there is a distracting insistence on the control of everything and everybody connected with research and all this requires "reports," "plans" and "requests in proper form." In the more advanced countries there has been sufficient experience with scientists and there are enough scientists in scientific administration to inhibit such bureaucratic *paperasserie*. In the underdeveloped countries, there has been too little experience of scientific research to permit a realistic image of how scientists really work and of what is compatible with their effectiveness. There is not enough of a culture of the scientific community to stand up against the tradition of the civil service and the distrustfulness of politicians.

Bureaucracy, beliefs in the magical powers of science and empire-building are usually accompanied by an insistence on secrecy. The jealous insistence on secrecy in underdeveloped countries is very often used as a means to protect scientific cranks and rogues who proliferate in countries without scientific tradition.

In the lower ranks of the small and fragmentary scientific community of the underdeveloped countries, there is a pronounced tendency to build "institutes" as private reserves, from which to attack and repel, as if from medieval castles, particular academic and scientific enemies.

All decision-making in government entails a political element and the higher the level of decision, the more pronounced is this political element and the more pronounced is the struggle to exercise influence and power. This is true of decision-making in science and science policy in every country. In underdeveloped countries these struggles for power sometimes become so acute that often the basic objectives of decision-making are lost in an atmosphere of political wire-pulling, lobbying and acrimony.

These vices are not, as I have already suggested, confined to the uppermost levels of science policy-making; they intrude into the "atmosphere" of laboratories, of the whole institutes, of branches of science and of the whole institutional system of science in some countries. They help to accentuate the attenuated creativity or diminution of the scientific workers of these countries.

VIII

There is a small number of first-class scientists in some underdeveloped countries, who, under the extremely difficult conditions existing prior to independence, when scientific research was scarcely encouraged, managed not only to survive scientifically but, through the strength of their character and their devotion to science, also managed to build first-class laboratories and institutes. The hard struggle for scientific existence they had to go through often left an imprint on their personalities, which made it difficult for some of them to adjust themselves to the new generation of scientists. Such personalities, however, as a rule hold very high the standards of fruitful scientific work and they often succeed in transmitting their views and beliefs to their pupils.

Yet, one often encounters other types of scientific personalities who have a more negative influence on scientific productivity. Some of these like "Diderot's monk" do as little work as possible, always speak well of their superiors, and let the rest of the world, including their fellow scientist monks, go to the devil. There are also "research politicians" who devote little time to their own research work but much time to laboratory politics, flattering their superiors, jockeying for posts and stipends for travel abroad.

Then there are the irreconcilable malcontents who complain incessantly about the conditions of their laboratories, the condition of science in their country, their colleagues' incompetence but do very little themselves.

Other frequently encountered types of scientists who help to hold science back in the underdeveloped countries are the "cranks" who have fixed ideas about solving one or more unsolved problems in science by methods which border on lunacy. With powerful political backing these cranks sometimes prosper for a long time, especially if they are, as is not rarely the case, "rogues." These are sometimes men of ability but they do not do research; their aim is to establish by hook or by crook "a reputation" out of all proportion to their real value as scientists.

All these corruptions of the scientific career can prosper, and have prospered, in advanced countries and in highly developed scientific communities. There, however, sooner or later, through the action of an open scientific community, they are more liable to exposure; in any case they very seldom attain any marked influence. In underdeveloped countries, however, such persons can become dominant in research institutions and can do much damage to the more gifted and more devoted individual scientists as well as to the scientific system as a whole.

These are all contributory to the lesser effectiveness of scientific effort in the underdeveloped countries. A wise president or prime minister will see to it that his adviser for scientific matters initiates studies which, although they are difficult, would measure and assess the research attainments of individual scientists, of whole institutes, the extent of utilisation of research results in industry, agriculture, medicine, etc., in his own country and compare it with similar indicators for other countries. Such inquiries would make more vivid and convincing the general impressions which are gained by less systematic observations concerning the harmful repercussions of "the lack of scientific tradition."

IX

There are three main causes for these obstructions to the effective use of the limited scientific resources of the underdeveloped countries. Each is interconnected with the other and together render difficult an efficient science policy. They are (1) the failure of such scientists as there are to carry and transmit the tradition of science and their consequent failure to constitute a scientific community; (2) the lack of appreciation of and demand for research by industry, agriculture and other sectors of society; (3) the government, a bureaucratic institution, lacking in experience and tradition in dealing with science is, and must be at the beginning, the only decision-maker on science.

But since the government alone has the funds and, to the extent that they exist, the powers to promote the development of science, it is up to the government and to its president or prime minister to remedy this situation. This can be done if the government, at least in its crucial parts, can change its character, and that being done, the first of two factors are made the major and most important concern of the research policy of the country:

(1) A scientific policy, to be effective, must build a scientific community with its own traditions, closely linked to the international scientific community and the universal standards of science.

(2) To foster the appreciation of and demand for research work and research results by the industry, agriculture and by the other major institutional sub-systems of the society.

The second of these objectives is a very complex problem depending on the economic and all the other national policies. Creating demand for science in industry, instilling in industry the need for inventiveness, the need to "make the better mousetraps" is a very difficult problem, which nevertheless must become a major objective of national policy. This is in general one of the key problems in most countries with highly and rigidly planned economies—economies in which no substitute has been found for the profit motive and the competitive effect of the internal or foreign market is not operative. What to substitute for the profit motive and competition as the drive for inventiveness in the economic life, which has to be increasingly based on continuous and long-range planning, is a problem which not only the underdeveloped, but the communist countries too, are beginning to face. The study of this problem and finding ways to stimulate the demand for research in agriculture, industry and all other economic sectors must be a special field of concern for those making national scientific policy. This is especially true for underdeveloped countries, where research institutes in applied science are often in fact "pure" research institutes, simply because their results are never demanded or used by their native agriculture or industry.

To attempt to solve the problem of making a scientific community with a scientific tradition, the political leaders of an underdeveloped country will have to turn to the social sciences for basic information, for particular studies and for empirical understanding on how to proceed.

Underdeveloped countries have scientists, but have no scientific community, if we define the latter, in the light of what exists in the developed countries, as an organised group with a developed system of beliefs, with a developed system of institutions for internal communication, as well as a system of communication for dealing with other social groups, and which is bound by certain traditional norms of behaviour for furthering their individual and collective work in science.

In an advanced country, the scientific community is large enough to permit differentiation with sufficient members in each special field to permit complex interaction with each other, and sufficiently different from each other to be able to stimulate each other; it must be organised into a system of training, research, publicity, standard-protecting institutions, which control admission by rigorous scrutiny of qualifications, and which represent and protect the community in its relations with other sectors of the society. It has its own culture, i.e. beliefs, values and vocabulary, its own heroes and models of conduct; it is aware of its history and its continuity with the past without being rigidly attached to that past. It has its own system of communicating and assessing the results of research and analysis and its own patterns of promoting the proficient and holding back those who do not meet its requirements. It has its own circles of face to face interaction and inter-individual communication. It is aware of itself at least in a vague way as something different from other sectors of the society, and it is recognised and acknowledged as a distinctive group by those other sectors of the society. It is linked with other scientific communities across political boundaries by personal contact, by mutual appreciation and by public communication and formal association as well as by a sense of fundamental affinity. It is aware of its identity within the national and world societies and has pronounced beliefs in the legitimacy of its role and calling. We can see how far the underdeveloped countries are from having a scientific community in this sense. Still that must be the objective of their policy.

In the underdeveloped countries scientists are relatively few in number, and they are often, as far as any particular field of research is concerned, dispersed over long distances. They suffer from isolation from each other and thus they do not have the benefits of the stimulation of the presence of persons working in closely related fields. They are in danger, a danger to which they too often succumb, of losing contacts with their colleagues in the international scientific community. They feel peripheral and out of touch with the important developments in science unless they can visit and be visited by important scientists from the more developed countries; they feel inferior and neglected because their own journals and organs of publication, where they exist at all, are seldom read by foreign scientists, seldom quoted in the literature and are indeed often neglected by their own colleagues at home. They have little contact with their colleagues in neighbouring underdeveloped countries.

They are in brief not fully-fledged members of the scientific community and their work suffers accordingly. Neither its scientific nor its practical value is what it could be, given the talent and training of many of the persons following scientific careers in underdeveloped countries. And the scientific community in the country being frail and anaemic, its relationships with the political order of its own country is poor too. Even were the political leaders willing to pay attention to it, its advice would be made unrealistic by the isolation of its advisers from the pulsations of the living scientific community. The strengthening of the internal structure of the scientific community of an underdeveloped country would improve the quality of its work, enhance its selfconfidence and increase the weight which it carries with the other sectors of its society.

This leads us back to the point made above. The fruitful pursuit of scientific truth and its application, once discovered, is not just a matter of talented individuals, well trained in foreign universities and supplied with the equipment they desire. These are very important, but the cultivation of science is a collective understanding and success in it depends on an appropriate social structure. This social structure is the scientific community and its specialised institutions. The underdeveloped countries can dispense with it no more than the developed countries can. The advantage of the latter is that they have inherited it and can develop it as the occasion demands, with the effort which is inherent in traditions which are already well established. The disadvantage of the underdeveloped countries is that they must develop it *ab ovo*. But develop it they must or they will have no scientific development and no economic and social development either.[8] Until science becomes an autonomously growing institution in the new states, all devolves on policy.

NOTES

1. Table 1 based on the following data: Dedijer, S., "Measuring the Growth of Science," *Science*, CXXXVIII (1962), 3542, pp. 781-788; Kramish, A., "Research and Development in the Common Market Vis-a-vis the U.K., U.S. and U.S.S.R.," P-2742 (Santa Monica: Rand Corporation, 1963). Data on Hungary from a Hungarian government publication.

2. Through more intensive contact and communication all countries throughout the world are slowly adopting and pursuing increasingly similar and more compatible national objectives. This seems to me to be a result of greater moral and intellectual consensus and the increased social and political participation of the increasingly educated population in all aspects of the life of the country.

3. At the Geneva conference this lesson was not so strongly emphasised and not so evident as the first one. The number of papers submitted to the section on science policy was smaller than the number submitted to any other section, amounting only to 2.8 percent of all the papers. This shows that the awareness of the importance of science policy is relatively embryonic among scientists and scientific administrators in the developed countries themselves, since the papers for the conference came largely from them.

4. Compiled from United Nations Conference on the Application of Science and Technology for the Benefit of Less Developed Areas, *List of Papers*, E./Conf. 39/ Inf.3, January, 1963; *Directory of Participants*, E./Conf. 39/Inf.7, February, 1963, also *Addendum* 1 and *Addendum* 2 to the *Directory of Participants*.

5. This has been done in some developed countries.

6. Dedijer, S., "Scientific Research and Development: A comparative Study," *Nature*, CLXXXVII (1960), 4736, pp. 458-461.

7. The social sciences can contribute not only inventories of what the country possesses, they can also increase and enrich the national self-consciousness, the mutual

awareness of the different sections of the population and thus aid in the transformation of a collection of scarcely integrated separate traditional societies united by a single government into a coherent modern society. This in turn, by unifying the culture of the country, would help to feed the reservoir of talent from which prospective scientists could be drawn.

8. The identification of the standards of judgement and conduct of scientists as a community is a very difficult task. Such norms are transmitted, mostly orally, through example. They are not easily articulated. Very rarely could one find examples of them described in the biographies or autobiographies of scientists. Nonetheless, the social sciences are now beginning, however rudimentarily, to analyse them. Progress is being made in the perception and articulation of the nature and content of these standards, and the community which is maintained by them and by which they are maintained. Even now social scientists are beginning to think in the social science idiom. Thus, for example, we find in *Science,* CXXXIX (1963), 3561, an editorial on "The Roots of Scientific Integrity," written practically in the language of sociology, where among other things it is said: "Part of the strength of science is that it has tended to attract individuals who love knowledge and the creation of it. Just as important to the integrity of science have been the unwritten rules of the game. These provide recognition and approbation for work which is imaginative and accurate, and apathy or criticism for the trivial and inaccurate. . . . Thus, it is the communication process which is at the core of the vitality and integrity of science. . . . The system of rewards and punishments tends to make honest, vigorous, conscientious, hardworking scholars out of people who have human tendencies of sloth-fulness and no more rectitude than the law requires." The social sciences are, however, only in their beginnings in this subject. The natural science policy-makers must draw from them what help they can in getting on with the task of supporting, fostering and encouraging the creation and development of their own scientific communities.

CHRISTOPHER K. VANDERPOOL

Center and Periphery in Science:
Conceptions of a Stratification of Nations and its
Consequences*

Throughout the world, one can identify centers of scientific activity in a variety of disciplines where research at the forefront of knowledge is transforming existing paradigms, or, in some cases, overthrowing them in a process of scientific revolution.[1] Usually these centers are also the major locations of prestigious scientific journals which attract articles from scientists in other countries, and national associations which draw foreign scientists to their meetings. In addition, scientists outside of the centers send their students to them for socialization under those scientists conducting strategic research in a field. In this way the centers of scientific activity exert a "pull" towards a vortex of prestigious scientific endeavors.

Outside of these centers lie the peripheral areas of science. Often major work in a field is being carried on here and, if such work is

Paper presented at American Association for the Advancement of Science Meetings: Southwestern and Rocky Mountain Division, April 21-24, 1971, Tempe, Arizona.
*This research report is part of a larger research project supported by NSF, NIMH, and the Hazen Foundation.

successful, the periphery may become a focal point of scientific research. But the likelihood of this occurring is low because the prestige of the centers begets power and resources in the forms of financial and social support and the influx of highly trained manpower. The poverty of the peripheral regions limits their potential for upward mobility in the international scientific system.

TABLE 1.   Level of Development of the Home Country and the Perceived Position of the Home Country in Relationship to the Leading Countries in the Respondents' Fields

PERCEIVED POSITION OF THE HOME COUNTRY	LEVEL OF DEVELOPMENT OF HOME COUNTRY	
	Developing Countries	Developed Countries
Among or close Behind the Leading Nations	36.1%	82.5%
Lagging Behind the Leading Nations	63.9	17.5
Total	100.0% (N=101)	100.0% (N=121)
	Q = −.786	

To find out whether or not scientists are conscious of a distinction between centers and peripheries of research in their fields, a sample of visiting foreign scientists from developed and developing countries[2] in the physical, biological and social sciences at several midwestern American universities[3] were asked to identify the nations in which research at the forefront of knowledge in their fields is being carried out. In addition, they were asked to discern the positions of their home countries relative to the leading countries in their areas of scientific inquiry.

The countries identified most frequently as centers of research by the visitors are the United States, the Soviet Union, France, Great Britain, and Japan. Furthermore, the United States is, in general, acknowledged by the visitors to be the leading country in their respective fields. The positions of the other four nations usually are seen as being interchangeable. As one Canadian scientist pointed out, "except for the United States, which is the highest country, there isn't much difference in the nations which are near her."

In terms of locating the position of their home countries in relationship to the ranking they presented, respondents either see their home country as being among or close behind the leading nations or they view it as lagging behind the leading countries in their fields. As Table 1 reveals, scientists from developing nations usually identify their nations as a peripheral area of scientific activity, i.e., lagging behind the leaders. Scientists from developed nations, on the other hand, rank their home countries as among or close behind the leading nations. In general, therefore, the greater the level of development of the home country, the higher the perceived position of the home country in relationship to the leading countries in a field.

The respondents were also asked if they thought their national scientific community will be upwardly mobile in the future in terms of their ranking of top countries in their fields. The majority of scientists, as Table 2 indicates, are optimistic about the future mobility of their home countries. What is interesting, however, is that scientists from developing nations are more likely to acknowledge the possibility of future mobility than scientists from developed nations.

Several reasons can be given for the relationship of level of development and perceived future mobility of the home country. Many scientists from developing countries are conscious of the overall "lowness" of the position of their home country, i.e., they see their country as so far behind the top nations in their fields that downward mobility is an impossibility. As a microbiologist from Greece noted, "my country is so distant from the leaders and so are other nations similar to mine, we can only move up. Down is where we are at." For others, there is an inherent optimism based on the types of scientific developments they see occurring in their home countries. These scientists point out that tremendous strides in self-improvement are under way and a better quality of scientist is becoming predominant in their home countries. Moreover, they often cite increased financial and social support for their fields by the public and power centers of their society. It is only a matter of time and the building of a critical mass of scientists before the gap between their nations and the leading countries is narrowed.

Some scientists share this optimism, but temper it with a cognizance of the possibilities of future mobility on the part of the top nations. They believe in the ability of their countries to climb the ladder of prestige and power in the scientific community, but also foresee the failure of their nations to achieve a position of equal rank with the current leaders in their fields because of the advancement in their fields taking place amongst the top nations. They feel that they will narrow the gap, but at the same time acknowledge that the distance will never be fully closed.

Those scientists from developing nations who do not see any future mobility for their home countries usually cite the lack of dedicated scientists and students, the absence of adequate equipment and facilities, low level of financial and social support given to their work, and the lack of interest in creativity. According to these scientists, no improvement in these conditions is likely to occur. Rather, the "malaise" of their scientific community will remain and, hence, increase the disparity between their nations and the top countries in their field.

TABLE 2. Level of Development of Home Country and Perceived Future Mobility of the Home Country in Relationship to the Leading Countries in the Respondents' Fields

FUTURE MOBILITY OF HOME COUNTRY	LEVEL OF DEVELOPMENT OF HOME COUNTRY	
	Developing Countries	Developed Countries
Yes	72.6%	50.5%
No	27.4	49.5
Total	100.0%	100.0%
	(N=101)	(N=121)
	Q = .445	

Scientists from developed nations, on the other hand, have a lower rate of perceiving future mobility on the part of their home countries than scientists from developing countries because they usually see their home countries as among or close behind the leaders. Some cannot visualize any circumstances in which their countries could experience downward mobility. Rather, they see their countries as "permanent" centers of scientific activities in their fields. Others make reference to the inability of nations behind them to catch up with their nations in the ranking system, because those nations who are not of an equivalent or better rank lack "quality" scientists, equipment, and support necessary to surpass their home countries. Finally, some of the scientists from developed nations acknowledge that even though their countries are high in the ranking system they outlined, that position will remain stable due to a lack of competition from the periphery and because of the further advancement of those nations above them. As one British chemist stated, "Britain will always be number two. The United States and Russia will always be ahead of us, but no nation can take second place from us in the foreseeable future."

The scientists from developing nations who see their countries as being mobile in the future usually identify the strengths of their national scientific communities as the sources of their mobility. Given their strengths, they visualize that many of the major breakthroughs in their fields will occur in their home countries. These successes will enhance the leadership position of their nations and increase their "centrality" in their areas of scientific inquiry. Those nations below them, as a result, will not be able to move close to or above their home countries.

A critical issue to raise in terms of this difference between scientists in developing and developed nations is the impact of the ranking system of nations on the scientists in this study. As has already been pointed out, these visitors usually regard the United States as the center or a center of scientific activity in their fields. Given that scientists from developing countries see their national scientific communities in their fields as far behind the leading nations and that scientists from developed nations cast their national scientific communities among or close behind the leaders, how does this perspective of the home country's place relative specifically to the United States affect the scientists in this study? Two areas of possible impact are the positions held by the scientists in the social system of work in the United States and the type of exchange networks which exist between their home countries and the United States.

Concerning the former, the scientists were asked to identify the nature of their current position. The positions they hold vary from research assistant to full professor. Generally the rank of instructor[4] and assistant to full professor can be regarded as "mainline" positions. Research assistant and associate, on the other hand, usually are considered to be "non-mainline," i.e., temporary. This distinction has been retained in the analysis and the respondents have been classified into mainline and non-mainline positions.

Beyond the temporary versus permanent distinction, mainline positions often are regarded as more prestigious than non-mainline positions. In addition to the prestige factor, there are monetary reimbursement differences between these positions. Usually, the financial rewards are higher for those scientists in instructor to professor ranks than for scientists in the research assistant or associate positions.[5] Moreover, the mainline positions entail greater involvement in the decision-making structures of departments and greater access to individuals in authority positions in the departments than non-mainline positions. Finally, the difference in positions may also involve varying definitions of roles and social identities on the part of the scientists in terms of their interaction with students and colleagues.

Table 3 relates the level of development of the scientists' home countries to the positions of the respondents in the social system of work in the United States. Here one finds that scientists from developing countries are more likely to hold non-mainline positions than scientists from developed countries. Therefore, the greater the level of development of the home country, the greater the tendency to occupy mainline positions in the United States.

TABLE 3.   Level of Development of the Home Country and the Position of the Respondents in the Social System of Work in the United States: Mainline vs. Non-Mainline Positions

POSITION OF THE RESPONDENTS IN THE SOCIAL SYSTEM OF WORK IN THE UNITED STATES	LEVEL OF DEVELOPMENT OF HOME COUNTRY	
	Developing Countries	Developed Countries
Mainline: (Instructor, Assistant, Associate or Full Professor)	13.8%	30.6%
Non-Mainline: (Research Assistant, Research Associate)	79.2	69.4
Non-response	7.0	
Total	100.0% (N=101)	100.0% (N=121)
	$Q = -.4313$	

In terms of the ranking of nations presented previously, these results indicate that scientists from the periphery occupy positions of low status in one of the high-ranking centers of scientific activity, the United States. Scientists from nations which are close behind or among the leaders in their field when travelling to another center hold positions which are of a high status. This implies that many of the scientists from developing nations do not receive the same level of reward, prestige, and role involvement as many of the scientists from developed nations receive.

The second area of interest related to the center-periphery ranking of nations is what types of exchanges do the scientists see as existing between their home country and one of the centers of scientific activity, the United States.[6] Interviewees were asked to identify any networks of exchange between their home countries and the United States in terms of communication of information in journals and exchanges of journals, transference of financial and other forms of resources, sponsorship of students, work contacts with scientists, and exchange of current news and gossip, including the availability of positions in the United States and the home countries. The major types of exchanges the scientists acknowledge are exchanges of resources, students, and journals. What is of interest to this discussion is the direction of these exchanges, i.e., are the exchanges reciprocal or non-reciprocal, and does perception of reciprocity vary with the level of development of the home country? Non-reciprocal exchanges have been divided into two types: (a) one-way exchanges from the United States to the home country and (b) one-way exchanges from the home country to the United States.

Table 4 (based on interview data) indicates that scientists from developing nations are more likely to view the network of exchanges to be non-reciprocal than scientists from developed nations. The latter scientists regard the exchanges as being reciprocal. In addition, more scientists from developing nations see the direction of exchanges of a non-reciprocal nature to be flowing out of their home countries to the United States rather than from this country to their nations. For those scientists from developed nations who see exchanges as non-reciprocal the one-way exchange is viewed as being from the United States to their home countries.

Cautiously interpreting these findings because of the small number of scientists involved in this analysis, one can conclude that the systemic linkages between the peripheral nations and the centers of scientific activity in a field are of a non-reciprocal nature and the direction of the

TABLE 4.   Level of Development of Home Country and the Type of Exchanges between the United States and the Home Countries of the Respondents: Reciprocal and Non-Reciprocal Exchanges

TYPES OF EXCHANGES BETWEEN THE UNITED STATES AND THE HOME COUNTRY	LEVEL OF DEVELOPMENT OF HOME COUNTRY	
	Developing Countries	Developed Countries
Reciprocal	50.0%	67.3%
Non-Reciprocal		
(a) One-way Exchanges from U.S. to the Home Country	23.5	15.4
(b) One-way Exchanges from Home Country to the U.S.	26.5	11.5
Non-response		3.8
Total	100.0% (N=30)	100.0% (N=52)

exchange can be from the center to the periphery or vice versa. Among the leading nations, exchanges are reciprocal. Two statements made by scientists yield an adequate description of this difference between center and periphery in exchange networks.

Describing the reciprocity of the systemic exchange linkage between his country and the United States, a scientist from West Germany said: "every week scientists in my department receive letters from Americans. Sometimes they even call each other on the 'phone for critical discussions on a research problem. We often tell them of promising students who are interested in their areas and we arrange for these students to work in the United States. The Americans also send students to us. We also receive journals from your country and we send ours to yours. So I must say in answer to your question on direction, it occurs both ways."

A Chilean cardiologist described the non-reciprocity of exchanges in the following manner: "to Americans, we are a scientific backwater region. No one bothers with us and no one knows that we even exist in the United States. Sure we get journals from the U.S., but Americans don't read ours. I'm not sure even if they receive them. When my colleagues and I write letters to Americans telling them of our interest in areas that they are working in, some never receive replies. If they do, as I did, the Americans always say 'what the hell are you doing in Chile?' Relatively few of our students go abroad to this country, and many do not return. We never had any American students, even though our work is quite good. Sometimes American cardiologists visit Chile for a vacation and, if they run into one of us, they are quite surprised at the dynamic research we are doing. My colleagues here don't want me to return home. But I will. I like the backwater region."

In summary, scientists from developing nations identify the positions of their home countries in a ranking of nations as low or peripheral to the centers of research in their fields. Scientists from developed countries, on the other hand, see their nations as centers or leaders in their fields. The peripheral or central rank of their home countries affects the type of position they occupy when they are employed in one of the centers of activity in their fields and the type of exchanges they see as existing between their home countries and the United States. These findings imply that scientists from the periphery do not receive the same level of support as do scientists from the center in the United States.

The evidence presented here suggests that future research in the sociology of science must take into account the differential status of nations in science and the functions of rank for the center and periphery. Such a line of inquiry may clearly demarcate zones of power and influence between national scientific communities, differential reward structures, and the processes which lead to the maintenance of a status position in the international scientific community. This also raises the question: how can peripheral nations become upwardly mobile in a system where existing power, privilege and prestige generate future power, privilege and prestige? Can peripheral nations be individually upwardly mobile or must they use a process similar to "sanskritization,"[7] i.e., group mobility, to change their position to that of a center? The latter avenue to mobility is suggested implicitly in ideas about developing regional centers of research in critical areas of science proposed by scientists, politicians, and agents of change in the developing world.

Another critical question is the following: If mobility is no longer possible in some fields because of the strength of the leading nations, might not an avenue of mobility for a peripheral nation be investment of financial and social support in another field in which there are no leaders? For example, in one of the interviews, a biologist stated that mainland China is now giving enormous support to the biological sciences in the hope of becoming the center of the biological revolution in the next century. In this way China may move ahead of the Western nations which have been in the past centers of research in the physical science revolution. Since both the physical and biological sciences generally require enormous expenditures to support adequate research and educational institutions, the majority of developing nations can never compete head on with the developed nations in these areas. The only hope for these nations is either the formation of coalitions in the form of regional science institutions, as has already been suggested, or investment in the social sciences or fields in the physical and biological sciences untouched by the centers, where the level of support needed to give birth to and sustain such institutions is still relatively small.

Research aimed at explorations in this domain of inquiry in the sociology of science could yield not only critical information on variations in the international scientific community, but also knowledge of international stratification systems and of such stratification phenomena as mobility patterns, caste and class formation, and status crystallization among nations.

NOTES

1. The literature on the sociology of science has not dealt specifically with the idea of center and periphery in science. The following works, however, implicitly touch upon this domain of inquiry: Stevan Dedijer, "Measuring the Growth of Science," *Science*, 138, November, 1962; pp. 781-788; Derek J. de Solla Price, *Little Science, Big Science*, N.Y.: Columbia University Press, 1963; Robert Gilpin, *France in the Age of the Scientific State*, Princeton: Princeton University Press, 1968; Frederick Harbison and Charles A. Myers, *Education, Manpower, and Economic Growth*, New York: McGraw-Hill, 1964; John Porter, "The Future of Upward Mobility," *American Sociological Review*, 33, 1968, pp. 5-19.

2. Level of development of home country was measured by Harbison's and Myers' classification of nations by levels of human resource development (op. cit. pp. 26-34).

Using a composite index of human resource development, Harbison and Myers ranked 75 nations. The index is composed of nine measures: 1) number of teachers per 10,000 population; 2) engineers and scientists per 10,000 population; 3) physicians and dentists per 10,000 population; 4) pupils enrolled at first level (primary) education as percentage of the estimated population aged 5 to 14 inclusive; 5) adjusted school enrollment for first and second level (secondary) education combined; 6) pupils enrolled at second level education as a percentage of the estimated population aged 15 to 19 inclusive, adjusted for length of schooling; 7) enrollment in third level (higher) education as a percentage of age group 20 to 24; 8) percentage of students enrolled in scientific and technical faculties in a recent year; and 9) percentage of students enrolled in faculties of humanities, fine arts, and law in the same year.

Harbison and Myers formulated four levels of development upon the basis of the scores of the nations. Level 1 is composed of those nations they consider as "underdeveloped" in human resource development. The range of scores in level 1 is from .3 for Niger to 7.55 for Sudan with countries such as the Ivory Coast, Congo, Haiti and Senegal falling between these scores. Level 2, "partially developed" nations range from Guatemala and Indonesia at 10.7 to Iraq at 31.2. Between these extremes lie such nations as mainland China, Turkey, Paraguay, and Pakistan. In level 3, "semi-advanced" countries, nations such as Czechoslovakia, Poland, India, South Korea, Cuba, etc. fall between the scores of 33.0 for Mexico to 73.8 for Norway. Sixteen nations are classified in level 4, "developed," with Denmark with a score of 77.1 at one extreme with the United States with a score of 261.3 at the other. The country whose score is closest to the United States is New Zealand at 147.3.

Since the range of scores between level 1 and level 3 is not as great as the range of scores in level 4, and since most of the scientists who come to this nation on visits are from levels 3 and 4 (see *Open Doors,* Institute for International Education, N.Y., 1968), countries in levels 1, 2, and 3 will be classified as developing and countries in level 4 will be considered developed. Hence, foreign scientists from the sixteen nations Harbison and Myers classified as advanced will be considered in this investigation as being from developed nations. Foreign scientists from the 59 other nations will be considered as coming from developing nations. The sixteen developed nations, in descending rank order, are the United States, New Zealand, Australia, Netherlands, Belgium, the United Kingdom, Japan, France, Canada, U.S.S.R., Finland, West Germany, Israel, Argentina, Sweden, and Denmark. Since no scientists from the United States are part of the population of this study, there are only fifteen nations from the developed level.

3. Eighty-two interviews were conducted and one hundred and forty mailed questionnaires were received from visiting foreign scientists at the Universities of Wisconsin, Illinois, Indiana, Minnesota, Michigan, Chicago, Iowa, and Michigan State, Ohio State, Purdue, and Northwestern Universities. The following table summarizes the characteristics of the respondents in this study:

TABLE A.   Characteristics of Visiting Foreign Scientists: Type of Science and Level of Development of Home Country of the Scientists

LEVEL OF DEVELOPMENT OF HOME COUNTRY OF THE SCIENTISTS	TYPE OF SCIENCE			
	Physical	Biological	Social	Total
Developed	53.5%	55%	58%	55%
Developing	46.5%	45%	42%	45%
Total	100% (N=95)	100% (N=115)	100% (N=12)	100% (N=222)

The figures in the above table correspond to a high degree with the national breakdowns of visiting foreign scientists in the United States, see: Open Doors, N. Y.: International Institute of Education, 1968.

4. Sometimes the position of instructor is regarded in some departments as a non-mainline position. Unfortunately, data on whether the rank of instructor in the departments studied was mainline or non-mainline were not available.

5. In some universities, the position of research associate can be quite prestigeful and financially rewarding. Data on such university differences were not gathered.

6. An attempt was made to identify the roles scientists perform in the exchanges between their home countries and the United States. Forty percent of the eighty-two interviewees did not respond to the question and two percent said they didn't know if they played any role. As a result, this question did not yield any results which could have given insight into differences between scientists in systemic linkage roles.

7. "Sanskritization" is a term derived from Hindu culture referring to attempts to achieve mobility at a group level in a society which prohibits individual vertical mobility. In India group mobility was possible if an entire sub-caste dropped all of the life-styles of their caste and adopted those of a higher caste, e.g., by becoming vegetarians, avoiding untouchables, etc.

PART V

The Third Culture of Science

The article by Norman Storer which introduces part five, "The Third Culture of Science," touches on many of the topics covered in the preceding sections and our introduction. He discusses the basic assumptions of science, the inherent "internationality" of basic science, "sources of energy and the structures through which it is manifested" in the "international scientific community," the norms of science, the emergence of modern science, and basic and applied research. He concludes his paper by considering the "prerequisites for self-sustaining national scientific communities;" these prerequisites are identified in terms of cultural, support, and management factors. His analysis is basically a normative one. He does offer some clear guidelines concerning the conditions which would have to prevail in order for us to reasonably approximate a "free science" in a "free (global) society." Our concluding essay, "The Third-Culture of Science," is a retrospect and a prospect on science in a global perspective.

NORMAN W. STORER

The Internationality of Science

INTRODUCTION

Even if one lives one's daily life in close contact with only a few trees, there may be some intellectual satisfaction in knowing the shape and essential character of the entire forest. This article is addressed to the broad topic of the relations between science as a particular type of human enterprise and the more time- and space-bound human groups called nations. Its aim is to suggest the outlines of a central conceptual model that may give order to our thinking about more specific aspects of this general problem. It will be largely analytical rather than therapeutic, descriptive rather than prescriptive.

We begin with a consideration of science as man's collective attempt to depict, accurately and economically, the "real world" about him in

Norman W. Storer, "The Internationality of Science and the Nationality of Scientists," *International Social Science Journal,* Vol. XXII, No. 1 (1970): 80-93. Reproduced with the permission of UNESCO.

terms of the causal relationships which account for the ever-changing panorama of observable phenomena. This will be followed by an analysis of the basic social dynamics of the world-wide scientific community as they are now understood, after which an attempt will be made to outline systematically the most important interface relationships between the innately non-national character of science and the inescapable nationality of its practitioners.

SCIENCE AND REALITY

To the extent that science is concerned with describing empirical reality, with the creation of valid, rigorous, and economical pictures of reality rather than with the provision of practical advice for solving immediate problems, it must begin with the assumption that there is but a single physical reality "out there" which can be discovered through the creative application of observation, intuition, logic and experiment. The unitary nature of this reality, or its essential coherence, provides a constant base-line against which the quality of scientists' achievements must ultimately be assessed. Mother Nature, as it were, is the final and totally impartial judge of what is true and what is false in science, and all must turn to her in the end for confirmation of their descriptions of reality.

In more concrete terms, this assumption of the singleness of reality implies that what is discovered about X-rays in Germany will be true also in Japan and in Peru, and will be as true in 1995 as it was in 1895. The location and date of a physical event are thus assumed to be irrelevant to the fundamental nature of the relationships involved. Roger Coates, in his 1713 preface to the second edition of Newton's *Principia,* expressed this assumption in the following words: ". . . if gravity be the cause of the descent of a stone in Europe, who doubts that it is also the cause of the same descent in America?"[1]

And if nature is indeed a single reality, it follows that the structure of knowledge which men develop to describe it will be built through a process in which certain generalizations must be established before certain others can be accepted as meaningful. Collectively, scientists could not have accepted the validity and significance of Avogadro's constant if Boyle's law had not preceded it, nor could Mendeleeff's periodic table have been constructed without prior knowledge of the nature of oxidation. Individual phenomena may be discovered almost at random, of course, as Mendel discovered certain regularities in the transmission of hereditary characteristics, but such discoveries are meaningless unless they can be logically related to other discoveries;

until there exists a context within which such discoveries can take on significance, they may well be ignored if not forgotten altogether.

Scientific knowledge is, after all, the sum of our descriptions of reality rather than the reality itself. These descriptions are cumulative and take on broader meaning only when they can be interrelated, so that the validity of each new addition to a body of knowledge rests as much on its logical "fit" with what is already known as it does on its immediate veracity as a description of a single observed relationship. While there is no rigid inevitability about the sequence in which individual phenomena are discovered, then, the fact that an organized body of knowledge can grow only in certain directions—because of its need to describe the structure of a single reality "out there"—means that scientific knowledge is subject to certain unalterable constraints in its development. These constraints refer to the order in which meaning must be extended rather than the order in which observations may be made, but they do mean that human values—ideological, religious, even aesthetic —are without influence in determining the final structure of scientific knowledge.

The virtual inevitability of the order in which scientists' understanding of natural phenomena develops, it must be noted, stands in propadeutic contrast to the much less inevitable sequence of technological advances. As Derek Price has pointed out, there is no necessary reason why we should have developed incandescent lights before fluorescent lights, or steam engines before internal combustion engines.[2] Practical application does not necessarily require precise understanding of the physical relationships involved, or mankind would never have practiced smelting, animal husbandry, or any of the other methods of controlling nature upon which the evolution of society has been based.

This point adds another dimension to the established distinction between basic and applied research in that it emphasizes the internationality of basic research in contrast to the more specifically local concerns of applied research.

If basic physical phenomena are universal, then, and the advance of science must generally follow a single path, it follows that the location of a scientist in space (if not in time) is irrelevant to his chances to aid in this advance—provided he has adequate access to information on the current state of-the-art in his field and the equipment necessary to engage in research at this frontier. And since the scientists who are interested in the study of the same phenomenon are located in different parts of the world, it follows also that the existence of national boundaries should be completely irrelevant to their concern for each other's work and their natural tendency to co-operate (even if antagonistically) in advancing the scientific understanding of this phenomenon.

Basic researchers, then, by nature form an international community. The social dynamics of this community will be taken up in the next section of this article, but it is important here to point out some of the consequences of such determined internationalism for this community. First of all, there is the scientist's need to develop some competence in one or more languages beside his own to understand the contributions of his foreign colleagues. Second, the internationalism of science reinforces the scientist's desire to quantify his work whenever possible. Certainly the greater precision which quantification affords facilitates the more rigorous organization of scientific knowledge, but the fact that mathematics is itself international in form means that linguistic barriers become less important as more scientific findings are expressed in mathematical form.

Finally, given this emphasis on the universality of basic science rather than the location and practical significance of its immediate manifestations, it is perhaps easier to understand the roots of what is known today as the "brain-drain." So long as the scientist can in principle study the same phenomenon anywhere in the world (geology and zoology are important exceptions to this rule), and this activity is of primary importance to him, nationality becomes less relevant to his behaviour than facilities and the proximity to other scientists, and he will have few inhibitions against going wherever these conditions can be maximized. To be sure, the decision to seek training in a more advanced nation usually precedes the decision to settle permanently in that country, but even with gradual introduction to a new society and the material rewards offered today by nations benefiting from the brain-drain, it cannot be denied that the psychic costs incurred in moving to an alien society are quite high. It must be the basic researcher's primary allegiance to science itself, and thus his interest in being more effectively able to participate in the international scientific community, rather than simply his desire for material gain which accounts for his willingness to move.[3]

Commitment to science is thus commitment to a perspective which takes little account of spatial and temporal location, because the foci of scientific attention are bound by neither. The nationalistic perspective that glorifies "home," "family," "citizenship," and sometimes "race," not to mention language and considerations of national power, is at odds with this perspective and must necessarily have less influence on the outlook and behaviour of the "true scientist" than do his interests in the advancement of pure knowledge.

In sociological terms, science is a "culture-oriented" institution:[4] a collective enterprise whose attention is focused on the development of a body of organized symbols rather than the achievement of goals intrinsic to locations in space and time; it is intellectual rather than material in its orientation. And since the intellectual world, the realm of abstract symbols having universal application, cuts across the material world on which nations are built, it is inevitable that a strong allegiance to the former entails a tendency to ignore the latter. No wonder that science ignores national boundaries; no wonder that "pure" scientists are eager to move to the centres of scientific activity often at the expense of the material concerns of their families and home countries.

Before going further with this line of reasoning, it will be useful to discuss the social arrangements through which a commitment to science *per se* can become viable, or through which the international community of science is able to continue as a realm of meaningful activity and a source of rewards justifying this commitment. After the basic social character of science has been established, we will be able to return to our consideration of the relationship between national interests and scientific interests.

Basic Dynamics of the (International) Scientific Community

To talk about dynamics is to talk about both sources of energy and the structures through which it is manifested. Here, we are concerned with the fundamental nature of the "energy" that keeps science going and with the ways in which it is channelled so that scientific activity can continue. Because science is essentially an intellectual activity rather than one that depends directly upon the exploitation of physical energy, we must agree in the beginning that this energy must be motivational in character; we bring it to mind when we ask why scientists want to engage in research and the other activities associated with it. And we must agree that its structuring is determined by the special set of norms and values that distinguish science from other sectors of society.

This definition of the problem has provided the framework for much of the "basic" research on science, beginning with Robert K. Merton's pioneering essays in the 1930s and continuing to the present day. It is unnecessary here to go into the history of the sociology of science, but it is of interest to note that what is now conceived as the central "energy" of science was not identified until after the norms that structure its flow had been determined. Merton suggested the four basic normative prescriptions which make up the "ethos" of science[5] in 1937—a description which has not been seriously challenged since—and presented his analysis of the central energy that underlies scientific activity[6] in 1957.

In retrospect, however, it is easier to begin with the nature of this energy, since the character of the goal toward which it is apparently directed has much to do with its guiding norms. Merton pointed out in 1957 that professional recognition, the celebration by colleagues of one's scientific achievements, is the single most appropriate and legitimate reward for achievement in science. Through the analysis of a series of disputes over priority in scientific discovery, ranging from the controversy between Newton and Leibniz over the discovery of the calculus to the less sensational but still meaningful resolution of the question of whether Darwin or Wallace had first hit upon the theory of evolution, Merton was able to demonstrate that the receipt of professional recognition, earned by first discovering something, is indeed of central importance to the scientist's motivation. Even if he is reluctant to admit it, the scientist yearns for indications that his work has been accepted by his colleagues as valid and significant—indications that may range all the way from being mentioned in a footnote to being awarded a Nobel Prize. This is not to say that all research is done simply in order to gain recognition, but without such feedback the desire to engage in research would quickly dwindle.

Why the scientist should want professional recognition is a question that has not yet been fully resolved. Two major hypotheses at present attempt to explain it. First, it is proposed that the scientist is trained to want recognition during his apprenticeship in science because it certifies that he has satisfied the demanding requirements of his role: he has advanced our knowledge of some aspect of reality.[7] A complementary hypothesis contends that the desire to create, to produce "meaningful novelty," is a basic human need and that the act of creation is not complete without the receipt of competent response to it from others.[8] The discovery of a regular relationship between physical phenomena is a type of creativity, especially since it must be described in words or mathematical equations if it is to take its place in the body of scientific knowlege, and the discoverer needs the affirmation of his peers that his creation is valid and meaningful. In science, positive response to the product of creativity constitutes professional recognition—and even a negative response is better than none at all. In other contexts a person may desire affirmation of the beauty, cleverness, or practical utility of what he has created, but the basic need for competent response seems to be the same in all fields of creativity.

Regardless of why the scientist desires professional recognition, it is possible now to assert that it is the normatively appropriate motive of the scientist—even though individuals may also find a variety of other rewards for engaging in research. This central assertion is supported by indirect evidence of several types, and the primary remaining objection to it is its apparent conflict with the idea that the scientist is disinterested, altruistic, and entirely unconcerned with fame.

Again, two different but complementary explanations have been offered for scientists' reluctance to admit their interest in receiving professional recognition. (Records of numerous priority conflicts, together with other data showing the large proportion of scientists who admit to occasional worry about "being scooped,"[9] are sufficient to dispose of the contention that they do not really care.) One explanation has it that there is another norm in science which calls for humility and works to make the scientist deny his interest in receiving any sort of reward for his achievements.[10] The other has it that, since professional recognition is worthless if not objective—it is supposed to represent Mother Nature's judgement of the validity and significance of a discovery, not of the discoverer—the scientist hesitates to admit his interest in professional recognition since this might lead his colleagues to bestow it on him as a favour, rather than as the expression of an impersonal evaluation of his work.[11]

The "energy" that underlies scientific activity being now identified, we can turn to a consideration of the rules which guide it, or which direct the relationships among scientists so that they can collectively continue to carry out their research, have it objectively evaluated, and receive enough of the reward of professional recognition to retain enthusiasm for their occupation. The four norms, first described by Merton, concern scientists' relations with each other rather than their attitudes toward empirical phenomena, and upon logical examination it can be seen that they make up the minimal set of such directives simultaneously able to promote the cumulative advance of knowledge while also sustaining the motivation of those engaged in this pursuit.[12]

These norms, it should be noted, are actually the sociologists' highly abstract names for distinctive clusters of behaviour preferences. No claim is made that scientists themselves are directly aware of these norms, or that they would call them thus if they were. They are discussed here in some detail for the sake of filling out the "model" of the scientific community upon which this analysis is based.

The first is "universalism" which describes scientists' tendency to assume that natural phenomena (when properly abstracted from their immediate concrete contexts) are everywhere the same, and also to make a complete distinction between the truth of what another scientist says and his personal characteristics. Thus, an oxygen molecule is supposed to combine with an atom of carbon to form CO_2 under the same conditions in Russia and in America, and a scientist should ignore the political, social, and religious characteristics of the man who discovered this in evaluating the discovery.

The second norm is "communality" referring to the complex of behaviour associated with scientists' unwillingness to keep their findings secret or to allow other scientists to do so. Instead, they insist upon absolute freedom of communication, upon defining a new discovery as a "gift" to the entire scientific community. If the norm of universalism works to keep the scientist's attention focused on his research rather than on the irrelevant characteristics of his colleagues, the norm of communality ensures that his own opportunities for scientific achievement will not suffer thereby. Not only is he supposed to have full access to the findings of others, he is also constrained to make his own findings freely available. This norm thus produces a kind of routine, reciprocal generosity which maximizes the speed and effectiveness of scientific advance.

The third norm is "organized scepticism." It is perhaps best summed up in the wry definition. "A scientist is a man who takes a quarrelsome interest in his neighbour's work." This norm describes every scientist's obligation to receive each contribution another scientist makes to knowledge in his field with critical scrutiny, to make known his evaluation of it, and also to be just as critical of his own work before he proceeds to share it with others. It thus encourages mutual policing of each other's work by scientists so that only high-quality research is accepted into the corpus of "certified knowledge" forming a discipline's body of scientific truth.

The fourth norm is that of "disinterestedness," which was at first simply a description of scientists' ambivalence toward the receipt of professional recognition. Merton originally defined this norm as operating to discourage scientists from openly seeking recognition, but it also seems to operate as a deterrent against using research to acquire any of the rewards society customarily bestows upon high achievement: wealth, influence, and fame. The norm thus serves both to insulate the scientist against the temptations society may offer were he to turn his research to the solution of practical problems, and to keep his attention properly focused upon the reward which his colleagues alone can give. In this way he is kept attentive to his colleagues' interests, which represent the current "needs" of a growing body of basic knowledge, and is motivated to work within the structure of his discipline and to maintain its momentum.

Actually, of course, these four norms are ideals or central tendencies in the behaviour of scientists, rather than precise descriptions of the way scientists really behave all the time. Yet by serving as guides to behaviour they make possible the continued, co-operative functioning of the scientific community and serve as standards by which scientists judge each other's role performance. In combination with the "energy"

derived from the scientist's interest in gaining professional recognition
—which can be obtained only in return for meaningful contributions to
knowledge—these norms have produced a viable, self-sustaining social
system which has been growing steadily on an international basis for
the past 300 years.[13]

Behaving generally in accord with the model presented here, scientists have produced a substantial body of certified knowledge which
serves both as the base-line against which current scientific achievements are evaluated, and also to generate further questions and thus
provide new opportunities to earn professional recognition through
answering them. Given this essential structure, the scientific community
needs only a benign social environment, providing material and moral
support for its work as well as a steady supply of new recruits, in order
to flourish.[14]

Because the social environment within which scientists must work
varies so widely from one part of the world to the next, however, we
cannot achieve an adequate understanding of science without paying
close attention to the nature of the various external influences which
impinge upon the work of the scientific community. In the following
section, these influences will be viewed as a necessary consequence of
the nationality of scientists.

The Nationality of Scientists

The fact that each scientist is an individual biological organism,
uniquely located in time and space, is of particular relevance to our
understanding of the scientific community. Every scientist is born into
a particular human group and his outlook has inescapably been shaped
by the culture of that group. Living in different parts of the world, under
different conditions and with their own histories, and being relatively
isolated from other groups—by design now, as often as by chance—
these groups have developed their own perspectives on man and nature,
their own goals and priorities for reaching them, and their own views
of how their members should behave toward one another. A scientist
is made, not born, and before he ever becomes a scientist he has been
almost indelibly stamped with the culture of his origins.

The largest human groups, characterized by sovereignty, a high level
of self-sufficiency and definite boundaries that separate them from
other similar groups, are nations. A scientist thus has a nationality
before he has a career in science, which means not only that he has
grown up with a certain set of values and speaking a certain language

but that his life will be greatly influenced by the material advantages (and disadvantages) that his nation bestows on all of its citizens.

Historically, science has never been free of these national concomitants. It has developed within nations rather than spontaneously on an international basis, for the simple reason that the material and cultural advantages offered their citizens by different nations, or by adjacent groups of nations, have varied widely throughout history. It is probably safe to say, in fact, that the range of differences between nations has been widening since at least the eighteenth century.

We tend to date the beginnings of modern science from 1543, and to think of it as first appearing in Western Europe during the sixteenth and seventeenth centuries (though it drew initially on the conceptual residues of earlier, incomplete starts in Babylon, India, Egypt and Greece). It developed and acquired momentum because conditions were ripe, and it is not difficult to identify these conditions.

Science could not develop without men intellectually drawn to the objective study of nature, who could afford the time for this non-utilitarian activity, and were able to communicate their findings to others. Further, of course, they needed the tolerance of society, if not its active support. This was not always forthcoming in Western Europe, but moods of intolerance were neither widespread nor long-lived, and the initial spark was never snuffed out.

By the end of the sixteenth century, the basic value-system of Western Europe was generally receptive to a neutral or even exploitative attitude toward nature.[15] Its economy had become productive enough to support an upper class with ample leisure and it was possible, for those who wished, to form private groups to promote their interests, often with royal encouragement, if not material assistance. It was possible, too, for these amateur scientists to communicate regularly with each other across substantial distances and thus to share in the advancement of basic scientific knowledge. There is no need here to recapitulate the history of modern science; it is sufficient simply to indicate that from its beginnings science was intimately associated with nationality. It has always been rooted in nations, and its international character has been an attribute of its goals rather than its origins.

After the original rise of science in Italy, England, and France, and as other nations acquired the material and cultural prerequisites for the growth of their own scientific communities, the essentially non-nationalistic character of science made it possible, indeed necessary, for citizens of the "newer" nations to look to their more advanced neighbours for training in scientific professions. Thus, the French and English originally looked to Italy for inspiration and help; as the centre of excellence moved north, Germans sought training in France and England. Then it was Germany's turn to serve as the centre for scientific training, particularly during the latter half of the nineteenth century, and both Americans and Russians looked to that country for their advanced training. In the twentieth century America has been perhaps the chief centre for training.[16] (Israel, of course, is a special case. After its establishment in 1947 it was able to muster a scientific community of the first rank within ten years, because of its attraction for scientists who had been born to diverse, primarily European, nationalities.)

During the second third of this century, particularly after the end of the Second World War, the practical value of science could no longer be ignored by the public. Despite its horror, the atomic bomb symbolized the awesome practical powers of science and the advantages accruing to a nation in both international relations and domestic affairs through the aid of science. Nations without such resources became desirous of developing them, and those which already possessed substantial scientific communities began to think more directly in terms of "scientific competition" than at any time in the past.[17]

Thus there developed the concept of science as a national asset—and of the significance of the national scientific community. This is a relatively new factor in world politics and is an important goal for many nations. But, as noted, the importance of science to national interests is based on the real or hoped-for practical benefits, rather than on the value of knowledge for its own sake.

At this point it will be necessary to return to a more theoretical consideration of the relationships between basic or "pure" science and research carried out in the service of specific, concrete human needs.

BASIC AND APPLIED RESEARCH[18]

The description of science given in the discussion of the dynamics of the scientific community is, of course, appropriate primarily to what is called basic research—the disinterested quest for new, universally valid knowledge sought without regard for its possible relevance to the solution of practical problems. Applied research, on the other hand, is oriented directly or indirectly to solving "real" problems. The importance of this distinction for our purposes is that the empirical problems which beset men as physical beings, in contrast to the theoretical questions which intrigue them as intellects, are usually localized in space and time. They are related to national rather than scientific interests.

A problem in one part of the world need not exist in other parts or at other times, so that a solution to it tends to lack the universality which characterizes answers to basic scientific questions. This means that if scientists' interests were to be guided solely by practical concerns, the consensus on what constitutes important scientific questions— which lies at the base of the international scientific community—would immediately be shattered. And since their problems would not only be peculiar to places and times but also identified by criteria which vary from one culture to the next, the scientists of a given nation would ordinarily be neither interested in nor able to contribute to the work of scientists in other nations.

Further, empirical problems do not arise in logical sequence as do the theoretical questions which occupy basic scientists, so that applied research has relatively little potential for building a cumulative, generalized body of knowledge. Only coincidentally would one scientist's research have meaningful implications for another's, so the chances of obtaining competent response would be drastically reduced. There is thus a considerably smaller scientific audience for achievements in applied research, and greatly diminished opportunity for the applied researcher to gain the kind of immortality offered by fundamental scientific discoveries.

This means that the applied researcher must look more to non-scientists for rewards than to his colleagues, and these rewards must be something other than the competent response to creativity which is the normatively appropriate reward. They are usually money and sometimes public adulation, neither of which requires any real understanding of what the scientist has done. Thus the applied researcher violates the norm of disinterestedness, which also makes it more difficult for him to participate fully in the activities of the scientific community. He is therefore perceived by basic scientists as threatening the moral rectitude and stability of their entire enterprise. For this reason there are appreciable pressures on the young scientist not to engage in applied research. The invidious distinction between basic and applied research is obvious to any advanced graduate student, ordinarily trained to seek the type of work carrying more prestige and the sort of position in which it can be carried out with maximum facility.

Yet society's interest in encouraging and supporting science must ultimately rest on the assumption that it will eventually be repaid in the form of solutions to pressing problems. Thus, despite the fact that basic research provides the specialized universe of discourse within which the applied scientist works, there are often demands on the scientist to apply his skills entirely to one or more of the problems his society has defined as in need of solution. To be sure, it may take twenty years or more before a basic discovery is embodied in the comprehensive conceptual framework guiding the applied researcher's definition of his problem, but it is clear that it would be impossible to build a viable scientific community on the basis of applied research alone, for reasons intrinsic to the very nature of science.[19]

No matter how affluent it is, then, a nation must realize that to demand only applied research from its scientists is to short-circuit the natural chain of events through which it will eventually profit from its support of science. It must recognize that these profits are not likely to be gained quickly, or from the basic researcher himself, but either through the eventual translation of his findings into practical application by someone else, or through the later activities of some of his students. His apparently "useless" research and teaching have a high probability of paying off some time in the future, if only society can afford to wait for this return on its investment.

PREREQUISITES FOR SELF-SUSTAINING NATIONAL SCIENTIFIC COMMUNITIES

Given both the dynamic imperatives of the international scientific community and the basis upon which nations are willing to support scientists internally, what can we say about the conditions under which a national scientific community may develop as an intellectually self-sustaining entity? There seem to be three major components to the answer, even though they are intricately interwoven and not easily separated by empirical observation.

The first of these is the cultural component, by which is meant the intellectual or non-material factors necessary for the existence of science. The second is the component of support, which covers the major material factors in this "social equation." Finally, there is the component of understanding, or the awareness of the nature of science and its essential needs; this must form the basis of effective national science policy and management in terms of day-to-day operations. To conclude, we shall examine each of these components in turn, hoping thereby to set forth a reasonably compact summary of the principal interfaces between the internationality of science and the nationality of its practitioners.

CULTURAL FACTORS

Required first, of course, is a general attitude towards the natural universe that legitimates or even encourages an objective effort to discover its secrets. Cultures which impute natural phenomena to the arbitrary or capricious acts of supernatural beings will find no reason to seek a systematic understanding of them through observation and experimentation. What caused lightning to strike a tree in one instance, for example, might not be duplicated the next time because the god's reason for bringing it about—the perceived cause of the lightning bolt—would be different. Thus, no dependable description of the sequence of observable physical events leading to this general type of event could ever be achieved.[20]

Beyond the existence of a favourable and confident attitude toward the systematic investigation of nature, a nation's culture must also support science positively. A scientific career must be honoured, not merely tolerated, else there will be no motivation to engage in such work. There must be at least one significant group within a society, be it government, the priesthood, or the lay intelligentsia, that supports science, before even a minimal number of its members will view it as a realistic career possibility.

Finally, the culture must approve of contact with the citizens of other nations and, at the more practical level, the provision of opportunities for contact both through travel and through training in languages of those more advanced in scientific research. This factor thus shades off into the realm of material support, particularly through adequate training in foreign languages as well as in science.

SUPPORT FACTORS

The amount of material support for science depends particularly upon the relative affluence of a nation, though the priorities by which its wealth is allocated to various activities are the critical factor in determining the development of a self-sustaining scientific community.[21] Adequate investment in higher education, which provides both new recruits to science and opportunities for employment of those already trained as scientists, is evidently essential. Ordinarily it is the prestige of a university education in general rather than of training in science in particular, and the prestige of being a "professor" or "doctor" rather than of being a scientist, which is of initial importance in attracting highly talented individuals to scientific careers. But if this is the route to be travelled in developing a national scientific community, then it must be followed.

In the absence of relevant data, we may speculate further that only after a nation has developed a "critical mass" of scientists, and provided the facilities to enable them to participate effectively in the work of the international scientific community, will it have the necessary foundation for a domestic scientific community and for the ultimate exploitation of this group to its own practical advantage. This "critical mass" of scientists, which might be defined as an audience of sufficient size and competence to provide adequate feedback—professional recognition—to its members so that they need not feel totally dependent upon feedback from abroad, cannot be determined by any hard-and-fast rule. International standards determine absolute competence, but the individual scientist's own standards determine how many competent local colleagues are needed to provide sufficient feedback to keep him satisfied with his position at home.

In the beginning, of course, a nation must send its students abroad for training, and thus run the risk that the best will be tempted to stay away because of the greater ease of participation in the international scientific community thus offered. The only effective answer is apparently to send more students abroad, on the assumption that eventually enough of them will return home to constitute the "critical mass"referred to above. Once this watershed is crossed—and the provision of research facilities by the home country must be taken for granted here—the gradual but progressive development of a national scientific community should occur.

MANAGEMENT FACTORS

Along with a nation's awareness of the practical value of a scientific community, the third component of successfully meeting this need comes into view. It relates to the conscious decisions made regarding domestic organization and support and the understanding of the nature of science (both basic and applied) which underlies them.

Perhaps the most important aspect of the national management of science, if this can indeed be thought of as a "national" policy, is the way in which science is organized in terms of positions within the social hierarchy. If scientists may aspire to relatively few positions of real influence (as was the case in Germany around the turn of the century)[22] then either fewer talented people will choose this career or they will migrate to other countries after completion of their training.

It should be noted here that the scientific community probably approximates the ideal of *laissez faire* much more closely than the economic institution of society ever did, so that it is vitally important that no structural impediments be placed in the way of talent wherever it appears. The original social class of a scientist should be totally irrelevant to his advancement in terms of reputation and position, provided his own abilities merit such advancement, and the extent to which irrelevant barriers are placed in his path affects the morale and motivation of the entire national scientific community.[23] The discovery of scientific truth is completely separate from the social characteristics of the discoverer, and for scientists' interest in research to be sustained, achievement must be fairly rewarded.

Finally, and at a higher level of national policy, the allocation of support for basic and applied research must be such as to strike a balance between the demands of the former (demands, really, for evidence that the nation values basic research for its own sake) and the more obvious pay-offs to come from the latter. The 10 per cent allocated in the United States for basic research in recent years, as a proportion of total research and development expenditures, is probably far too low for a developing nation, but again we have no accepted rule by which to determine the best division of support between basic and applied research for a nation at a given level of development.

Regardless of the lack of formulae for governmental science policies, however, the general relationship between basic and applied science as described above must be observed if a nation is to build its own self-sustaining scientific community. And as a nation's understanding of the fundamental nature of science increases, it must inevitably be more successful in its efforts to develop its scientific capacity for domestic progress and international influence.

NOTES

1. Roger Coates, "Preface," in the Motte translation of Isaac Newton, *Principia,* 1713.

2. Derek J. de Solla Price, *The Difference Between Science and Technology,* Detroit, Thomas A. Edison Foundation, 1968.

3. cf. Stevan Dedijer, "Why Did Daedalus Leave?," *Science,* Vol. 133, No. 3470, 30 June 1961, p. 2047-52.

4. Talcott Parsons, "The Institutionalization of Scientific Investigation," in: Bernard Barber and Walter Hirsch (eds.), *The Sociology of Science,* p. 7-15, New York, Free Press, 1962.

5. Robert K. Merton, "Science and Democratic Social Structure," in *Social Theory and Social Structure,* rev. ed., p. 550-61, New York, Free Press, 1957.

6. Robert K. Merton, "Priorities in Scientific Discovery: A Chapter in the Sociology of Science," *American Sociological Review,* Vol. 22, No. 6, December 1957, p. 635-59.

7. Warren O. Hagstrom, *The Scientific Community,* p. 9, New York, Basic Books, 1965.

8. Norman W. Storer, *The Social System of Science,* p. 57-74, New York, Holt, Rinehart & Winston, 1966.

9. Warren O. Hagstrom, *Competition and Teamwork in Science,* Madison, Wisconsin: University of Wisconsin Department of Sociology, July 1967, 20 p. (mimeographed).

10. Robert K. Merton, "The Ambivalence of Scientists," in: Norman Kaplan (ed.), *Science and Society,* p. 112-32, Chicago, Rand McNally, 1965.

11. Storer, op. cit., p. 103-6.

12. Merton, "Science and Democratic Social Structure," op. cit.; also Bernard Barber, *Science and the Social Order,* Chapter 4, Glencoe, Ill., Free Press, 1952; Storer, op. cit., p. 76-86.

13. Derek J. de Solla Price, *Little Science, Big Science,* New York, Columbia University Press, 1963.

14. Norman W. Storer, "The Coming Changes in American Science," *Science,* Vol. 142, No. 3591, 25 October 1963, p. 464-7.

15. Robert K. Merton, "Science, Technology, and Society in Seventeenth-Century England" *Osiris* (Bruges, Belgium), Vol. 4, 1938, p. 360-632.

16. Joseph Ben-David, *Fundamental Research and the Universities,* Paris, OECD, 1968, p. 29-53.

17. Vannevar Bush, *Science, the Endless Frontier,* Washington, National Science Foundation, 1960; originally published 1945.

18. Storer, *The Social System of Science,* op. cit., p. 106-15.

19. See, for instance, Peter Thompson, "TRACES: Basic Research Links to Technology Appraised," *Science,* Vol. 163, No. 3865, 24 January 1969, p. 374-5.

20. Thomas R. Odhiambo, "East Africa: Science for Development," *Science,* Vol. 158, No. 3803, 17 November 1967, p. 876-81.

21. Derek J. de Solla Price, "Nations Can Publish or Perish," *International Science and Technology,* No. 70, October 1967, p. 84-90.

22. Ben-David, *Fundamental Research and the Universities,* op. cit., p. 33.

23. See for instance, Richard L. Meier, "Research as a Social Process: Social Status, Specialism, and Technical Advance in Great Britain," *British Journal of Sociology,* Vol. 2, No. 1, March 1951, p. 91-104.

SAL P. RESTIVO
CHRISTOPHER K. VANDERPOOL

The Third-Culture of Science

Roots of the Third-Culture of Science

The third-culture of science is the cultural (including intrascientific) patterns created, shared, and learned by scientists of different societies who are in the process of relating their societies (or sections thereof) to each other. The concept of a scientific third-culture links the cross-societal activities of scientists to (1) the development of an "international scientific community," and (2) the development of cross-societal linkages which ideally foster increasing communication and cooperation among the world's peoples (Useem, Donaghue, and Useem 1963; Useem 1967, 1971).

From the perspective of a social system of science, the ability of scientists from different societies to engage in meaningful scientific interaction depends on at least three factors: (1) there must be media of communication, such as specialized languages, which scientists everywhere have access to, can learn, and will use; (2) scientists must share a view of "reality," and of the nature and goals of scientific inquiry; and (3) the role of the scientist must be comparable in different societies.

Generalized and codified knowledge, classification schemes, standards and mechanisms for evaluating research, measurement procedures, and common problems form the core of the communication system in science (Coblans 1964, p. 93). Actual and potential consensus among scientists, and the convergence of interests in scientific activities rest on (1) the relative constancy of an accepted "grammar of learning" (Woolf 1964, pp. 2-3), and (2) the universals of human experience, including the existence of a discoverable "reality." This need not, and should not, imply a static "thingness out there" that all human beings at all times and in all places experience, or can experience similarly. David Bohm, for example, arguing from the assumption that there exists "an absolute, unique, and objective reality," conceives that reality as an "inexhaustible multiplicity and diversity" of things, an "infinite totality of matter in the process of becoming." In principle, "any *given* kind of thing is . . . knowable." Scientific activity is a process of approaching "the absolute" in terms of better and better approximate analyses of more and more things and their interrelations, and discovering the limitations of specific concepts and laws in increasing detail (Bohm 1971, pp. 168-70).

To the extent that the scientific process and the general increase in cross-societal movement of scientists and other persons is unimpeded, we can expect the transactions between scientists and "reality" to generate a dynamic system of values, norms, and beliefs (concomitant with the growth of scientific knowledge) which ideally transcends conventional cultural boundaries.

An important determinant of whether and to what extent the process of science develops internationally or globally is the social definition of scientific activity in different societies. Scientific communication among scientists in different societies is facilitated or obstructed by the degree to which social definitions of scientific activity converge or diverge. The specific patterns of cross-societal interaction tend to reflect the convergence of these social definitions. Interaction patterns also follow the various activities scientists engage in (e.g., research, teaching, administration, consultation, "gate-keeping," publishing). For example, researchers tend to interact with other researchers and to support organizations that serve as foci and contexts for scientific communication (Kaplan and Storer 1968). One of the main stimuli to the movement of scientists across societal boundaries, and to the recurrence of "brain drains" is the failure or inability of many societies to support scientific research.

Historically, there has been a general development and expansion of the third-culture of science. This process is sketched in the following section. It is important to avoid reading this history as a unilinear, unidirectional, and inevitable process that proceeds continuously, unimpeded, uninterrupted, and irreversibly. In our concluding sections, we discuss some of the obstacles to the development and expansion of the third-culture of science.

THE THIRD-CULTURE OF SCIENCE

The physical mobility of persons engaged in scientific activities has been a noteworthy aspect of the history of science. Large numbers of outstanding Greek scholars in the late pre-Christian era apparently migrated (Dedijer 1968, pp. 13-14), and until about 300 B.C., the main flow of this migration was to Plato's Academy and Aristotle's Lyceum in Athens. As a consequence of the efforts of Ptolemy Lagi (323-285 B.C.) and his son, Ptolemy Philadelphus (285-247 B.C.), Alexandria succeeded Athens as a scientific center; among the scholars who lived and worked in third-century Alexandria were Zeno the Stoic, Epicurus, Euclid, Eratosthenes, Archimedes, and Aristarchus of Samos (Albright 1957, pp. 339-40). As Sarton has noted (1959, pp. 9-10), the movement of superstitions was greater than that of scientific ideas. And the movement of scientific ideas to the East was not great: in proportion to the Asiatic population, "the Greek emigrants were too few in pre-Christian times and too little interested in science and scholarship to affect and change Eastern minds. . . ." (Sarton 1959, p. 11). No continuous diffusion of science occurred, and men who possessed scientific ideas were by no means mobile in great numbers.

PRE-INSTITUTIONALIZED

Nevertheless, centers for scientific activity which attracted scholars from widely separate areas emerged and flourished in East and West prior to the beginnings of the scientific revolution in sixteenth century Europe. These early contacts among persons engaged in scientific activities were sporadic; a specialized scientific role had not yet emerged. Scientific activities were rooted in and controlled by ruling political, military, and religious elites. This period can be classified as the *pre-institutionalized* phase of the third-culture of science. Cultural (including intra-scientific) patterns were being created, shared, and learned by "scientists" of different societies who were in the process of relating their societies or sections thereof to each other. The process of developing a language of science and a common value system among scientists from different parts of the world had begun.

By the fourth century A.D., Rome had become the center of so great a flow of students from Gaul and other provinces that special decrees were issued to govern their conduct (Haarhof 1920, p. 241). The most famous of the early academies in China was the Academy of the Gate of Chhi, founded in the fourth century B.C. (Needham 1969, p. 243). In Nalanda, India, a Buddhist school attracted Asiatic pilgrims between the fifth and twelfth centuries A.D. (Moskerji 1947, pp. 563-64; Altekar 1948, pp. 123, 125); a center for higher learning was organized at Gundi Sapur in East Persia early in the sixth century A.D. (Dedijer 1968, p. 17); and in A.D. 639, the emperor T'ai Tsung established a center in China which attracted "barbarian" students (Galt 1951, p. 328; Marin 1901, p. 378). In 1259, at an astronomical observatory at Maragha in Azerbaijan, Spanish, Egyptian, Persian, Chinese, and Mongol astronomers assembled to relate their discoveries and observations (Needham 1949, p. 6). Baghdad was also an important scientific center in the centuries preceding the Middle Ages. Not only centers, but also individuals attracted scholars, Abelard being an outstanding example (Dedijer 1968, p. 20).

From the thirteenth to the fifteenth century, the vanguard of the West's scientific revolution was visible in the migrations of European intellectuals to the great universities. At Bologna and Paris, foreigners seem to have outnumbered native students (Dedijer 1968, p. 21).

INSTITUTIONALIZING

With the emergence of modern science in Western Europe from 1500 on, and the institutionalization of the scientific role, an increase in the scale of international science occurred. The third-culture of science entered an *institutionalizing* phase. Scientists attacked the shibboleths of knowledge and pursued recognition for their scientific activities, often with missionary zeal. From the sixteenth to the nineteenth century, scientists stressed the potential of science for transforming society. Many of their arguments were "scientistic." Advocating a materialistic and empiricist explanatory orientation to the physical, natural, and social world, scientism is characterized by "a particular understanding of the power of science . . . a critique of tradition, and . . . a form of substitute religion" (Kwok 1965, p. 23). The "idea of progress" became a predominant theme in scientific ideology. Progress and the inevitability of social and scientific evolution, for example, were major and recurring themes in the Presidential addresses of the British Association for the Advancement of Science from its founding in 1831 to the death of Queen Victoria (Basalla, et. al. 1970).

During this institutionalizing phase, the first scientific associations and research institutions were established, and science began to enter the university curriculums (Ben-David 1971). The part-time gentleman scientist began to be replaced by the full-time scientist. Many of the associations and research institutions were established or supported by the governments of several nations in response to scientism, and to the enormous military and technological potential of scientific inquiry. These organizations became focal points of cross-national communication among scientists. The participation of scientists and scholars in the activities sponsored by scientific organizations (e.g., The Royal Society, and the Academie des Sciences) stimulated the emergence of scientific work roles, specialized languages, and a value system ideally shared by all scientists.

INSTITUTIONALIZED

The contemporary third-culture of science, though heterogeneous and changing, is characterized by enough established patterns to warrant the description *institutionalized* phase. This phase began to emerge in the early nineteenth century. The first of the annual international scientific congresses was held in 1860 (Needham 1949, p. 6). In 1919, an international scientific federation, the International Research Council (IRC) was formed (Baker 1971, p. 1). The IRC was originally a federation of scientists from Belgium, Canada, France, Italy, Japan, New Zealand, Poland, Portugal, Romania, Serbia, the United Kingdom, and the United States; today, 62 nations are represented in the IRC (Baker 1971, pp. 8-9).

In addition to international non-governmental organizations such as IRC, international governmental organizations have emerged, e.g., the Organization for Economic Cooperation and Development (OECD), the European Organization for Nuclear Research (CERN), and the Intergovernmental Oceanographic Commission (OIC) (Crane, 1971). A variety of international professional associations, such as the International Sociological Association, have been founded. The major national professional associations have, as a consequence of the increase in communication, transportation, and exchange links among the world's scientists, become international in nature.

In the institutionalized stage of the third-culture of science, international activities include (1) the organization and support of international congresses, conferences, and symposia; (2) the establishment of international scientific journals, and abstracting and translation services; (3) exchange of scientists programs; (4) facilitation of exchange programs through the compilation of personnel and program directories, public announcement of opportunities for participation in exchange programs, and financial support for individuals and programs; (5) exchange of data, instruments, and techniques; and (6) systematization of scientific nomenclature (Revelle 1963, p. 116).

International science is dependent upon the support of scientific activities within nations. Without national recognition of science as an important part of national welfare, the range of international exchanges would be quite narrow. With the emergence of this recognition on an increasingly international scale has come a population explosion in the scientific community, the growth of viable research and education institutions where scientific knowledge is created and transmitted, and a rapid increase in the number of scientific journals, abstracting services, and forums for informal scientific communication (Storer 1966, p. 138). Derek J. DeSolla Price called attention some time ago to exponential growth rates in the scientific community (Price 1963). At the same time, political and economic barriers prevent, and in some cases retard, the development of international scientific cooperation. There are economic restrictions on the passage of instruments across national boundaries, restrictions on the geographical areas in which research can be carried out, and passport and visa restrictions which often involve "conference-killing" delays (Revelle 1963, p. 126; Rose and Rose 1970, p. 181).

Another factor affecting patterns of international scientific activity is variations in levels of national development (Dedijer 1967, pp. 155-56). Scientists from former colonial nations are often slighted by the governments and scientists of the colonial mother countries. In addition, scientists from developed countries do not actively seek contacts with scientists from developing countries because they view science in the developing nations as a "scientific back-water region." The scientist from the developing country, on the other hand, frequently does not receive the level of material support needed to participate fully in international scientific associations and organizations.

The international scientific communication network is affected by variations among scientific disciplines. The rate of growth of research and the changing nature of the research enterprise within a discipline, shifts in the world distribution of research output, and fluctuations in national research policies affect the forms of direct and indirect communications among scientists (Crane 1971; Dedijer 1967, p. 144). The establishment of research institutions often changes the location of transnational involvement (Ben-David 1971). When new paradigms emerge threatening the existing world-view for explaining natural and social phenomena and a period of scientific revolution is set in motion (Kuhn 1962), scientists distant from the centers of science often will resist adoption of the new paradigm. Yet scientists have adapted to these shifts and changes in the past and what was considered to be an innovation or drastic alteration has become part of "normal science" in the third-culture of science.

Of all the scientists who are members of the international scientific community, only a portion are affected by these internal and external barriers to international cooperation among scientists. Only the highly involved leaders and informal leaders of a scientific discipline are directly affected by these barriers (Hagstrom 1965; Rose and Rose 1971, p. 180). The highly involved leaders are those scientists who participate in all the communication channels within a scientific discipline. They publish frequently and receive formal recognition for their leadership in an area of scientific inquiry. Their contacts with scientists abroad are equally frequent. They correspond weekly and sometimes daily with their foreign colleagues. They visit these colleagues and are, in return, visited by them. In addition, they are actively involved in a variety of international scientific associations and organizations.

The informal leaders, on the other hand, rely heavily on informal contacts with their colleagues abroad. In their "invisible colleges," those informal communication networks established by scientists interested in similar domains of investigation (Storer 1968, pp. 20-21), the informal leaders keep up with developments in their field and make known their own research progress.

A third type of scientist, the "scientific statesman," is also affected at times by these barriers. These scientists, e.g., Linus Pauling in biology and Kenneth Boulding in economics, spend a great deal of time with specialists in other disciplines and non-scientists at home and abroad. Their prestige in their respective fields enables them to have recurrent contacts with non-specialists as experts without jeopardizing their reputations among colleagues. Having made contributions to their own fields in the past and now contributing to knowledge outside of their areas of specialization, they reduce contacts within their specialty and increase their contacts with the wider scientific and non-scientific communities.

The restrictions placed on international exchanges affect the involvement of the leaders and statesmen in the third-culture of science. Scientists who do not participate as frequently at the international level are also affected by these restrictions. The filtering down of information through the formal and informal networks is hampered, and thus the state of science in general is adversely affected.

THE THIRD-CULTURE OF SCIENCE AND PRESTIGE

Throughout this essay, we have viewed international interaction among scientists as consisting of exchange of information, the development of a language of science, and the sharing of a common value system. But the organization of scientific communication systems involves other exchanges. It also consists of an exchange of information for social recognition and prestige (Hagstrom 1965, p. 13). The information that is carried along the variety of networks is evaluated by the recipients of the research reports, journal articles, and pre-prints. Through such evaluation, the source of the information acquires recognition as a basic contributor or non-contributor to scientific knowledge.

Since the third-culture of science is an arena of action whereby scientists receive prestige, certain benefits accrue from the competition for prestige (Reif 1965, p. 142). High standards of accomplishment are maintained buttressing the value system of science. Prestige is given to those scientists whose work has been proven fruitful in advancing the endeavors of their colleagues in similar domains of inquiry. Critical evaluation of the results of their discoveries by fellow researchers leads to the avoidance of mistakes and oversights in research, forges new directions for future investigations, and identifies new approaches to the examination of physical and social phenomena. As a result, the competitive prestige system often advances the state of knowledge in a scientific discipline.

But such competition at the international level also has its drawbacks. It subjects scientists to strains and conflict between the values inherent in science for the advancement of scientific knowledge and personal self-aggrandizing values (Reif 1965, p. 142). The pressures which result can lead to the abandonment of serious reflection and insight on a scientific problem and to the adoption of a routine of rapid and frequent production of research notes, articles, and monographs in a search for recurrent acknowledgement of the scientist's contributions to a field. Moreover, the communication networks that carry formal and informal messages across societal boundaries may break down because secrecy is considered necessary to avoid having a discovery "scooped" by others and simultaneous discoveries being made (cf. Merton 1968).

Watson's account of the finding of the double-helix structure of the DNA molecule is illustrative of such a crack in the information flow between scientists (Watson 1968). Linus Pauling was working on the same problem in California as were Watson and his colleagues in England. The latter group obtained information regarding the state of Pauling's research. This information revealed that Pauling was directing his efforts towards a "dead-end" which Watson and his colleagues had abandoned. Rather than notifying Pauling of his misdirection, Watson and his colleagues celebrated the assurance of the Nobel prize they would subsequently receive (cf. Merton 1963; Gaston 1971).

Secrecy and competition is frequently overlaid with a coating of nationalism. Receiving international recognition is not only a blessing for a scientist's career, but also is an honor bestowed on his country.

Hence the competition for prestige is, on the one hand, a question of whether scientist A beats scientist B to a discovery and, on the other, of whether nation X is further along than nation Y in its scientific and technological progress. The addition of "nationalism" to the competition for prestige amplifies the tendency to disrupt the international flows of scientific information.

Nationalism is only partly stimulated by governments searching for increments in their nations' ranking in the international system of power and prestige, and scientists acting in accordance with such a search. Scientists also use the competition for power at the international level to enhance their own status positions in their home countries. Only those professions which enjoy prestige in the eyes of the members of a given society are likely to receive the degree of support, deference, and moral backing needed for systematic and sustained growth of the profession (Coser 1965, p. 33). One way of obtaining such benefits is to argue that "if we don't accelerate our effort in this field, we will likely lose the race with nation X."

Similar arguments have enabled scientists in this country to receive support that they would not have otherwise obtained. For example, the Sputnik crisis encouraged the president of the United States to appoint a full-time science advisor to his staff to advise him on the state of science in America and abroad (Greenberg 1967, p. 139). In the absence of crisis neither government nor other scientists demanded active participation of scientists in Washington.

The National Science Foundation also has in the past partially defined its interests in terms of the competition for prestige at the international level and the social benefits to the citizens of the United States derived from basic research. A former director of the National Science Foundation summarized the federal government's objectives in providing funds for science and technology in the following manner: "The objectives of federal support for science and technology are the very best the social structure can produce. . . . The National Science Foundation's mission is general and may be described as 'the advancement of science in the national interest' " (U.S. Congress 1961, pp. 4-5). As long as the international system is dominated by "scientific nation-states" (Gilpin 1968, p. 25; King 1964, p. 115) national interests will continue to take precedence over the third-cultural interests of the scientists.

PROSPECTUS ON THE THIRD-CULTURE OF SCIENCE

Bernal wrote that while science has been "from the start international," especially and most fully from the eighteenth century on, "the present century has marked a definite retrogression" (1939, p. 191). More recently, Rose and Rose have commented that "while the individual scientist might maintain his universalistic ethic, it has increasingly become a reality for only an elite among scientists" (1970, pp. 180ff). Even in the full flowering of internationalism, which Bernal associates with the eighteenth and nineteenth centuries, there were technical and geographical constraints on international activity. Then, as in the past, the communication and mobility patterns implied by internationalism were a reality "to only a relatively small proportion of those who regard themselves as—and are regarded as—scientists (Rose and Rose 1970, p. 181). Perhaps, as it has been suggested, internationalism "remains a predominant myth just because it is precisely [the scientific elite] who tend to write about the philosophy and ethics of science as an institution and activity" (Rose and Rose 1970, p. 181).

The concept of "third-culture," as formulated by the Useems (1967, 1968, and 1971) is rooted in the idea of an increasingly complex and interdependent world. Third-cultures arise out of the cross-cultural movements of individuals and have the potential for increasing the scale of human activity and consciousness. The significance attributed to cross-cultural experiences as a stimulus for the growth of international awareness and cooperation is, however, problematic (e.g., Angell 1967, p. 129; Restivo 1971, pp. 178-87). Cross-cultural scientific activity, like other human activities, can increase in scale without facilitating cooperation among individuals and nation-states. The emergence of a supranational system of social activities can stimulate a concern among its participants for their system and their roles independent of national and/or international systems in general. Scientists may, for example, become committed to a transnational profession without necessarily becoming committed to or aware of the possibility for a world society (Restivo 1971, p. 180).

The coming changes in the third-culture of science, to paraphrase Norman Storer, *may* involve enlarging the outlooks, heightening the self-confidence, and imparting "a zest to innovate" among scientists. It may mean contributions "to the ecumenical collectivity of science, enlarged understandings of what we mean when we talk about discovering knowledge useful for mankind" (Useem 1971, pp. 24-25). But they may also be directed by the dysfunctions of professionalization and bureaucratization. We refer to today's third-culture of science as "institutionalized." Certain patterns have emerged; others are emerging which may or may not be consistent with the patterns which define this institutionalized stage. Institutionalization can, where no counterprocesses emerge, lead to rigidity, stagnation, and retrogression.